Sustainable Nuclear Power

Academic Press
Sustainable World Series

Series Editor
Richard C. Dorf
University of California, Davis

The *Sustainable World* series concentrates on books that deal with the physical and biological basis of the world economy and our dependence on the tools, devices, and systems used to control, develop and exploit nature. Engineering is the key element in developing and implementing the technologies necessary to plan for a sustainable world economy. If the industrialization of the world is to continue as a positive force, the creation and application of environmentally friendly technologies should be one of the highest priorities for technological innovation in the present and future.

This series includes titles on all aspects of the technology, planning, economics, and social impact of sustainable technologies. Please contact the editor or the publisher if you are interested in more information on the titles in this new series, or if you are interested in contributing to the series.

Current published titles:

Technology, Humans and Society: Towards a Sustainable World, edited by Richard C. Dorf, 500 pages, ISBN: 0122210905, published 2001

Wind Power in View: Energy Landscapes in a Crowded World, edited by Martin J. Pasqualetti, Paul Gipe, Robert W. Righter, 234 pages, ISBN: 0125463340, published 2002

Coal Energy Systems by Bruce G. Miller, 526 pages, ISBN: 0124974511, published 2004

Hydrogen and Fuel Cells by Bent Sørensen, 400 pages, ISBN: 0126552819, published 2005

Sustainable Nuclear Power

Editors

**Galen J. Suppes
Truman S. Storvick**

AMSTERDAM • BOSTON • HEIDELBERG • LONDON
NEW YORK • OXFORD • PARIS • SAN DIEGO
SAN FRANCISCO • SINGAPORE • SYDNEY • TOKYO

Academic Press is an imprint of Elsevier

St. Philip's College Library

Elsevier Academic Press
30 Corporate Drive, Suite 400, Burlington, MA 01803, USA
525 B Street, Suite 1900, San Diego, California 92101-4495, USA
84 Theobald's Road, London WC1X 8RR, UK

This book is printed on acid-free paper. ∞

Copyright © 2007, Elsevier Inc. All rights reserved.

No part of this publication may be reproduced or transmitted in any form or by any means, electronic or mechanical, including photocopy, recording, or any information storage and retrieval system, without permission in writing from the publisher.

Permissions may be sought directly from Elsevier's Science & Technology Rights Department in Oxford, UK: phone: (+44) 1865 843830, fax: (+44) 1865 853333, E-mail: permissions@elsevier.co.uk. You may also complete your request on-line via the Elsevier homepage (http://elsevier.com), by selecting "Customer Support" and then "Obtaining Permissions."

Library of Congress Cataloging-in-Publication Data
Application submitted

British Library Cataloguing in Publication Data
A catalogue record for this book is available from the British Library

ISBN 13: 978-0-12-370602-7
ISBN 10: 0-12-370602-5

For all information on all Elsevier Academic Press publications visit our Web site at www.books.elsevier.com

Printed in the United States of America
07 08 09 10   9 8 7 6 5 4 3 2 1

Working together to grow libraries in developing countries

www.elsevier.com | www.bookaid.org | www.sabre.org

ELSEVIER   BOOK AID International   Sabre Foundation

Contents

List of Figures	ix
Preface	xv
Organization of the Book	xvi
Acknowledgments	xviii

1 Introduction   1
   Energy in Today's World   1
   Energy on Planet Earth   2
   What Are the Right Questions?   4
   Sustainable Nuclear Power   7

2 The History of Energy   9
   Energy   9
   Nature's Methods of Storing Energy   10
   Man's Interaction with Nature's Stockpiles and
      Renewable Energies   13
   The Industrial Revolution and Establishment of
      Energy Empires   16
   Environmental Impact   25
   Environmentally Responsible Nuclear Power   31
   References   32

3 Energy Reserves and Renewable Energy Sources   33
   Fossil Fuel Reserves   33
   Cosmic History of Fossil Energy Reserves   42
   Nuclear Energy   48
   Recent Solar Energy   58
   Ethanol and Biodiesel from Agricultural Commodities   62

vi Contents

    Emergence of Nuclear Power   71
    References   72

4 Emerging Fuel Technologies and Policies Impacting These Technologies   **75**
    Politics of Change in the Energy Industry   75
    Cost of Feedstock Resources   77
    Case Study on Investment Decisions and Policy Impacts   82
    Taxes and Social Cost   95
    Corporate Lobbying Retrospect   99
    Diversity as a Means to Produce Market Stability   102
    The Details Are Important   104
    Environmental Retrospect   108
    Efficiency and Breakthrough Technology   109
    Farm Commodities and Land Utilization   112
    Global Warming   114
    Diversity and the Role of Nuclear Power   116
    References   117

5 History of Conversion of Thermal Energy to Work   **119**
    Use of Thermal Energy   120
    The Concept of Work   121
    Early Engine Designs   125
    Turbine-Based Engines   134
    Fuel Cells   147
    Recommended Reading   156
    General References   156
    References   157

6 Transportation   **159**
    Transportation Before Petroleum Fuels   159
    Petroleum Fuels: Their Evolution, Specification, and Processing   161
    Alternative Fuels   168
    Vehicular Fuel Conservation and Efficiency   176
    References   184

7 Production of Electricity   **185**
    History of Production   185
    Production of Electrical Power   196
    Recommended Reading   200
    References   200

Contents

8 Energy in Heating, Ventilation, and Air Conditioning — 201
  The Heating, Ventilation, and Air Conditioning
    Industry — 201
  Air Conditioning — 208
  Heating — 211
  Peak Load Shifting and Storing Heat — 214
  The Role of Electrical Power in HVAC to Reduce
    Greenhouse Gas Emissions — 217
  Example Calculations — 218
  References — 221

9 Electrical Power as Sustainable Energy — 223
  Sustainability and Electrical Power — 223
  Expanded Use of Electrical Power — 224
  Increased Use of Electrical Power in Transportation — 226
  Increased Use of Electrical Power in Space Heating — 238
  Increased Use of Electrical Power for Hot Water
    Heating — 245
  Topics of National Attention — 245
  Example Calculations — 246
  Recommended Reading — 247
  References — 248

10 Atomic Processes — 249
  Energies of Nuclear Processes — 249
  Chart of the Nuclides — 254
  Nuclear Decay — 259
  Conditions for Successful Nuclear Fission — 260
  Transmutation — 270
  Nuclear Fusion — 273
  Radiological Toxicology — 275
  References — 282

11 Recycling and Waste Handling for Spent Nuclear Fuel — 283
  The Nuclear Energy Industry — 283
  Recycling and Green Chemistry — 284
  Why Reprocess Spent Nuclear Fuel? — 284
  Discovery and Recovery — 289
  Reprocessing: Recovery of Unused Fuel — 299
  Waste Generation from Reprocessing — 312
  Report to Congress — 314
  References — 315

12 Nuclear Power Plant Design — 319
Advances in Thermal Efficiency — 321
Steam Cycles in Commercial Operation — 327
Generation IV Nuclear Power Plants — 331
Lessons from History — 341
Challenges in Nuclear Power Plant Design — 344
Implementation Strategies and Priorities — 347
Recommended Reading — 350
References — 350

13 For-Profit Industrial Drivers — 353
Levelized Cost Approach — 353
Capital Costs — 355
Case Studies — 363
Costs of Reprocessing — 364
Advocates for Nuclear Power — 367
Transportation and Nuclear Power — 369
Expanded Use of Nuclear Power in Residence and Commercial Applications — 372
Approaches to Long-Term Handling of Spent Nuclear Fuel — 373
Fuel Costs and Energy Options — 376
Comparison to Other Studies on Economics of Nuclear Power — 378
Concluding Comments — 379
References — 381

Index — 383

Online companion site:
http://books.elsevier.com/companions/0123706025

List of Figures

1-1 The legacy of 30 years of commercial nuclear power in the United States, including 30 years of fission products that are of little value and sufficient stockpiled fissionable fuel to continue to produce electrical power at the same rate for another 4,350 years.   7
2-1 Typical composition of oil.   20
2-2 Year 2000 oil flow in quadrillion Btu.   22
2-3 Recent U.S. oil prices.   23
2-4 U.S. imports of petroleum.   23
2-5 World oil reserves by region. Estimates of Canadian reserves by *Oil and Gas Journal* in 2003 are much higher than in previous years. Most likely they include easily recovered oil sands.   24
2-6 U.S. energy consumption by source.   24
2-7 Estimate of U.S. energy reserves.   25
3-1 Summary of world and U.S. fuel reserves in Btus.   34
3-2 Impact of atomic mass number on permanence of atoms. H is hydrogen, He is helium, Li is lithium, C is carbon, O is oxygen, F is fluorine, Ar is argon, Fe is iron, Kr is krypton, Sn is tin, Gd is gadolinium, Pu is plutonium, Bi is bismuth, and U is uranium.   44
3-3 The history of energy.   47
3-4 Escalating chain reaction such as in a nuclear bomb.   52
3-5 Controlled steady-state chain nuclear fission such as in a nuclear reactor.   53
3-6 Approximate inventory of commercial spent nuclear fuel and fissionable isotopes having weapons potential (Pu-239 and U-235). The solid lines are for continued operation without reprocessing, and the dashed lines are for reprocessing (starting in 2005) to meet the needs of current nuclear capacity.   56

4-1	Alternative benchmark technologies.	81
4-2	Summary of tax breakdown on $28 barrel of synthetic crude.	90
4-3	Summary of price contributions on a gallon of gasoline on $2.01 per gallon of unleaded regular gasoline.	95
4-4	Summary of electrical power-generating capacity by fuel sources for electrical power generation in the U.S. (1999).	102
4-5	Carbon dioxide emissions by sector.	115
5-1	On Earth, most energy comes from the sun and ultimately becomes heat. This is a fascinating story of trial and error with the successful inventions providing the many devices we use every day.	120
5-2	Illustration of how pistons perform work.	126
5-3	Condensing steam used to move a piston.	131
5-4	Use of high- and low-pressure steam to power a piston.	132
5-5	A basic steam turbine power cycle.	138
5-6	A boiling water reactor (BWR) and steam power cycle.	140
5-7	A pressurized water reactor (PWR) and steam power cycle.	141
6-1	Crude oil fractions and market demands.	163
6-2	Summary of energy losses in use of fuel for automobile travel. Effectiveness and Impact of Corporate Average Fuel Economy (CAFÉ) Standards.	177
6-3	Average fuel economy of motor vehicles.	178
7-1	A simple generator.	190
7-2	A generator with armature rotating inside magnetic field.	190
7-3	A transformer.	194
7-4	Increases in thermal efficiency electrical power generation during past century.	197
8-1	A heat pump operating in heating and cooling modes.	212
8-2	Impact of space heating on baseload for electrical power generation.	214
8-3	Example of 24-hour electricity demand during July for Fort Jackson in South Carolina.	214
8-4	Example PCM device and tank for active climate control.	216

List of Figures xi

9-1	Energy consumption in the United States. Distribution by energy source only includes sources contributing more than 2% of the energy in each category.	224
9-2	Top technologies for providing beneficial impact to California's economy while reducing fuel consumption. Beneficial impact is reported in projected 2001 dollars for the time period from 2002 to 2030.	226
9-3	Simplified presentations of parallel and series HEV designs.	227
9-4	Comparison of PHEV and BEV designs. The PHEV has an engine and smaller battery pack. The BEV does not have a backup engine.	228
9-5	Comparison of net present cost for operating a conventional vehicle (CV), hybrid electric vehicle (HEV), plug-in HEV with a 20-mile range (PHEV-20), and a battery electric vehicle (BEV) with 200-mile range. Present values are based on a 7-year life cycle, $1.75 per gallon gasoline, and 6 ¢/kWh electricity.	229
9-6	A city BEV. The city BEV is a compact vehicle that has a maximum range of 60 miles. It is a niche market vehicle that can meet the needs of "some" commuter needs.	231
9-7	Miles traveled with typical automobile each day and implied ability for PHEVs to displace use of petroleum.	231
9-8	Fuel cell technologies and their applications.	236
9-9	Ratios of costs for heating with fuel versus heating with electrical heat pump. Ratios based on COP of 2.0 and heater efficiency of 90%. Shaded regions show where heat pump is more cost effective.	239
9-10	Dependence of a typical heat pump performance on outside temperature.	241
9-11	Comparison of heat pump using air heat sink to ground source unit.	243
10-1	A neutron-induced fission of U-235.	251
10-2	Excerpt from the Chart of the Nuclides.	254
10-3	Presentation format for stable isotopes in chart of nuclides.	255
10-4	Presentation format for unstable isotopes in chart of nuclides.	256

10-5	Skeleton of complete chart of nuclides illustrating stable nuclei.	256
10-6	Energy level diagram for Nickel-60.	258
10-7	Decay tracks for fertile collisions with Th-232 and U-238.	263
10-8	Plot of binding energies as function of mass number. Higher values reflect more stable compounds. The values are the binding energy per nuclei release of energy if free protons, neutrons, and electrons combine to form the most stable nuclei for that atomic number.	264
10-9	Typical neutron absorption cross section vs. neutron energy.	268
10-10	Laboratory fusion.	275
11-1	Mass balance for reprocessing. Mass is mass of heavy metal and fission products of heavy metal.	285
11-2	Mass balance of once-through fuel use as practiced in the U.S. The fission products are the "waste." The years indicate the years of available energy if used at the same rate as used in once-through burns. Both France and England have immobilized the concentrated fission products in glass for long-term storage.	286
11-3	Recovery of unused fuel is the first phase of fuel reprocessing. Casings are physically separated as the initial step in future nuclear waste management—separation of structural metals in the bundles and recovery of fissionable fuel.	287
11-4	Schematic of overall process to minimize hazardous waste from nuclear power. The masses in tons represent total estimated U.S. inventory from commercial reactors in 2007—about 50,000 tons. The volumes are based on uranium density—the actual fission product volumes would be about twice the values indicated.	288
11-5	Full use uranium providing centuries of energy. Thorium is also a fertile fuel that can be used in this closed cycle.	290
11-6	Conversion of U-238 to Pu-239.	293
11-7	Block flow diagram of PUREX reprocessing of spent nuclear fuel.	300
11-8	UREX block flow diagram.	303
11-9	Pyropartitioning process to recover heavy metals.	308

List of Figures   xiii

11-10	Impact of advanced nuclear fuel reprocessing.	315
12-1	Steam cycle operating at 33% thermal efficiency.	321
12-2	Evolution of thermal efficiency in a steam cycle. Higher-temperature steam turbine operation was key.	322
12-3	Staged expansion.	323
12-4	Accuracy of empirical model for power cycle thermal efficiency.	324
12-5	Steam reheat in power cycle.	326
12-6	Comparison of efficiency projections of different models. The Joule and modified Joule models assume a feed temperature of 313 K.	326
12-7	Projected thermal efficiencies as a function of maximum steam temperature and a low temperature of 313 K.	327
12-8	Boiler, superheater, and steam reheat in a pulverized coal power plant.	328
12-9	Boiling water reactor (BWR).	329
12-10	Pressurized water reactor (PWR).	330
12-11	Supercritical water-cooled reactor.	333
12-12	Very high-temperature reactor.	334
12-13	Gas-cooled fast reactor.	336
12-14	Sodium-cooled fast reactor.	337
12-15	Lead-cooled fast reactor.	338
12-16	Molten salt reactor.	339
12-17	Level of risk as the level of materials development complexity increases.	347
13-1	Distribution of energy consumption in the U.S. by sectors with electricity separated as its own sector.	373
13-2	Energy reserves available as a direct and indirect result of 30 years of operation of nuclear power plants in the United States. The mass is all commercial-heavy metal for 30 years of operation. The years are the period a power plant could continue to operate using the uranium that was mined to provide the first 30 years of operation.	374
13-3	Spent nuclear fuel stored on-site at a nuclear power plant if reprocessing started in 2007.	375
13-4	Extrapolation of recent rates of spent fuel accumulation from U.S. commercial facilities. Based on the rate of generation between 1990 and 2002, 30 years of stored spent nuclear fuel will be reached in 2007.	378
13-5	Uranium/plutonium in its "sustainable" phase.	380

Preface

After 30 years of commercial U.S. nuclear power production, there is sufficient fissionable fuel stored at power plants to meet all electrical power needs for the next 150 years. For the 31 states that have these reserves, this nuclear fuel is the most accessible and inexpensive new energy source in each of those states. This book is about sustainable nuclear energy and what proven nuclear fission technology these reserves can deliver.

In addition to the fissionable fuel stored at power plants, an additional 630 years of fissionable depleted uranium is stored at other locations. Combined, these reserves can meet all electrical power needs, as well as the vast majority of transportation, residential, and commercial (building) energy needs for the next 400 years. Proven technology can deliver this energy while disintegrating the waste into harmless elements—the responsibility of not leaving a legacy of nuclear waste is implicit and attainable.

While the technology is proven, a vision is needed to identify paths to commercialization that are economically sustainable. It is this vision and discussions of select empowering technologies that make this book unique.

Our purpose for writing this book is to help you, our reader, better understand energy sources, how energy is used to meet transportation and residence needs, and how nuclear power is one (if not the only) means to provide abundant and sustainable power to the world.

The book's contents are designed to incrementally build upon the science and engineering foundations of those engineers and scientists who do not have a background in nuclear science or engineering. It is hoped that this book will help empower engineers and scientists to make the needed advances.

The sustainability that nuclear fission can bring includes more than electrical power. Nuclear power can bring sustainability and extended prosperity through applications impacting transportation and space heating using grid electricity as a conduit. This extended topic has been the topic of multiple reports to Congress and Congressional Research Service reports. Where possible, these reports are cited and directly quoted.

The text is designed to be understandable and useful to interested citizens and legislators. They will ultimately empower the government to change policies and allow nuclear fission to be safer, more available, and free of waste-handling issues.

Organization of the Book

The chapters of this book are intended to be self-contained. This results in duplication of topics, but it should be easier to read where you have special interest.

Chapters 1, 2, and 3 cover the history of energy, the reserves, and some of the renewable resources available to us. Energy's history on earth starts with the sunlight that helped form wood and living organisms. Wood and living organisms are the raw materials stored by nature and transformed by geological processes over millions of years to give us coal, petroleum, and natural gas. Wood warmed the cradle of civilization. Coal was probably first used 2,000 years ago, and, more recently, it powered the industrial revolution, starting in about 1700. Liquid fuel is easiest to use in engines and served to power the modern automobile. The first fixed-wing aircraft flew using this same liquid fuel just after 1900. Liquid petroleum fuels made the 20th century the petroleum age, with the midcentury addition of natural gas pipeline distribution.

Chapter 4 evaluates alternative energy sources and technologies. We depend exclusively on petroleum fuels for transportation, and any interruption of our supply of imported petroleum can become an instant economic and social problem. Coal is the main source of fuel for electric power production. Here, the competition from other energy sources keeps the price of electricity fairly stable. Known and emerging technologies can stabilize energy prices and create security from unemployment and military conflict.

The story of energy through the 19th and 20th centuries concentrates on the work produced by hot gases expanding in engines (machines designed to do work). The science and technology of the development of these machines are summarized in Chapter 5. Chapters 6, 7, and 8 describe the technologies that provide transportation, electricity, and the equipment we use to heat and cool our homes and workplaces.

Prior to the sun's radiation touching Earth, atomic energy was shaping the universe. Man tapped into the power of nuclear energy near the end of World War II with two thunderous explosions over Japan. Two city centers were leveled and thousands of people evaporated.

This unfriendly introduction to nuclear energy has produced the attitude among many that everything nuclear should be banned. Even the name nuclear magnetic resonance imaging (an important medical diagnostic tool) had to be changed to calm patient anxiety. Attitude withstanding, nature's nuclear energy touches us in the form of the sun's radiation and geothermal heat every day. And when confronted with depleting oil and coal reserves, we cannot ignore the huge energy reserves available through nuclear energy.

Today, there are over 100 nuclear power plants in the United States producing 18% to 20% of the electricity we use every day. A few pounds of "nuclear fuel" can replace thousands of tons of diesel or coal fuel, allow a submarine to cruise underwater for months instead of hours, and provide electrical power without the air pollution associated with burning coal, petroleum, and vegetation.

The source of this nuclear energy goes back to the time when atoms were formed, long before our solar system existed. All of the atoms that we find in the gases, liquids, and solids on Earth were assembled in and among the stars from the particles and energy that make up our sun and the rest of the Milky Way galaxy. The history of energy starts—and ends—with nuclear energy.

Chapters 1 through 8 provide a case for nuclear power's ability to substantially accommodate electrical, transportation, and residential power needs. Nuclear processes are put in their accurate context as natural processes that are vital to earth's ecosystem, including the warmth of the sun's radiation and the nuclear fission's role in maintaining earth's molten core and creating a habitable ecosphere.

Chapter 9 is on the use of electrical power grid as the conduit through which nuclear power can deliver not only electrical energy needs, but also the energy to sustain transportation and space heating for homes and businesses. Technologies like plug-in hybrid electric vehicles and heat pumps empower nuclear energy to eliminate the import of both petroleum and natural gas.

Chapters 10 through 13 focus on the nuclear science, reprocessing spent nuclear fuel to eliminate waste, and economics.

Chapter 10 examines atomic processes. An understanding of these processes allows nuclear power to be harnessed. An improved understanding and commitment allow transmutation of wastes to all but eliminate waste in the immediate future and, with a realistic outlook, to fully eliminate waste issues before they become a burden on society. The burden will be on the power producers; it will be limited and sustainable. Chapter 11 describes the technology and approaches to eliminating nuclear waste.

Nuclear processes provide the sources of heat used to power the heat engine known as the power cycle. There are unique aspects of the power plant with which most engineers are not familiar.

Heat transfer in nuclear processes is more complex. In addition to convective, conductive, and radiative heat transfer, the impact of neutrons with molecules of the working fluid are able to directly heat the working fluid. Heat transfer in nuclear processes is covered in Chapter 11.

Chapter 12 completes the general design details of the many nuclear power plant options. The plant design ultimately provides passive and high safety. Proper plant designs can greatly increase the amount of energy produced relative to the waste generated.

All too often, texts on the introduction of technology fail to evaluate the technical barriers to commercialization alongside nontechnical barriers. For nuclear power to provide a sustainable solution to today's energy needs, it must be commercialized in a sustainable manner. This translates to overcoming both the technical and nontechnical barriers to commercialization. Chapter 13 identifies and prioritizes the barriers to commercializing the next generation of cleaner, safer, and more efficient nuclear power plants. A comparison to those barriers impacting the commercialization of alternative liquid fuels in the United States reveals some common barriers, but, in general, the barriers are very different. In both cases national policies present the greatest barriers, and there is an opportunity for our leaders to set the course for overcoming these barriers.

Acknowledgments

Discussions with colleagues led us to write about the strong interaction between science and technology, economics and legislation, in our modern society. We thank the hundreds of students who have been patient with our presentations and taught us that the KISS (Keep It Simple, Stupid) theory of teaching really works.

Complex solutions are best formulated by answering a series of well-stated simple questions.

The hospitality of the Chemical Engineering Department shown to a professor in retirement (TSS) served as an inspiration and made this book possible.

The companion Website containing government reports, additional exercises, case studies, and teaching tools provided by the authors to complement the text can be found at http://books.elsevier.com/companions/0123706025.

CHAPTER 1

Introduction

Energy in Today's World

The remarkable improvement in the standard of living in the United States during the 20th century is unprecedented in world history. An almost total transformation from an agrarian to an urban society occurred during this period. Work once done by people and animals is now performed by machines powered by petroleum or electricity. Both the quality and duration of life have improved.

Petroleum fuels altered the way we grow and distribute food, where we live, the location and configuration of manufacturing, and even the way we entertain ourselves. Air travel makes it possible to reach any point in the world in less than a day.

An abundant, reliable source of electricity revolutionized the factory and multiplied worker productivity. Electricity in the home allowed for refrigeration, lighting, indoor climate control, and complete home entertainment centers.

The creative and inventive use of energy is the foundation of modern society. With the emergence of the 21st century, we are challenged to improve this foundation while faced with challenges like global warming, importing over $200 billion in crude oil each year, overreliance on natural gas resources, and a momentum of established infrastructure that resists needed changes. How we respond to these challenges will dramatically impact our future.

At the onset of the 21st century, the United States has abundant and available energy, but insufficient fossil fuel reserves exist for the rest of the world to imitate U.S. consumption. Also, an overreliance on imported petroleum and natural gas may quickly climax with economic turmoil as nations position themselves for these limited resources. A theme of sustainable energy emerges—energy technology options that can be practiced worldwide through the 21st century and beyond. Within the theme of worldwide sustainability, the petroleum option soon wanes, as petroleum cannot even provide present transportation fuel needs without frequent price fluctuations.

Coal is often cited as offering 600 years of reserves, but most of those reserves are of such low quality that they are not considered recoverable. If the United States were to use domestic recoverable coal reserves to meet all U.S. energy needs, the coal would be depleted in 90 years. If coal were to replace petroleum and natural gas on a world scale, the reserves would be depleted in a few decades. Today, ready-to-use uranium fuel costs about $0.62 per delivered MBtu of heat (assumptions on U-235 content and 3.4% fuel burnup) as compared to coal at $1.29. The price of coal is stable, but on a decade timescale it will begin to increase. On the other hand, if new reactor technologies were to increase the burn of uranium from 3.4% to 34%, the price of new uranium fuel could decrease to $0.062 per delivered MBtu (assuming same fuel rod composition). Nuclear alone emerges as a proven source capable of providing abundant energy to the world through the 21st century and beyond.

This narrative is written to provide an introduction to sustainable energy science and technology that may help engineers, scientists, and policy makers better meet the challenges of providing an abundant supply of sustainable energy. An overview of energy options provides a case for nuclear power, and a review of transportation energy options shows how abundant electrical power can also provide sustainable and abundant energy for transportation. We will examine current nuclear technologies and approaches to nuclear power that allow treatment of nuclear waste for safe, long-term storage.

The story starts with the concept of energy and how all forms of energy have a common origin. These common features include an origin in the energy of the atom. All energy on earth originated from atomic energy. Through the years, energy has degraded and has been stored, and we have learned to use it.

Energy on Planet Earth

The sun's light was yesterday's atomic energy. The energy stored in wood and vegetable oils was yesterday's sunlight. Yesterday's wood and vegetable oils are today's coal and crude oil. Yesterday's coal and crude oil are today's natural gas. An examination of these natural energy stockpiles and their history provides a basis for subsequent sections on technologies that use these energy reserves. Chapters 2 and 3 describe how energy reserves were formed and the quantity of these reserves.

Nature used time to transform sunlight to wood, oil, coal, petroleum, and natural gas. Today, man can transform these reserves in a matter of hours. Relatively simple processes for converting petroleum into gasoline have evolved into technologies that allow coal to be taken apart and put back together at the molecular level. Fuel cells can convert the chemical energy of hydrogen or methane directly to electricity without combustion.

Understanding the advantages and disadvantages of nature's various energy reserves requires an understanding of engines and power cycles. The text on gasoline engines explains how these machines work. Likewise, processes for converting coal into electricity that took centuries to develop can be quickly explained.

At the start of the 20th century, suitable liquid fuels were rare, and the proper match of a fuel with an engine was an art. Today, we can move vehicles or produce electricity from energy originating in petroleum, coal, natural gas, wood, corn, trash, sunlight, geothermal heat, wind, or atomic energy. Each can be used differently. Natural gas, for example, can be used directly in spark-ignition engines, converted to gasoline fuel, converted to diesel fuel, converted to hydrogen fuel, or used to produce electricity. The countries that have actively advanced these technologies now have the highest standards of living (see box, "Instruments of Change").

With these multiple energy sources and hundreds of ways to use them, is there one homemade technology-ready alternative to replace petroleum? Is at least one alternative cost competitive with more than $200 billion in the crude oil that we import each year? If we were able to make this one transformation, many of our international problems and fluctuations in the economy would be replaced with increased national security and a more robust economy.

Instruments of Change

It took man centuries to develop the metals used for today's energy machines. At the start of the 20th century, with the materials (primarily steel) in place, high-performance machines were developed in a matter of years. Having the right materials available was necessary for the advances, but it alone was not sufficient.

One of the biggest differences between the time period before and after the year 1900 was the role governments played in fostering technical development. Largely starting with the aircraft and motorized vehicles of World War I, federal appropriations to improve military capability have driven progress in the machines for energy conversion. Governments provided the funding and dedication to fully develop machines to help meet national objectives such as winning a war. The benefits extended far beyond the war.

The steam turbine designed to power warships in the early 1900s produced improved turbines for domestic electric power production. Gas turbines used to drive air compressors in military jets at the end of World War II were used in commercial jets after the war. The early 1940s Manhattan Project produced the atomic bomb, and about ten years later, the same nuclear science produced the first nuclear reactor that replaced the diesel engine in submarines.

Taking concepts out of the laboratory and into public use typically comes with a high price tag. Today we benefit tremendously from the developments undertaken by governments in the 20th century. First-world governments recognize the need for basic research. Support for basic research is essential to maintain their leadership status.

One of the greatest challenges of capitalism in the 21st century is to continue world-impacting technologies without the motivation of world conflict to take the technologies forward. In what could be referred to as a social experiment of the late 20th century, governments have relied more on major companies to develop technology. Major corporations seldom assume the cost of long-term research. Progress has slowed in many areas. This social experiment appears to be failing.

What Are the Right Questions?

The process for unlocking the potential of technology starts with asking the right questions. Both history and science have a story to tell. In 1940, Germany was converting coal into the highest-quality diesel and jet fuel, and it was able to sustain this industry (aside from allied bombing) using coal that was considerably more costly than the vast, rich reserves of today's Wyoming coal. Wyoming has vast supplies of coal in 40-foot-thick seams just feet below the surface. It could be harvested for a few dollars a ton as quickly as it can be loaded into trucks.

Synthetic fuel production, as an alternative to crude oil, was sustainable in Germany in 1940. Why is it not sustainable today with cheaper coal, 60 years of scientific and technological advance, and pipeline distribution that does not rely on costly petroleum tanker shipment from the other side of the world? Originally, the German synthetic fuel process was designed to produce refinery feedstock. Can the synthetic fuel industry leapfrog the competition by producing a fuel that can be directly used in engines? If the refinery could be bypassed, the cost advantages of synthetic fuels are advanced over petroleum alternatives at current prices.

South African synthetic fuel (known as Fischer-Tropsch) facilities were able to sustain production of synthetic oil from coal while in competition with world crude oil prices at $10 per barrel in the late 1990s. Canadian syncrude facilities are reported to be producing petroleum from oil sands at $20 per barrel. The oil sand reserves are about the same size as world reserves of petroleum. Today, Canadian oil sands are used instead of imported oil—the technology is sustainable and profitable.

Why have South Africa and Canada been able to incubate these industries during the past few decades, while the United States failed and is to this day without a significant synthetic fuel industry to replace crude oil imports that exceed $200 billion per year? Lack of competitive technology is not at fault.

Repeatedly, U.S. voters have given the mandate to foster cost-competitive alternatives to imported petroleum. Do U.S. policies foster the development of replacements for petroleum, or do U.S. policies lock in competitive advantages for petroleum over alternatives? When you get past the hype of fuel cells, ethanol, and biodiesel, a comparison of U.S. tax policies on imported crude oil relative to domestic fuel production reveals practices that favor crude oil imports. These and similar policies are the economic killers of the technology that can eliminate the need to import fuels and create thousands of quality U.S. jobs.

In the arena of alternative fuels, both the liquid fuel distribution infrastructure and the refinery infrastructure are controlled by corporations with a vested interest in gasoline and diesel. Because of this and other barriers to commercialization in the United States, the most likely options to succeed are those that do not rely on a new fuel distribution infrastructure. These two options are electrical power and natural gas, and, of these, natural gas imports are rapidly rising.

Natural gas provides a limited advantage over petroleum, and recently the price of natural gas per unit of energy has exceeded the price of gasoline. Electrical power provides a domestic alternative that does not rely on a new fuel distribution infrastructure—a reliance on diverse indigenous energy supplies creates stability in prices and reliability in supply. Electricity is the one option that can substantially replace petroleum as a transportation fuel. Of the options to produce electrical power, nuclear stands out due to its abundance, and its fuel supply provides electrical power without the generation of greenhouse gases.

The utility of electrical power is extended to automobiles with "plug-in" hybrid electric vehicles (PHEVs). PHEVs are able to use electrical power and replace all imported oil without producing air pollution. Use of PHEVs could reach cost parity with conventional gasoline vehicles in a matter of months if development and production of the technology were made a national priority. In a decade of evolution the average consumer could save $1,000 to $2,000 over the life of a vehicle using these technologies rather than conventional gasoline engines.

Having missed the entry positions on technologies like Fischer-Tropsch fuels and Canadian tar sands, is PHEV technology now an opportunity? If PHEV technology is the right opportunity at the right time, is it also the last real opportunity before other nations challenge U.S. economic might?

Through PHEV technology, sustainable electrical power generation has far-reaching consequences beyond the continuation of abundant and reliable electrical power. Sustainable electrical power generation is the key to sustainable transportation and an end to reliance on imported petroleum. This narrative presents a case for nuclear power to meet this opportunity.

Of all the energy technologies, nuclear energy is probably the most misunderstood. Nuclear energy can be produced safely, and we understand the technology well enough to minimize the risk of even the worst-case accident. However, the handling of nuclear waste is a thorn in the side of the nuclear industry.

From the fundamental perspective, the issue of nuclear waste is rather ironic. One question is obvious: Can the really nasty, fission products in spent nuclear fuel be separated from the bulk of the waste—the bulk being considerably more benign and quite valuable? The answer is surprising.

Sustainable Nuclear Power

Figure 1-1 summarizes the legacy of 30 years of nuclear power production in the United States. While much attention has been paid to the radioactive waste generated by commercial nuclear power, the fact is that 30 years of fission products from all the U.S. facilities would occupy a volume less than the size of a small house. On the other hand, the inventory of stockpiled fissionable material in the form of spent fuel and depleted uranium could continue to supply 18% of the electrical power to the United States for the next 4,350 years. This fuel inventory is a most valuable resource and represents material that has already been mined, processed, and stored in the United States.

Reprocessing spent nuclear fuel emerges as the key to sustainable, abundant, and cheap electricity. Reprocessing is removing the really nasty fission products in spent nuclear fuel... the bulk being considerably more benign and valuable. The removal of the fission products is easier (its chemistry) than concentrating the

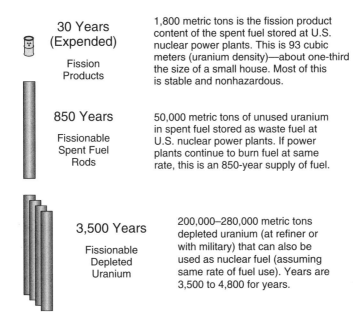

FIGURE 1-1. The legacy of 30 years of commercial nuclear power in the United States, including 30 years of fission products that are of little value and sufficient stockpiled fissionable fuel to continue to produce electrical power at the same rate for another 4,350 years.

fissionable uranium isotope (isotope enrichment) used to convert natural uranium into fuel-grade uranium. The energy inventory summarized by Figure 1-1 is available through chemical reprocessing of spent nuclear fuel and use in Generation IV nuclear reactors. The technologies can actually destroy the nuclear waste generated by the first generation (Generation II) of commercial nuclear reactors. This electrical power can be generated with little to no greenhouse gases while reducing the quantity of spent fuel stored at the power plant facility. Fission products make up 3.4% of spent fuel, and only 0.3% to 0.5% of that represents long-term radioactive waste.

Technologies are available that allow nuclear power to meet every aspect of sustainability. The abundance of uranium that has already been mined will produce energy much longer than scientists can reasonably project new energy demands or sources.

The technologies that enable electricity to meet the demands of transportation and heating markets are covered in Chapters 7, 8, and 9. In Chapter 10 the key concepts behind nuclear science are introduced. Reprocessing technology is covered in Chapter 11, and nuclear reactor technology is covered in Chapter 12. Finally, in Chapter 13, the economics of current generation and Generation IV nuclear power production are covered with the goal of identifying key technologies that will deliver the potential of nuclear power.

CHAPTER 2

The History of Energy

As you sit in a climate-controlled room, light available at the flip of the switch, eyes roaming the page, lungs breathing in and breathing out, heart beating, and electrical pulses of your brain contemplating the words on this page, it is easy to take the *energy* that powers your world for granted. Without energy there would be no light, no book, no life. Mankind is surfing on a wave of energy that was initiated at the dawn of the universe.

Energy

The past one hundred years are like the blink of an eye in the life of humanity, and yet, within this blink, scientists have unraveled the history of energy. This story goes hand in hand with the history of the universe. Following energy back in time takes you to the origin of the universe.

Your body is powered by the energy stored in the chemical bonds of the food you eat. The energy in this food is readily observed by taking a match to a dried loaf of bread and watching it burn. Both your body and the fire combine oxygen and the bread to form water and carbon dioxide. While the fire merely produces heat in this reaction, your body uses the energy in a very complex way to move muscles and produce the electrical energy of your nervous system to control motion and thought.

Both your body and the fire use the chemical energy stored in the starch molecules of the bread. This energy is released as chemical bonds of starch and oxygen are converted to chemical bonds in water and carbon dioxide. Even the molecules your body retains will eventually revert back to carbon dioxide, water, and minerals.

The energy in the chemical bonds of the food came from photosynthesis, which uses the energy from the sun to combine carbon dioxide and water to produce vegetation and oxygen. While the oxygen and carbon stay on the earth and cycle back and forth between vegetation and the atmosphere, the sun's radiation has a one-way ticket into the process where it provides the energy to make life happen. Without this continuous flow of energy from the sun, our planet would be lifeless.

The radiation that powers the photosynthesis is produced by the virtually endless nuclear reaction in the sun. In this process, hydrogen atoms combine to form helium. When hydrogen atoms join to form more stable helium, the total mass is reduced. The lost mass is converted into energy according to Einstein's equation, $E = mc^2$. Enough mass was formed during the birth of the universe to keep the stars shining during the past 15 to 20 billion years.

The presence of different elements in our planet, solar system, and galaxy reveals energy's history. All forms of energy on Earth originated at the birth of the universe. Our life and the machines we use depend on energy's journey, catching a ride as the energy passes by. We are literally surrounded with energy in hydrogen, uranium, and chemical bonds with limits of our use of this energy largely determined by our choices and, in some cases, our pursuit of technology to better utilize these resources.

Nature's Methods of Storing Energy

All forms of energy, whether nuclear, chemical energy in coal, chemical energy in petroleum, wind, or solar, are part of energy's journey that started with the birth of the universe. In our corner of the universe, the energy output of the sun dwarfs all other energy sources. Nuclear fusion in the sun releases massive amounts of energy. The only way this energy can escape from the sun is in the form of radiation. Radiation output increases as temperature increases. Somewhere along the journey, the sun came into a balance where the sun's radiant energy loss tends to decrease the temperature at the same rate as the nuclear fusion tends to increase the temperature. In this process, the outward force of the constant nuclear explosions is balanced by the sun's gravitational force to form a nearly perfect sphere.

Before life evolved on Earth, the sun's energy reached Earth. If you close your eyes and look at the sun, your face receives the warmth while the top of your head receives little. The radiation causes Earth's equator to be warmer than the poles. These temperature differences cause wind and ocean currents. See the box "The Nature of Wind."

> **The Nature of Wind**
>
> The principal is easy to understand. At warm locations like the Texas coast, warm water rises in the oceans and warm air rises in the atmosphere. The space is filled from the flow of cooler water or air coming in from the sides. The cooler fluid is now warmed and keeps the process going.
>
> Meanwhile, at colder locations like Greenland, the warmer air high in the sky or on the ocean surface is cooled and displaces the cooler fluid below. The cooler fluid moves outward to areas where fluids are rising.
>
> Water tends to amplify wind patterns, since water vapor, evaporated by the sun, is even lighter than heated air at the same temperature. When water vapor cools, it can become a thousand times more dense by condensing into rain or snow. This can happen on massive scales like the Gulf Stream or on smaller scales, like the flow of air and moisture that keeps the skies clear of clouds on the Costa De Sol in southern Spain.

Before life existed on Earth, the sun's radiation formed water vapor and caused it to rise from the oceans. This water vapor caught the wind and was blown to the mountains, where it cooled to form rain. The high elevation of this water in the mountains gave it energy to flow downhill. Rocks and gravel dissipated this energy on its journey back to the oceans. The potential energy from water's height in the mountain is converted to thermal energy that is reflected by a slight increase in the temperature of the water as is progresses from the mountain to the ocean.

The first primitive organic life appeared on Earth about 3 billion years ago, with photosynthesis first occurring about 1 billion years later.[1] The vast majority of the vegetation fell to the ground and decomposed, combining with oxygen and going back to carbon dioxide and water. Some fell to the floor of swamps, where oxygen could not reach them as fast as they piled up. These deposits were buried deeper and deeper, making it even more difficult for oxygen to reach them and convert them back to carbon dioxide and water. After a sufficiently long time, the vegetation rearranged into more stable deposits that we call coal. Different types of coal developed, depending on the depth, temperature, and moisture of the deposits. This preservation process was particularly effective in swamps, where the water reduced the rate at which oxygen could reach the fallen vegetation.

In the seas, much of the surface was inhabited by bacteria called phytoplankton (small, floating, or weakly swimming animals or plantlife in water). The cells of these phytoplankton contained oils that are in some ways similar to the corn oil used to cook french fries. When these phytoplankton died, most of them were converted back to carbon dioxide and water by the oxygen dissolved in the water or by feeding animals. Some were swept to ocean depths, where there was little oxygen, and they accumulated. Some of these bacteria deposits were buried by silt. The passage of time and the pressure from the overburden of water and silt transformed these deposits into petroleum oil.

In the turmoil of erosion, volcanoes, and general continental drift, large deposits of coal and oil made it back to the surface, where, in contact with oxygen, they oxidized back to water and carbon dioxide. Other deposits persist for us to recover. Still other deposits were buried deeper, reaching higher pressures and higher temperatures, due to Earth's geothermal heat. There, the coal and petroleum converted to a combination of natural gas and high-carbon deposits of hard coal or carbon in the form of graphite.

Over tens of millions of years, the sun's energy, working with the life on Earth, formed energy deposits of coal, petroleum, and natural gas. Currently, yesterday's radiation is available as vegetation such as wood, corn, and palm oil. Today's radiation is available as sunlight, wind, ocean currents, and the hydroenergy of water in high-altitude rivers and lakes.

The legacy of the universe is all around us. Compared to our consumption of energy, the fusion energy available in the hydrogen of the waters of the ocean is almost endless. Uranium available in the soil and dissolved in the ocean can produce energy by fission. All atoms smaller than iron could be fused to form iron, while all atoms larger than iron could undergo fission (splitting) to form iron. Both processes involve the nucleus of atoms releasing vast amounts of energy.

The geothermal heat of the Earth originated at the birth of the universe. The cosmic forces at the beginning formed the atoms that collected, formed rocks, and became Earth with a molten center. If the heat were left unreplenished, the core of the Earth would have long since cooled. Adding to the heat of colliding masses, uranium and other larger molecules are constantly undergoing nuclear rearrangements (including fission) from Earth's surface to its core. The fission energy release occurs one atom at a time, but the energy adds up. The released heat maintains molten magma from Earth's core to near the surface. On the surface we see this energy released as volcanic eruptions and geysers.

It is important to recognize that nuclear conversions have played a vital role in the evolution of life on Earth and continue to maintain Earth's molten core. A natural nuclear reactor actually formed in Oklo, Gabon (Africa), about 2 billion years ago. This occurred due to the concentration of U-235 in ore at Oklo: The high concentrations caused a fission chain reaction of the U-235, leading to lower than normal U-235 concentrations and trace plutonium found in those deposits today.

We did not introduce nuclear processes (or even nuclear reactors) to Earth. We merely learned to control nuclear processes and harness the energy. We have options on where we can tap into energy's journey to power our modern machines, and halfway through the 20th century, nuclear power became one of the options to meet rapidly increasing energy demands.

Man's Interaction with Nature's Stockpiles and Renewable Energies

Primitive man was successful in tapping into the easily available and easily usable forms of energy. He lived in warmer climates where the solar warmth protected him from the cold. Even the most primitive animals, including early man, recognized the need to nourish their bodies with food.

As the use of fire developed, man was able to move into colder climates, where the energy in wood was released by burning campfires. Animal fat and olive oil were soon discovered to be useful sources of fuel to feed the fire for heat and light. These fats and oils were observed to burn longer and could be placed in containers or wrapped on the end of a stick to create a torch—hence, they are early endeavors into fuel processing. Whale blubber was added to animal fat and olive oil for food and fuel.

The wheel and axle were another early step in developing energy technology. The wheel and axle assisted man to use his physical energy to move heavier loads. The cart was made more effective by using domesticated animals. For stationary applications, water wheels and windmills converted the hydraulic and wind energies into shaft work for many applications including pumping water and grinding grain. Wind energy powered ships to explore new lands, establish trade, and expand the fishing industry.

Machines using wind and hydraulic energy made it possible for one person to do the work of many—freeing time for them to do other tasks. A most important task was educating the young. Time was also available for the important tasks of inventing newer and better machines. Each generation of new machines enhanced man's ability to educate, invent, discover, and add to leisure time.

Societies prospered when they used the freedom created by machines to educate their youth and to create new and better machines. Inventions and discoveries extended to medicines that conquered measles and polio. The benefits of modern society are available because of the effective use of energy and the way energy-consuming machines enhanced man's ability to perform routine tasks, freeing time for education, discovery, and innovation. Civilization emerged and prospered.

History shows that civilization evolves based on technology. For man, the "survival of the fittest" is largely the survival of the culture most able to advance technology. In modern history, while Hitler's technology dominated the World War II battlefields, Germany was winning the war. As the Allies' technology surpassed Germany's technology, the Allies began to dominate the battlefields, leading to victory.

If you drive through the Appalachian Mountains, you will observe how coal seams (varying from an inch to over a foot thick) once buried a few hundred feet in the Earth are now exposed on cliffs. The upheaval that created mountains also brought up deposits of coal and oil. At cliffs like these, man first discovered coal. Coal was considerably easier to gather at these locations than firewood, and eventually coal replaced firewood. Marco Polo observed "black rocks" being burned for heat in China during his 1275 travels.[2] Coal's utility caught on quickly. Between 1650 and 1700, the number of ships taking coal from Newcastle to London increased from 2 to about 600.[3] In 1709 British coal production was estimated to be 3 million tons per year. Benjamin Franklin noted in 1784 that the use of coal rather than wood had saved the remaining English forests, and he urged other countries to follow suit.

When oil was found seeping from the ground, it could be collected and used to replace an alcohol-turpentine blend called camphene (camphene being less expensive than whale oil).[4] Eventually, mining and drilling techniques were developed to produce larger deposits of coal and oil found underground.

From an historic perspective, energy technology has tended to feed upon itself and make its utility increase at ever-faster rates. Large deposits of coal made it possible for just a few people to collect the same amount of fuel that it took an entire community to gather a few centuries earlier. Easy and efficient gathering of fuel freed up more time and resources to develop new and better machines that used the dependable fuel supply.

Prior to the 19th century the decisions to proceed with newly demonstrated technology were easy because the benefits were obvious. The vast amounts of virgin wilderness dwarfed the small tracts of land that had been devastated by poor mining practices, and an energetic entrepreneur could simply go to the next town to build the next generation of machines as local markets were dominated by local businesses.

At the end of the 19th century, vast tracts of land or oceans were no longer barriers to the ambitions of the people managing corporations. The telegraph allowed instant communication and steam engines on ships and locomotives helped them reach their destinations in a matter of days. Suddenly, budding entrepreneurs could no longer travel to the next town to get outside the influence of existing corporations. In energy technology, companies became monopolistic energy empires.

The growth of local businesses into corporations with expanding ranges of influence made their products quickly available to more people. The benefits were real, but the problems were real, as well. One problem was that innovation was being displaced with business strategy and influence, determining which technologies would be developed.

For energy options to be commercialized today, both technical and nontechnical barriers must be overcome. The nontechnical barriers generated by corporations and their far-reaching political influence can be even greater than the technical barriers. These nontechnical barriers must be understood and addressed.

The Industrial Revolution and Establishment of Energy Empires

The Standard Oil Monopoly

The Standard Oil monopoly of the early 20th century demonstrated what happens when a corporation loses sight of providing consumers a product and becomes overwhelmed with the greed for profit.

After the civil war, men swarmed to western Pennsylvania to lay their claims in a "black gold rush." John D. Rockefeller was among these pioneers. Within one year of discovering the potential for drilling for petroleum, the price went from $20 per barrel to 10 cents per barrel. Rockefeller realized that the key to making money in oil was not getting the oil out of the ground but rather refining and distributing the oil.[5] Standard Oil crossed the line when it changed its corporate philosophy to one of profiting by stifling the competition through monopolistic control of refining and distribution of all petroleum products.

The Atlantic and Great Western Railway controlled the cheap rail transit in the western Pennsylvania region, and this controlled the oil market. Rockefeller prevented his competition from using this railroad to sell their oil.

This was a change in paradigm for the energy industry. A company controlled the market by controlling access to the commodity. While nations and shipping fleet owners had done this in the past, this was different. This was a company operating in a free country aggressively moving to eliminate all competition.

Artificially inflated oil prices were just like taxes without representation. When previously faced with a similar situation, the people united in the American Revolutionary War. Here, the adverse impacts of the oil monopoly were difficult to quantify, unlike a tax on tea, and there was no precedent to show the way to reasonable remedies.

Competitors of Standard Oil were stifled. Some of the competition sold out. High consumer prices—higher than the free market would bear—were the result of this monopoly. Technology and innovation were also stifled. Corporate success was determined by controlling access to the products, not by the best technology.

In the past, improving technology benefited both the consumer and the company. When business savvy replaced innovation, the actions of the company were at odds with what was good for the consumer. The creative innovation and technology were now forced off the highway of ideas onto the back roads where progress was slow.

For 32 years Standard Oil profited from its monopoly on oil refining. In May 1911, U.S. Supreme Court Chief Justice White wrote the decision that mandated that Standard Oil must divest itself of all its subsidiaries within six months. In 1974 assets of the descendants of John D. Rockefeller were estimated to have the largest family fortune in the world, estimated at $2 billion.[5]

During the 20th century, pioneers had reached the end of habitable frontier. The steam engine and telegraph provided the means by which companies could extend their influence across the globe. One can argue that the international nature of today's mega corporations elevates them to a status as great as the nations they claim to serve. One can further argue that a mega corporation can be a friend or an enemy to a society in the same sense that a neighboring country can be a friend or an enemy.

In 1942 Senator Harry S. Truman led an investigating committee on treasonous prewar relationships between General Motors, Ethyl Corporation, Standard Oil, and DuPont in collaboration with the German company I. G. Farben. Company memos document corporate agreements designed to preserve the corporations no matter which side won World War II. Corporate technology exchanges compromised the competitive edge held by the United States as it entering World War II, including leaded gasoline technology (critical for high-octane aircraft fuels) and noncompetitive stances on synthetic rubber technology. At the same time, British intelligence called Standard Oil a "hostile and dangerous element of the enemy."[6,7,8] Continuation of this behavior led to anticompetitive-related antitrust hearings on leaded gasoline technology against these American companies in 1952.

With technology and innovation taking a backseat to business interests, politics and energy technology became perpetually intertwined. The larger companies were formally pursuing their agendas, even when these agendas were in conflict with public interest and involved collaboration with the enemies of our nation's closest allies.

A corporation that profits by providing consumer products more efficiently and at a lower cost is significantly different than a corporation that makes profit by controlling the supply or price of a consumer commodity or product. The Rockefeller Oil monopoly demonstrated that the profits were greater for the business strategy that is in conflict with national benefit. In the end, the only punishment was the mandated divestment of the Rockefeller Oil subsidiaries. The Rockefeller family emerged as the wealthiest family in the world.

What was the real precedent set by the Standard Oil monopoly? Was it that monopolies will not be allowed, or was it that great fortunes can be made and kept even if you are caught? The consequences are that business practices and not technical merit tend to have more and more impact on which technologies become commercial and ultimately benefit the public. With the introduction of the corporate lobbyist, technical merit is debatably in at least third place in this hierarchy.

Innovation in a World of Corporate Giants

There is little doubt that obstructed commercialization of technology stifles technology innovation. Companies and individuals have little incentive to build a nuclear-powered automobile, since the government would not allow this vehicle to be used on the highways (for good reason). Restricted commercialization can be good by redirecting efforts away from projects that endanger society. Restricted commercialization can be bad when the motivation is to maintain a business monopoly and to stifle competition.

History has shown little evidence of the impact of unrestricted entrepreneurism because modern history has been dominated by business rather than technical innovation. The true potential of unrestricted entrepreneurism is rarely seen. World War II is the best example in recent history illustrating what happens when we focus on developing the best technology available and the machines to get a job done. Within a ten-year period, the following technologies were developed:

- Nuclear bomb
- Jet aircraft
- Radar
- Transistors
- Intercontinental rockets
- Guided missiles
- Synthetic oil produced from coal
- Mass production of aircraft and tanks
- Swept wing and flying wing aircraft
- Stealth submarine technology
- Plastics industry, including synthetic rubber, nylon, and synthetic fiber

Many of the commercial advances between 1946 and 2000 occurred because decisive technology was developed during World War II. Some 20th-century accomplishments that fall into this category are nuclear power, jet air travel, landing a man on the moon, guided missile technology, transistor-based electronics and communication, stealth aircraft, including the B1 bomber, and the modern plastics industry.

The technological developments of World War II illustrate what can be achieved when technology and commercialization become a national goal. If technology has inherent limits on the good it can provide; we are far from reaching these limits. We are limited by mankind's pursuit, by the willingness of environmental groups to allow new technologies to become part of our societal infrastructure, and by the willingness of corporations or governments to invest in development and commercialization.

At the beginning of the 21st century, politics and energy technology are hopelessly entangled. Nine out of the top ten of the Global 500 companies are in energy or energy technology (e.g., the automotive industry), and business savvy trumps innovation in these companies.

The United States rose to superpower status in the 1940s when the national focus was on developing and commercializing strategic technologies. History has shown powerful countries fall when the national focus switches from advancing technology to maintaining the steady flow of cash to corporations and well-connected individuals (maintaining corporate status quo). The czars of Russia or aristocrats of Rome are two of many examples where common people were driven to revolt against a system dominated by who you were rather than what you had contributed.

Germany's synthetic oil production from coal is one strategic technology that did not become commercial in the United States after World War II. This same technology is currently commercial in South Africa because there was a different philosophy toward investing in this infrastructure. Today in South Africa the industry is self-sustaining and the technology is being sold for use in other countries. These production facilities were designed and constructed by U.S. engineering firms.

Canada started developing its oil sand resources in the 1960s, even though the technology could not undercut the price of crude oil. Because of continued and dedicated development, the oil sand oil now costs $20 per barrel to produce as compared to crude oil at over $50 per barrel (year 2005) on the world market.

The Oil Economy

The 21st-century U.S. civilization as we know it would be impossible without crude oil. We get over 90% of all automotive, truck, train, and air transport fuels from crude oil, as well as the majority of our plastics. The plastics are used to make everything from trash bags and paints to children's toys. If it is a solid device that is not paper/wood, ceramic/glass, or metal, it is probably plastic.

Crude oil is a good fuel. Figure 2-1 illustrates how this natural product can be separated by boiling point range to provide gasoline, a middle fraction (kerosene, jet fuel, heating oil, and diesel fuel), and fuel oil (oil used for boilers or large diesel engines for ships and electrical power plants). The natural distribution of crude oil into these three product classifications varies with the source of the crude oil.

Modern refining processes convert the crude oil into these three product categories and also provide chemical feed stocks. The modern refining process breaks apart and rearranges molecules

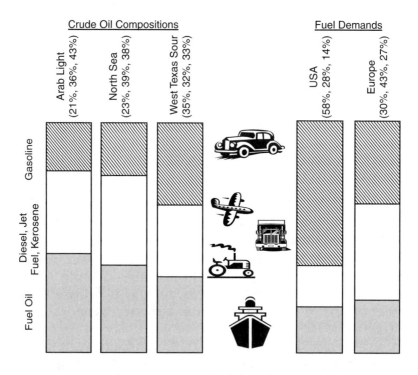

FIGURE 2-1. Typical composition of oil. (Keith Owen and Trevor Coley, *Automotive Fuels Reference Book*, 2nd ed. Warrendale, PA: Society of Automotive Engineers, 1995.)

to produce just the right amounts of gasoline, diesel, and fuel oil. In the United States, this typically means converting most of the "natural" fuel oil fraction and part of the middle fraction to increase the amount of gasoline.

Figure 2-2 is a detailed description of crude oil processing for the United States and shows import versus domestic production and other commercial applications. The figure also illustrates how complex crude oil processing is when it comes to providing commercial product demands.

Figures 2-3 and 2-4 show the extent of U.S. oil imports and how prices fluctuate. Since the year 2000, the United States has been spending over $100 billion per year to import crude oil (estimated at over $200 billon per year in 2005). Cheap oil in 1997 and 1998 brought a prosperous U.S. economy; more expensive oil in 2001 and 2002 added to the economic slump.

As illustrated in Figure 2-2, the United States imports well over half of the crude oil it processes. One can argue whether the world will have enough oil for the next 100 years or only the next 25 years. One thing is certain: The useful and significant domestic oil production in the United States would last less than ten years (about 7.6 years if there were no imports), and it will continue to decline. Over half the world's known oil reserves are in the Middle East. Figure 2-5[9] shows the breakdown, with Saudi Arabia, Iran, and Iraq with the greatest reserves.

Energy Sources

Petroleum provides more than 90% of vehicular fuels in the United States, but in addition petroleum represents 53% of all energy consumed in the United States, as shown in Figure 2-6. The energy stocks used in the United States are in sharp contrast to U.S. reserves (see Figure 2-7), and this will ultimately lead to energy crises.

If the world continues energy consumption at its present rate, it has approximately 3.6^{i} years of petroleum (to supply all energy needs), 17 years of coal, 46 years of natural gas, and millions of years of uranium (assuming full use of uranium and ocean recovery).[10] If the United States were the sole consumer of world energy reserves, world petroleum would last 75 years toward meeting all the U.S. energy needs, coal 500 years, natural gas 1,000 years, and uranium tens of thousands of years.

---

^{i} For world oil reserves of 5.3e18 Btu and world total energy consumption of 1.5e18 Btu/yr.

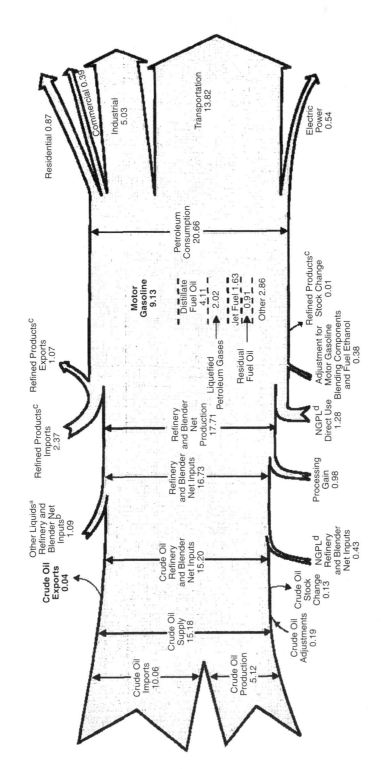

FIGURE 2-2. Year 2000 oil flow in quadrillion Btu. (U.S. DOE Public Domain Image. http://www.eia.doe.gov/pub/oil_gas/petroleum/analysis_publications/oil_market_basics/default.htm, May 17, 2002.)

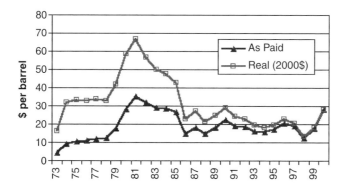

FIGURE 2-3. Recent U.S. oil prices. (U.S. DOE Public Domain Image. http://www.eia.doe.gov/pub/oil_gas/petroleum/analysis_publications/oil_market_basics/default.htm, May 17, 2002.)

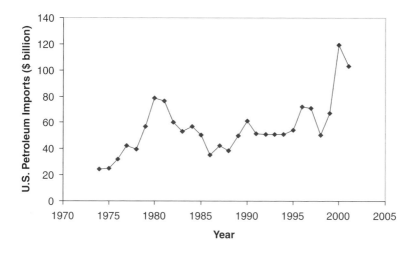

FIGURE 2-4. U.S. imports of petroleum. (http://www.eia.doe.gov/emeu/mer/txt/mer1-6, May 17, 2002.)

In the year 2000, the United States imported 53% of its petroleum to satisfy a consumption of 19.7 million barrels per day. If total U.S. demand for petroleum had to be met with known reserves in the United States, we would run out in about three years. The strategic petroleum reserve (most stored in salt caverns) would last a total of 31 days.[11]

It is where consumption exceeds availability that technology makes the difference. There is little doubt that the world's demand for petroleum is rapidly exceeding the availability of petroleum.

24 Sustainable Nuclear Power

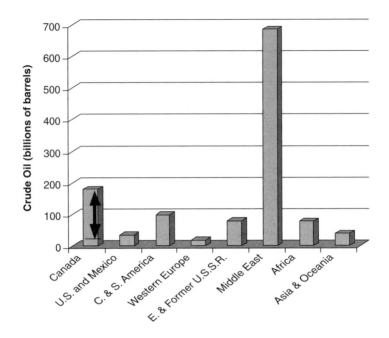

FIGURE 2-5. World oil reserves by region. Estimates of Canadian reserves by *Oil and Gas Journal* in 2003 are much higher than in previous years. Most likely they include easily recovered oil sands. (PennWell Corporation, *Oil and Gas Journal*, Vol. 100, No. 52 (December 23, 2002).)

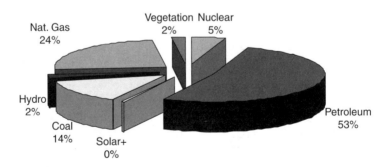

FIGURE 2-6. U.S. energy consumption by source.

The use of petroleum as currently practiced is not sustainable. From the position of available energy reserves (see Figure 2-7), coal and nuclear are the obvious choices to replace petroleum. The right combination of technologies can make the difference between security or vulnerability, between cheap energy or economic recession due to restricted oil supply.

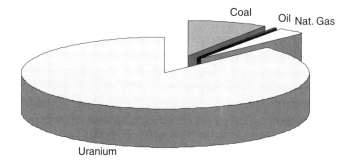

FIGURE 2-7. Estimate of U.S. energy reserves.

Environmental Impact

Environmentalism has a rich tradition of keeping industry under control. History bears witness to the devastation caused by deforestation. As early as 6000 B.C., the collapse of communities in southern Israel were attributed to deforestation.[12] In southern Iraq, deforestation, soil erosion, and salt buildup devastated agriculture by 2700 B.C.[13] The same people repeated their deforestation and unforgiving habits in 2100 B.C., a factor in the fall of Babylonia. Some of the first laws protecting timbering were written in 2700 B.C.[14]

Advances in citywide sanitation go back to at least 2500 B.C. and can be attributed to people uniting in an effort to improve their environment against the by-products of civilization. In 200 B.C. the Greek physician Galen observed the deleterious acid mists caused by copper smelting. Lead and mercury poisoning was observed among the miners of the A.D. 100 Roman Empire, and high levels of lead may have been a factor in its fall. The bones of aristocratic Romans reveal high levels of lead, likely from their lead plates, utensils, and, in some instances, food.[15]

Poor sanitation during the Dark Ages, including raw sewage and animal slaughter wastes littering the cities, contributed to both the Bubonic Plague and cholera. In these ancient cases, the factors that allowed civilization to attain its magnificence also presented new or recurring hazards. In this early history, the lives saved by the benefits of agriculture and metal tools far outweighed those lives lost or inconvenienced due to adverse environmental impacts.

The dark smoke of coal burning became evident as a significant problem in the 13th century. In 1306 King Edward I forbade coal burning in London.[16] Throughout history, the tally of deaths attributed to air pollution from heating and other energy-related technology accumulated. The better-documented of these cases are reported prior to government regulations that finally brought the problems under control.

On October 26 and 31, 1948, the deaths of 20 people along with 600 hospitalizations were attributed to the Donora, Pennsylvania, smog incident. A few of the other smog incidents in the next few years include 600 deaths in London from "killer fog" (1948); 22 dead and hundreds hospitalized in Poza Rica (Mexico) due to killer smog caused by gas fumes from an oil refinery (1950); 4,000 dead in London's worst killer fogs (December 4–8, 1952); 1,000 dead in a related incident in London in 1956; 170–260 dead from New York's smog (November 1953); and in October 1954 most of the industry and schools in Los Angeles were shut down due to heavy smog conditions (a smart, proactive measure made possible by the formation of the Los Angeles Air Pollution Control District in the 1940s, the first such bureau in the United States). In the 1952 London incident, the smoke was so thick that a guide had to walk ahead of buses, with all of London's transportation except subway traffic coming to a halt on December 8, 1952.

In 1955 the U.S. Congress passed the Air Pollution Research Act. California was the first state to impose automotive emission standards in 1959, including the use of piston blow-by recycle from the engine crankcase. The automakers united to fight the mandatory use of this modification that cost seven dollars per automobile. Subsequent federal legislation has been the dominant force on changes in U.S. energy infrastructure during the last 25 years.

The late 1960s has been characterized as an environmental awakening in the United States. Prior to 1968, newspapers rarely published stories related to environmental problems, while in 1970 these stories appeared almost daily.[17] Sweeping federal legislation was passed in 1970 with the Clean Air Act establishing pollution prevention regulations, the Environmental Policy Act (EPA) initiating requirements for federal agencies to report the environmental ramifications of their planned projects, and the establishment of the Environmental Protection Agency. The Clean Air Act was amended in 1990 specifically strengthening rules on SOx and NOx (sulfur and nitrogen oxides) emissions from electrical power plants to reduce acid rain. This legislation ultimately led to the closing of some high-sulfur coal mines.

Beginning with 1968 automobiles, the Clean Air Act (CAA) required the EPA to set exhaust emission limits. Ever since, the EPA has faced the task of coordinating federal regulation with the capabilities of technology and industry to produce cleaner-running vehicles. Since the precontrol era, before 1968, automotive emissions (gasoline engines) of carbon monoxide and hydrocarbons have been reduced 96%, while nitrous oxide (NOx) emissions have been reduced 76% (through 1995, the result of 1970 CAA).[18] The phasing out of lead additives from gasoline was a key requirement that made these reductions possible.

On February 22, 1972, the EPA announced that all gas stations were required to sell unleaded gasoline with standards following in 1973. Subsequent lawsuits—especially by Ethyl Corp., the manufacturer of lead additives for gasoline—ended with the federal court confirming that the EPA had authority to regulate leaded gasoline. Leaded automotive gasoline was banned in the United States in 1996. In 2000, the European Union banned leaded gasoline as a public health hazard.

The removal of lead from gasoline was initially motivated by the desire to equip automobiles with effective catalytic converters to reduce the carbon monoxide and unburned fuel in the exhaust. The lead in gasoline caused these converters to cease to function (one tank of leaded gas would wreck these converters). The influence of energy corporations was obvious when the U.S. Chamber of Commerce director warned of the potential collapse of entire industries from pollution regulation on May 18, 1971. This has been viewed as a classic example of *industrial exaggeration*.

Corporate influence was again seen in 1981 when Vice President George Bush's Task Force on Regulatory Relief proposed to relax or eliminate U.S. leaded gas phaseout despite mounting evidence of serious health problems.

Since the banning of lead in gasoline, scientific communities are essentially in unanimous agreement that the phasing out of lead in gasoline was the right decision. In addition to paving the way for cleaner automobiles, these regulations have ended a potentially greater environmental disaster. All the lead that went into automobiles did not simply disappear—it settled in the soils next to our highways. Toxic levels of lead in the ground along highways continue to poison children and contribute to mental retardation even today. (See the box "Lead (Pb) and Its Impact.")

In perspective, the air quality in our cities is good and generally improving. While the federal government monitors emissions and works to reform emission standards, the public and media tend to follow other issues more closely. Issues such as oil spills and global warming make the news.

> **Lead (Pb) and Its Impact, as Summarized by the EPA**
>
> *Health and Environmental Effects*: Exposure to Pb occurs mainly through inhalation of air and ingestion of Pb in food, water, soil, or dust. It accumulates in the blood, bones, and soft tissues. Lead can adversely affect the kidneys, liver, nervous system, and other organs. Excessive exposure to Pb may cause neurological impairments, such as seizures, mental retardation, and behavioral disorders. Even at low doses, Pb exposure is associated with damage to the nervous systems of fetuses and young children, resulting in learning deficits and lowered IQ. Recent studies also show that Pb may be a factor in high blood pressure and subsequent heart disease. Lead can also be deposited on the leaves of plants, presenting a hazard to grazing animals.
>
> *Trends in Pb Levels*: Between 1988 and 1997, ambient Pb concentrations decreased 67 percent, and total Pb emissions decreased 44 percent. Since 1988, Pb emissions from highway vehicles have decreased 99 percent due to the phaseout of leaded gasoline. The large reduction in Pb emissions from transportation sources has changed the nature of the pollution problem in the United States. While there are still violations of the Pb air quality standard, they tend to occur near large industrial sources, such as lead smelters. Between 1996 and 1997, Pb concentrations and emissions remained unchanged.
>
> Source: http://www.epa.gov/oar/aqtrnd97/brochure/pb.html.

The *Exxon Valdez* oil tanker spill (March 1989) is one of the most onerous oil spill incidents. This oil tanker ran aground in Price William Sound, Alaska, spilling 11 million gallons of petroleum. It is all the more infamous because of the costly remediation/penalties (over $1 billion in fines, with Exxon claiming $3.5 billion in total expenditures) that Exxon was required to perform as a result of this incident. Five billion dollars in punitive damages was also awarded against Exxon, but this remains to be collected after almost a decade.[20] In 1992 the supertanker *Braer* spilled 26 million gallons of crude oil in the Hebrides islands. Both of these incidents are dwarfed by the *Amoco Cadiz* wreck off the coast of France in 1978 with a spill of 68 million gallons.

In view of the *Amoco Cadiz* incident and the cumulative tens of thousands who died in London's killer fogs, it is easy to understand the increased environmental consciousness in Europe compared to that of the United States. For example, the European governments are aggressively addressing potential global warming issues, while the U.S. government tends to withdraw from international cooperation on the issue. Neither the United States nor European governments dispute the fact that carbon dioxide levels are increasing in the atmosphere. They do have varying opinions on the implications of these increasing carbon dioxide emissions.

On June 23, 1988, NASA scientists warned Congress about possible consequences from global warming with potential effects of drought, expansion of deserts, rising sea levels, and increasing storm severity. On December 11, 1997, the Kyoto Protocol was adopted by President Clinton (a Democrat) and 121 leaders of other nations. The Republican-dominated U.S. Congress refused to ratify the protocol. More recent comments by President Bush concerning the Kyoto Protocol sound like the comments and actions of the U.S. Chamber of Commerce director and former Vice President George H. W. Bush on the phaseout of lead from motor gasoline.

What the Future Holds

The complexity of U.S. politics that forced the breakup of Standard Oil in 1911 is now dwarfed by the manner in which energy and international politics are coupled. Operation Desert Storm and the Gulf War in 1991 put all doubt aside. The U.S. government is willing to protect oil supplies with direct military action.

Our transportation systems, fuel oil and propane heating, and plastic materials have unquestionably saved more lives than have been lost in the conflicts fought to keep oil flowing. However, if alternatives like nuclear power or biomass can maintain the benefits of cheap transit, heating fuels, and plastics, these alternatives should be pursued rather than entering into military conflict. Canada and South Africa have certainly demonstrated that crude oil can be cost effectively replaced with coal and oil sands.

Today's politics and government policies inhibit technology. They can also inhibit national prosperity. Understanding this requires learning about energy science, energy conversion technology, and economic/profitability analysis. It requires evaluating the facts and resisting the temptation to look for simple answers. Vested corporate and political interests are ready to capitalize on ignorance providing their public answers.

The summary of energy reserves in Chapter 3 shows that there is no energy shortage. Spikes in gasoline prices, electricity prices (California), and natural gas prices (throughout the United States) that occurred during the first two years of the 21st century cannot be justified based on a shortage of energy resources. Essentially all recent historical spikes in oil prices can be attributed to a lack of competition. It appears that in some cases the consumer was a victim of corporate strategies that created the appearance of shortages.

Two mechanisms produce shortages: government regulations that effectively prevent new capacity from coming on line (or define a fuel that limits potential suppliers) and a lack of diversity in energy feed stocks. Both mechanisms prevent the free market from establishing the price. In electrical power generation, increased use of natural gas presents the opportunity for gas suppliers to limit electrical power diversity, and this increases prices. For vehicular travel, an overdependence on petroleum fuels has already created national and consumer vulnerability.

A number of technologies are available to address today's greatest energy problems, but they challenge the current status quo of the energy industry. On electrical power generation, 100% nuclear fission (reprocessing of spent nuclear fuel) can address problems with both nuclear waste and electrical power diversity.

On petroleum, U.S. tax strategies actually give foreign producers a $7–$15 per barrel competitive advantage (see Chapter 8). The Canadian oil sand industry and U.S. synthetic petroleum (infant) industry are capable of meeting U.S. oil consumption needs at current prices (greater than $25 per barrel). More importantly, if given an equitable tax strategy compared to imported crude oil, these industries could create new-source competition that would tend to stabilize fuel prices at levels lower than current prices fluctuating above $30 per barrel.

Finally, an increasing overlap of electrical power and vehicular fuel energy networks would provide additional diversity to stabilize gasoline prices. The use of electrical power for automobile transportation has an important advantage over Fischer-Tropsch fuels and Canadian oil sands. Electrical power distribution bypasses the refinery and liquid fuel distribution infrastructure that is dominated by oil corporations.

The concept of market overlap is simple. When consumers have the ability to select between powering their vehicles with electrical power or gasoline, the price of gasoline will stabilize. If the fuel cell researchers deliver what they promise, rechargeable fuel cell systems could leapfrog rechargeable batteries for electric cars in commuting applications. Rechargeable fuel cells also have distinct advantages when used in hybrid cars. This topic is covered in greater detail in Chapter 9.

Environmentally Responsible Nuclear Power

The nuclear processes that occur in nuclear fission reactors were not invented by man. These processes have been occurring on the earth since its formation. What is new (within the past century) is our understanding of nuclear processes and the controlled use of nuclear fission to produce electrical power. There is nothing to be gained by avoiding the use of nuclear power, but there is much to be gained by the responsible use of nuclear fission to provide the energy.

When compared to the thousands who have died in the smog produced from coal or the thousands who have died in military conflicts designed to control the flow of petroleum, commercial nuclear power in the United States has proven to be the safest and most environmentally friendly energy source. The statistics shown in Table 2-1 reported by Hinrichs[19] further show that the environmental impact of nuclear power is a small fraction of the impact of coal power. The lessons of history are clear.

George Santayana once said, "Those who fail to learn the lessons of history are destined to repeat them." In energy technology, history's lessons are that more people will die from coal and petroleum utilization—from occupational accidents, military confrontations, and pollution generation—than from nuclear. The environment will also suffer more with coal and petroleum—from oil spills to the buildup of carbon dioxide in the atmosphere. Responsible nuclear power (as in historical U.S. commercial nuclear power) should continue to be safer and more friendly to the environment provided engineers and scientists continue to apply the lessons they have learned in regard to safely designing and operating nuclear power facilities.

TABLE 2-1
Impacts as summarized in 1986 of coal versus nuclear for a 1 GW power plant operating for one year.

	Coal	*Nuclear*
Occupation Health Deaths	0.5–5	0.1–1
Occupational Health Injuries	50	9
Total Public and Worker Fatalities	2–100	0.1–1
Air Emissions (tons)	380,000	6,200
Radioactive Emissions (Curies)	1	28,000

On the topic of nuclear waste, history shows that waste should not be buried (as was irresponsibly done with chemical waste in the early 20th century). Rather, processes should be designed to first minimize waste, and the waste that is generated should be treated to minimize its volume and contained to prevent diffusion into the environment.

References

1. B. P. Tissot, and D. H. Welte, *Petroleum Formation and Occurrence*. New York: Springer-Verlag, 1978, p. 5.
2. http://www.runet.edu/~wkovarik/envhist/2middle.html.
3. Ibid.
4. http://www.runet.edu/~wkovarik/envhist/4industrial.html.
5. Anthony Sampson, *The Seven Sisters*. New York: Bantam Books Inc., 1981, p. 27.
6. http://www.runet.edu/~wkovarik/envhist/7forties.html, May 20, 2002.
7. Stevenson, William, 1976, A Man Called Intrepid. New York: Harcourt Brace Jovanovich.
8. Borkin, Joseph, 1978. The Crime and Punishment of I. G. Farben. New York: Free Press Source for Fig. 2-4 http://www.eia.doe.gov/emeu/mer/, May 17, 2002.
9. Also see http://www.eia.doe.gov/emeu/international/reserves.xls.
10. Consumption at 7.1e16Btu/yr in U.S. and oil, coal, and gas reserves of 5.3e18, 2.5e19, and 6.9e19. World consumption of 1.46e17, 1.18e17, and 1.4e18 for total of 1.4e18 → use 1.5e18 for world. From http://www.eia.doe.gov/pub/international/iealf/table11.xls.
11. http://www.eia.doe.gov/neic/quickfacts/quickoil.html, May 17, 2002.
12. Grove, 1995.
13. Perlin, 1991.
14. http://www.radford.edu/~wkovarik/envhist/, May 19, 2002.
15. Noel De Nevers, *Air Pollution Control Engineering*. New York: McGraw Hill, 1995, p. 2.
16. http://www.runet.edu/~wkovarik/envhist/2middle.html.
17. *Changes in Gasoline III*, published by Technician's Manual, Downstream Alternatives, Inc. P.O. Box 190, Bremen, IN 46506-0190, 1996, p. 8.
18. http://www.runet.edu/~wkovarik/envhist/10eighties.html, May 20, 2002.
19. R. A. Hinrichs, and M. Kleinbach, *Energy—Its Use and the Environment*, 3rd ed. New York: Brooks/Cole, 2002, Table 14.6.
20. http://www.sitnews.us/1004news/100604/100604_knowles_exxon.html.

CHAPTER 3

Energy Reserves and Renewable Energy Sources

During industrial expansion, we can rapidly deplete available resources. This has occurred for petroleum on U.S. soil. Proven, recoverable oil reserves in the United States would only power our thirst for oil for three years if cut off from oil imports. Technology can meet the challenges of dwindling U.S. oil reserves but only by switching to the abundance of other energy forms and reserves. This energy is available in three forms: fossilized solar energy, nuclear energy, and recent solar energy.

Fossil Fuel Reserves

The "fossil" designation of certain fuels implies that the fuel energy content originates from prehistoric vegetation or organisms. Fossil fuels are the most commonly used energy source to drive our machines. Unlike wind and sunlight, which are dispersed in low concentrations across the surface of the Earth, commonly used fossil fuels tend to be concentrated at locations near the Earth's surface. Where these are easily accessible, we are able to collect them with great efficiency. Fossil fuel sources include the following:

- Coal
- Petroleum
- Heavy oil
- Oil sands
- Oil shale

- Methane hydrates
- Natural gas

In Wyoming, some coal seams are 40 feet thick and less than 100 feet underground. In the Middle East, hundreds of barrels per day of crude oil can flow from a single well under its own pressure. Each source provides abundant energy.

Coal,[1] petroleum,[2] and natural gas[3] are accessible fossil fuels and easy to use (see box, "Petroleum and Gas"). By far, they are today's most popular fuels. Figure 3-1 summarizes the known accessible reserves of these fuels in the entire world and in the United States. World recoverable reserves for coal, natural gas, and petroleum are $2.5E+19$ (25 billion billion), $6.9E+19$, and $5.3E+18$ Btus.[4] For coal, the total estimated reserves are about a factor of ten higher than the estimate for recoverable reserves.[5]

In the year 2000, the United States consumed 19.7 million barrels of petroleum per day (see Chapter 2) or $3.8E+16$ Btus per year. This consumption would deplete known U.S. petroleum reserves in about 3 years and estimated U.S. petroleum reserves in about 7.6 years. These statistics are summarized in Table 3-1.

Petroleum represents about 53% of the total annual U.S. energy consumption.[i] The U.S. total energy consumption would

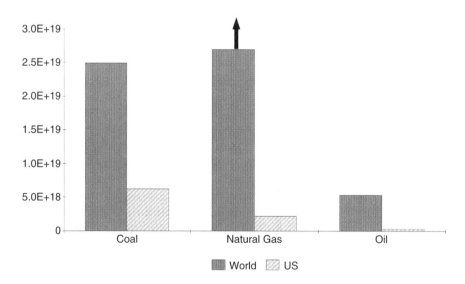

FIGURE 3-1. Summary of world and U.S. fuel reserves in Btus.

---

[i] The total U.S. energy consumption is about $7.1E+16$ Btus per year.

TABLE 3-1
Comparison of energy reserves and rates of consumption.

Energy Description	Amount (Btus)	Amount (as indicated)
World Recoverable Coal Reserves	2.5E 19	
World Recoverable Natural Gas Reserves	6.9E 19	
World Recoverable Petroleum Reserves	5.3E 18	
U.S. Petroleum Consumption (year 2000)	3.8E 16	19.7E 6 barrels
U.S. Petroleum Consumption divided by U.S. Known Petroleum Reserves		3 years
U.S. Total Recoverable Coal Reserves Divided by U.S. Total Energy Consumption		90 years

deplete U.S. estimated reserves of petroleum in four years. Natural gas would last 30 years, and coal would last 90 years. Coal can be used for much more than electrical power production (see box, "Strategic Technologies"). World natural gas reserves would last 1,000 years if they were used only to meet U.S. consumption. Figure 3-1 illustrates the relative magnitude of these reserves.

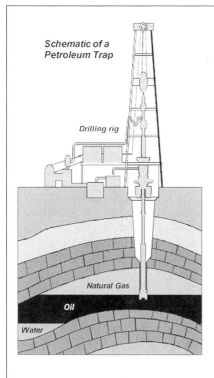

Schematic of a Petroleum Trap

Petroleum and Gas—from the Ground to the Refinery

The inserted image[6] illustrates a typical petroleum reservoir and drilling used to recover that petroleum. A rock cap has kept the reserve isolated from atmospheric oxygen for millions of years. Since the petroleum is lighter than water, it floats above water aquifers. Petroleum gases, including natural gas, are the least dense material in the formation and are above the oil. Drilling and inserting a pipe into the petroleum reserve allow recovery. Oil reserves recovered by conventional land drilling applications are typically several hundred feet to about one mile deep.

Estimated energy reserves in heavy oil, oil sands, oil shale, and methane hydrates dwarf known reserves in coal, natural gas, and petroleum. One evolutionary route to form these three reserves includes the advanced stages of petroleum decay. Petroleum contains a wide range of hydrocarbons, ranging from the very volatile methane to nonvolatile asphaltines/tars. When petroleum is sealed securely in rock formations, the range of volatility is preserved for millions of years or converted to methane if buried deeper, where it is converted by geothermal heat.

> ### *Strategic Technologies of the 21st Century*
>
> A few technologies stand out as logical extensions of current technology that have both economic and strategic value. These technologies generally require greater coordination than first generation technologies, and they are based on adding value to abundant indigenous resources. Uranium reprocessing is a strategic technology for electrical power generation.
>
> Coal and biomass qualify as abundant local resources that could be better utilized. The solid fuel refinery uses these feed stocks to supply an array of conversion and synthesis processes, including electrical power generation, production of liquid fuels, and production of chemicals like ammonia. The synthesis gas pipeline is based on much the same concept as the solid fuel refinery, but in the synthesis gas pipeline a pipe network is fed mixtures of hydrogen and carbon monoxide from several source locations. Likewise, the pipeline is used as a feed stock for chemicals and fuel by many processes. Two technologies stand out to tap into these resources (see "Option 1" and "Option 2" boxes).

If the rock overburden is fractured, either due to erosion of rocks above the formation or simply due to the weak or porous nature of the rock, the more volatile components of petroleum escape. This leaves less-volatile residues in the forms of heavy oils, oil sand, and oil shale.

Heavy oils are volatile-depleted deposits that will not flow at reservoir conditions but need assistance for recovery. Oil sand liquids are heavier than heavy oils—typically not mobile at reservoir conditions, but heat or solvents can make the oil flow through the porous rock. Oil shales are usually immobile and present in rocks that do not allow oil flow. Unlike the oil in oil sands that can be removed in situ or with low amounts of solvent and heat, the oil in shale tends to be very difficult to remove.

This makes oil shales more difficult to recover than oil sand. Unlike coal, which is a concentrated fossil fuel, oil shale is best characterized as a relatively nonvolatile oil dispersed in a shale. World reserves are estimated to be 600 to 3,000 times world crude oil reserves.[7] Lower estimates specific to the western United States place reserves at two to five times known world oil reserves.[8]

Option 1: Synthesis Gas Pipeline

In the solid fuel refinery, the synthesis gas generation can be separated from the other processes. This represents a potential added cost to recovering the heat from the synthesis, which is not required in a solid fuel refinery. The synthesis gas pipeline has this drawback, plus the expense of the pipeline, but the benefits are many.

Reduced transportation costs (pipeline versus railroad) are an advantage, as well as the ease of disposal of coal ash at the mining location as landfill to replace space created by removing the coal. Sulfur removal from the coal would be easier using this approach. This would allow the reopening and increased use of several minefields across the country that contain high-sulfur coal.

The synthesis gas will allow electrical power generation from coal approaching 50% thermal efficiency, which is about 10% better than direct firing coal. Furthermore, customers are likely to use this option, since they would not have the burden of building a gasification facility. The synthesis gas would have all the advantages of natural gas, but it would presumably be less costly on an equal energy basis.

A synthesis gas pipeline would allow smaller companies and entrepreneurs to enter the energy big business. Opportunities would exist both for production of synthesis gas (potentially from biomass or municipal solid waste) and use of the gas from the pipeline in a large array of processes. The pipeline would make it possible for companies to enter business with smaller investments and easy feedstock acquisition.

Heavy oil reserves in Venezuela are estimated to be from 0.1 trillion barrels[9] to 1 trillion barrels.[10] These heavy oils are generally easier to recover than oil sands and much easier to recover than oil from oil shale. The United States, Canada, Russia, and Middle East also have heavy oil reserves totaling about 0.3 trillion barrels[11] (lower side of estimates). In all, the heavy oil reserves are estimated to be slightly greater than all the more-easily recovered conventional crude oil reserves.

Surface reserves of oil sands have been mined and converted to gasoline and diesel since 1969 in Alberta, Canada. Production costs are about $20 per barrel.[12] These supply about 12% of Canada's petroleum needs. The sands are strip-mined and extracted with hot water. Estimated reserves in Alberta are 1.2–1.7 trillion barrels with two open pit mines now operating.[13] Other estimates put oil sand reserves in Canada, Venezuela, and Russia at about 3, 3, and 0.6 trillion barrels. Estimates approximate 90% of the world's heavy oil (and oil sand) to be in Western Canada and Venezuela.[14] Cumulatively, oil sand reserves are six to ten times proven conventional crude oil reserves. (Conventional crude oil reserves are reported at 1 trillion barrels.)

Option 2: Solid Fuel Refinery

A solid fuel refinery capitalizes on the strengths of several processes to overcome the weaknesses of each process to produce much higher final conversion efficiencies. For example, electrical power generation does not effectively use low-temperature steam and flue gases, but it produces large quantities of these gases.

For Fischer-Tropsch synthesis, conversion of the last bit of hydrogen and carbon monoxide to liquid fuels is considerably more expensive than the initial conversion of fresh feed. In a solid fuel refinery, the residual carbon monoxide and hydrogen are suitable for driving a gas turbine to generate electricity. These residual gases left from liquefied fuel production are used for electrical power generation that is more efficient than a coal-fired power plant.

Ammonia is an important chemical produced in a solid fuel refinery because of the energy intensity of the ammonia production process and its value as a fertilizer. Building one big gasifier to feed several processes brings an important economy of scale for this important step.

Energy Reserves and Renewable Energy Sources 39

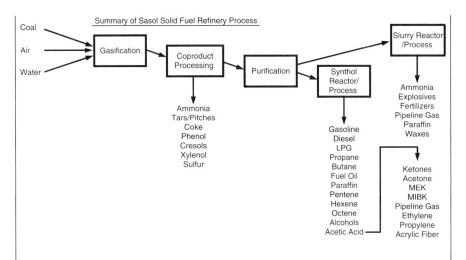

Solid fuel refineries are a reality in South Africa, as illustrated by the flow diagram. A large array of chemicals are produced. The overall process makes the best use of coal resources for supplying many useful products.

If methane escapes from an underground deposit (due to porous rock or erosion of overburden) and comes in contact with a combination of increased pressure, water, and cold temperatures, methane hydrate is formed. Methane hydrate is ice that contains methane and is stable below the freezing point of water as well as at temperatures slightly warmer than the freezing point of water at high pressure. Conditions are right for the formation of methane hydrates on the sea floor (under a few hundred feet of seawater, where the temperature is relatively constant at the temperature of maximum water density, 4°C; this is true even in the Caribbean) and in the Arctic permafrost. In addition to formation mechanisms involving petroleum decay, methane is commonly formed directly from biomass—both geological and recent biomass methane can form hydrates.

Methane hydrate reserves are not presently recovered. Countries like Japan have great interest in the potential of this technology because of the lack of natural fossil fuel reserves in the country and the large coastal water areas. In the United States, methane hydrates have received the attention of congressional hearings where reserves were estimated at 400 million trillion cubic feet (200,000 trillion cubic feet of gas (Tcf) in reserves under the jurisdiction of the United States). In the most conservative interpretation, these hydrates have enough energy to last maybe 5,000 years.

Natural gas emissions from these reserves occur naturally, so the methane greenhouse effect from this source will occur regardless of whether or not we use the energy. If we burn the natural gas from these emissions, the resulting carbon dioxide would have about one-tenth of the greenhouse effect of the equivalent methane release. The release of methane from methane hydrates on the sea floor may have contributed to the end of many of the ice ages. During ice ages, lower sea levels reduced the pressure on the sea floors. This lower pressure would lead to methane release.[15]

The trick to reducing natural methane emissions is to mine and recover those reserves that are most likely to release naturally. Most hydrate mining research involves changing the temperature and pressure at the solid reserve location to cause the methane hydrates to melt or sublime and then to recover the methane that is evolved. Experts at the congressional hearing agreed that *Alaska's North Slope was the most likely candidate for initial research because of its relative easy access (compared to the deep-water Gulf of Mexico) and in-place infrastructure.*[16]

In some instances, natural gas reserves are below ground and in equilibrium with methane hydrate reserves. When the natural gas is recovered, the methane hydrates melt, resupplying the gas in the reserve for easy recovery.

Table 3-2 summarizes the energies available in recoverable fuels. All of these fuels are in concentrated deposits, with the exception of uranium. The uranium availability includes recovering uranium from sea waters, which is possible but costly compared to today's prices. The numbers approximate the magnitudes of the different reserves relative to conventional crude oil reserves.

Estimating energy reserves has historically been inaccurate. For example, in the 1980s when the oil-producing countries shifted from a mentality of "creating the perception of oil shortfalls" to setting production quotas based on "countries reported reserves," the reported reserves increased dramatically. Likewise, reported coal reserves decreased by a factor of 10 from 1980[17] to 2002[18] due to redefinitions of recoverable reserves and the influence of oil companies on U.S. policy and U.S. Department of Energy positions.

Qualitatively, the image portrayed by Table 3-2 is correct. The costs of fuels reflect this. Petroleum ($45–$75 per barrel) and natural gas cost $9.00–$15.00 and $6.00–$12.00 per MBtu, respectively.

TABLE 3-2
Relative abundance of recoverable fuels.

U.S. Petroleum	1/20 X
World Petroleum	1 X
World Recoverable Coal	5 X
World Recoverable Heavy Oil and Oil Sands	10 X
World Recoverable Natural Gas	15 X
World Recoverable Oil Shale	>500 X
World Methane Hydrates	>5,000 X
World Recoverable Uranium	50,000 X

Coal costs about $1.20–$1.40 per MBtu. Uranium costs $0.62 per MBtu when 3.4% is fissioned or $0.062 per MBtu if a fission of 35% is assumed.

The following uses of these reserves are consistent with recent trends:

- Major oil corporations will likely progressively tap into oil sands and heavy oil as the reserves of petroleum are depleted. The corporations will likely be able to meet petroleum needs for several decades in the progression; however, this progression will likely occur to maintain corporate profit, will result in huge trade deficits, and may be at the expense of continuous military activity.
- Natural gas use in the United States is likely to follow the course of petroleum. The depletion of U.S. reserves will lead to increasing reliance on imports.
- Coal will continue increased use in electrical power generation. However, its high rate of carbon dioxide generation and limited

recoverable reserves will dampen its expansion relative to 1970s estimates that stated centuries of abundance. New technology is likely to bring a greater portion of the total coal reserves into the "recoverable" coal reserves category.
- Energies in oil shale are unlikely to be realized due to the high energy and cost of recovery.
- Methane hydrate recovery is uncertain because not enough is known about safe methods for its recovery, and the extent of reserves of "recoverable" methane hydrates depends on this technology.

Impacting technologies likely to develop in the next 30 years include 100% nuclear fission of nuclear fuel for abundant electricity, hybrid vehicles that are partially rechargeable with grid electricity, and chemicals and fuels made from biomass. Closed cycle nuclear fuel cycles that use all of the energy in uranium (including U-238) include reprocessing of spent nuclear fuel and the concentration of fission products so that the volume of waste is 100 times less than if the fuel rods were directly placed in repositories. These are the technologies that can change the rules, can deliver sustainability, and can promote economic prosperity by eliminating the need to import oil and natural gas.

Cosmic History of Fossil Energy Reserves

Similar to the fossilized bones of a dinosaur, fossil fuels are the remains of plants and microscopic organisms. Their compositions reveal a history going back hundreds of millions of years; however, the history of the energy in these fuels does not start there. This history goes back to the origin of the universe. An understanding of the dominant role nuclear energy has played in the history of energy helps us understand how nuclear energy will always be part of the energy mix we use.

What Is Permanence?

The Permanence Scale in Figure 3-2 is the binding energy in MeV for each nucleon (a proton or neutron) in the nucleus of an atom. The most stable atoms are those with the highest binding energy per nucleon. The nucleon-binding energy is plotted against the atomic mass number. The maximum binding energy per nucleon occurs at mass number 54, the mass number for iron. As the atomic mass number increases from zero, the binding energy per nucleon also increases because the number of protons (with positive charge) increases, and protons strongly repel each other. The binding energy of the neutron (no charge) holds the nucleus of the atom together. The atomic mass number for each atom is approximately the mass of the protons and neutrons in the nucleus, since the electrons (equal to the number of protons) have a mass that is about 1/1,837 that of a nucleon.

Above a mass number of about 60, the binding energy per nucleon decreases. Visualize the large atom nucleus as many protons try to get away from each other, with a much larger number of neutrons holding that nucleus together. The common isotope of iron-56 (about 92% of the mass of natural iron) has 26 protons and 30 neutrons. The common isotope of uranium-238 (about 99.27% of the mass of natural uranium) has 92 protons and 146 neutrons. The iron nucleus is very stable. The uranium-238 isotope does spontaneously decay, but it will take about 4.5 billion years for one-half of a lump of pure U-238 to decay, so it is practically a stable isotope.

An additional comment: The nucleon-binding energy increases as the atomic mass decreases from 260 to 60. This is the primary reason for the energy release in the fission process that makes nuclear reactors work. Since the forces in the nucleus of large atoms are so carefully balanced, a small energy addition (a low-energy neutron entering the nucleus) will cause the large forces acting between the protons to become unbalanced, and the nucleus comes apart (flies apart), releasing lots of energy. These large atoms are the fuel for nuclear reactors. The strong binding energies of the smaller atoms lead to the permanent end point of the natural isotope decay processes.

The arrays of different elements in our planet, the solar system, and the galaxy reveal their history. Hydrogen is the smallest of the atoms assigned an atomic number of one. Physicists tell us that at the birth of the universe, it consisted mostly of hydrogen. Stars converted hydrogen to helium, and supernovas (see the box "Supernovas") generated the larger atoms through atomic fusion.

Atoms are identified based on the number of protons (positively charged subatomic particles). Since both hydrogen and helium are smaller atoms than the nitrogen and oxygen in the air, the Goodyear blimp (filled with helium) and the Hindenburg zeppelin (filled with hydrogen) floated in air.

Helium has 2 protons, lithium has 3 protons, carbon has 6 protons, and oxygen has 8 protons. The number of protons in an atom is referred to as the "atomic number" of the atom. Atoms are named and classified by their atomic number. Atoms having between 1 and 118 protons have been detected and named (see the box "Making New Molecules in the Lab"). The atomic mass is the sum of the mass of neutrons, protons, and electrons in an atom—the atomic mass and the atomic spacing determines the density of materials.

Protons are packed together with neutrons (subatomic particles without a charge) to form an atom nucleus. There are more stable and less stable combinations of these protons and neutrons. Figure 3-2 illustrates the permanence of nuclei as a function of the atomic mass number (the atomic mass is the sum of the protons and neutrons). Helium 3—abbreviated He,3—is shown to have a lower permanence than He,4. Two neutrons simply hold the two protons in He,4 together better than one neutron in He,3. In general, the number of neutrons in an atom must be equal to or

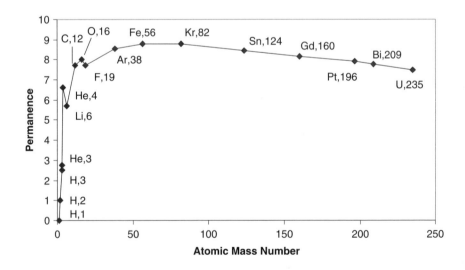

FIGURE 3-2. Impact of atomic mass number on permanence of atoms. H is hydrogen, He is helium, Li is lithium, C is carbon, O is oxygen, F is fluorine, Ar is argon, Fe is iron, Kr is krypton, Sn is tin, Gd is gadolinium, Pu is plutonium, Bi is bismuth, and U is uranium.

greater than the number of protons, or that atom will disintegrate into more stable combinations of protons and neutrons.

The wealth of information in Figure 3-2 explains much about the chemistry of our planet. For example, why do hydrogen atoms combine to form helium instead of breaking apart to form hydrogen atoms? Figure 3-2 illustrates that the "permanence" is greater for He,4 than for hydrogen (H,1). In nuclear reaction processes, atoms tend to move uphill on the curve of Figure 3-2 toward more stable states. In Chapter 10, the concept of atomic stability will be discussed in greater detail, and the term "binding energy" will be defined and used in place of "permanence."

Supernovas—The Atomic Factories of the Universe

Astronomers have observed "lead stars" that produced heavier metals like lead and tungsten. Three have been observed about 1,600 light-years from Earth. To paraphrase a description of the process:

Stars are nuclear "factories" where new elements are made by smashing atomic particles together. Hydrogen atoms fuse to create helium. As stars age and use up their nuclear fuel, helium is fused into carbon.

Carbon, in turn, is fused into oxygen, and the process continues to make heavier elements until a natural limit is reached at iron. To make elements heavier than iron, a different system is needed that adds neutrons to the atomic nuclei. Neutrons are a kind of atomic "ballast" that carry no electric charge.

Scientists believe there are two places where this can occur: inside very massive stars when they explode as supernovas and more commonly, in normal stars right at the end of their lives before they burn out.

To make atomic transitions to more permanent/stable atoms, extreme conditions are necessary (see the box "Supernovas"). On the sun, conditions are sufficiently extreme to allow hydrogen to fuse to more stable, larger molecules.[19] In nuclear fission reactions in a nuclear power plant or in Earth's natural uranium deposits, large molecules break apart to form more stable smaller molecules.

46 *Sustainable Nuclear Power*

Each atomic event is toward more stable combinations of protons and neutrons, and this process releases energy.

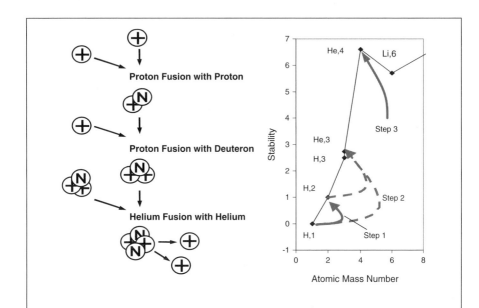

From Hydrogen to Helium

The most abundant atom in the universe is hydrogen. Hydrogen is the fuel of the stars. The diagram[20] illustrates how four protons interact to produce a new stable molecule: helium. Considerable energy is released in the process—about 25 MeV of energy for every helium atom formed (570 million Btus per gram helium formed).

Figure 3-3 is the starting point for qualitative understanding the history of energy. Nuclear reactions are where the history of energy begins. The story of the history of energy goes something like this: Once upon a time, long ago—about 15 billion years—there was a big bang. From essentially nothingness, in an infinitely small corner of space, protons and helium were formed. Carbon, iron, copper, gold, and the majority of other atoms did not exist.

Energy Reserves and Renewable Energy Sources 47

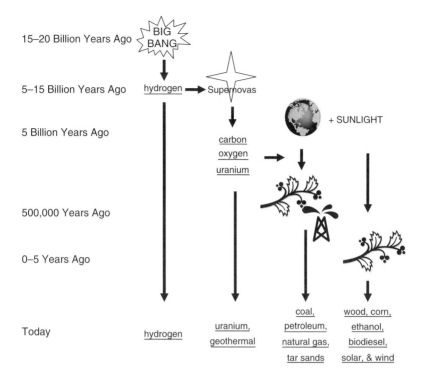

FIGURE 3-3. The history of energy.

Unimaginably large quantities of hydrogen and helium clustered together to form stars. The most massive of these stars formed supernovas. Fusion conditions were so intense in these supernovas that atoms of essentially all known atomic numbers were formed. Hence, carbon, oxygen, iron, copper, gold, and the vast array of atoms that form solid objects around us were formed.

Uranium was also formed along with atoms larger than uranium. The largest of these atoms rapidly fell apart to form more stable molecules. Uranium and plutonium have intermediate stability. They could be induced to fall apart but were stable enough to last for billions of years without spontaneous decomposition.

The spinning masses continued to fly outward from the big bang. As time passed, localized masses collected to form galaxies, and within these galaxies, solar systems, stars, planets, asteroids, and comets formed.

Making New Molecules in the Lab

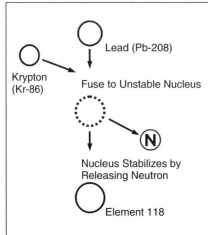

On small scales, scientists are able to make new molecules similar to the way in which supernovas combine two smaller molecules to form a larger molecule. As illustrated in the figure, krypton and lead combine to form a compound nucleus. This nucleus is very unstable and rapidly degrades to a more stable atom indicated as Element 118.[21]

This work was performed at the Lawrence Berkeley National Lab. During this synthesis, two atoms are joined to actually form a less stable molecule. This is possible in a particle accelerator that puts kinetic energy (high speed) into the krypton. This high speed provides the extra energy needed to fuse the nuclei. The atomic rearrangement resulting in the release of a neutron helps lock in a final element that is stable. High atomic number atoms are unstable and not found in nature.

It is here where energy and the universe as we know it began to take form. We are just beginning to understand the processes of the stars and supernovas to tap into the vast amounts of binding energy in the atom. The atomic binding energy available in one pound of uranium is equivalent to the chemical binding energy present in 8 million pounds of coal.[ii]

Nuclear Energy

In principal, nuclear energy is available in all elements smaller than iron through nuclear fusion and all elements larger than iron through nuclear fission. Iron is on the top of the permanence curve so it is one of the most abundant elements. While all other atoms

---

[ii] 12,000 tons per day (Ch. 4, Steam Turbine Section) = 365 days/year × 40%/ 30%/(750 kg × 1 ton/∼1,000 kg).

can degrade to iron, iron does not degrade. When most of the nuclear energy of the universe is expended, it will consist mostly of iron and elements of similar atomic number (all being stable).

In general, the largest atoms are the most likely—given enough time—to undergo nuclear decays such as the release of an alpha particle (a helium atom). Atoms larger than uranium (see[22]) have undergone fission to the extent that they can no longer be found on Earth. The amount of U-238 on Earth today is slightly less than half of what was present at Earth's formation. The amount of U-235 on Earth today is less than 1% of what was present at Earth's formation.

Of interest to us is the ability to perform these nuclear processes in a controlled and safe manner, because the nuclear binding energy can be used to produce electricity. The energy released as protons, neutrons, and atoms combines and rearranges in the progression to higher "binding energy." We are able to use nuclear fission on a practical/commercial scale with one naturally occurring element: uranium. We could perform fission on elements larger than uranium, but these are not readily available. In the H-bomb, we have demonstrated an ability to tap the energy of fusion for massive destruction, but use of fusion for domestic energy production is much more difficult. Practical nuclear fusion methods are an area of active research.

The only practical nuclear energy sources today are nuclear fission of uranium in nuclear reactors and the recovery of geothermal heat produced by nuclear decay under the surface of Earth (occurring continuously). Uranium is the primary fuel for both of these processes.

At 18.7 times the density of water, uranium is the heaviest of all the naturally occurring elements (the lightest is hydrogen; iron is 7.7 times the density of water). All elements (as defined by the number of protons in the nucleus) occur in slightly differing forms known as *isotopes*. These different forms are caused by the varying number of neutrons packed with the 92 protons in uranium's nucleus. Uranium has 16 isotopes, only two are stable. Most (99.3%) of natural uranium is composed of uranium-238 (U-238, 238 is the sum of neutrons and protons) and U-235, about 0.71% of natural uranium.

U-235 is slightly less stable than U-238 and when enriched to 3% to 8% can be made to release heat continuously in a nuclear reactor. Enriched to 90% U-235 and the sudden release of large amounts of energy becomes a nuclear bomb. We have mastered the technology to perform both of these processes. The U-238 decays slowly with about 2 kilograms of uranium decaying to 1 kilogram of uranium and slightly less than 1 kilogram of fission products in about 4.5 billion years. About half of the U-238 present when Earth was formed (and >99% of the U-235) has decayed, keeping the Earth's interior a molten metal core.[23]

U-235 decays faster than U-238. We are able to induce the fission of U-235 by bombarding it with neutrons. When one neutron enters the U-235 nucleus to form U-236, it breaks apart almost instantly because it is unstable. It breaks apart to form the nucleus of smaller atoms plus two or three neutrons. These two or three neutrons can collide with U-235 to produce another fission in a sustained chain reaction.

Nuclear fission occurs when the nucleus of an atom captures a neutron and breaks apart expelling two or three neutrons. The U-235 continuously undergoes fission (fission half life of 1.8E 18 years, alpha-decay half life of 6.8E 8 years) that proceeds slowly because there is so little U-235 in the metal; most of the emitted neutrons are lost with only a few producing fission. The exception was the natural nuclear reactor that formed in Oklo, Gabon (Africa), about 2 billion years ago. This occurred when the concentration of U-235 in ore at Oklo was high enough to cause a chain fission of the U-235 leading to lower-than-normal U-235 concentrations and trace plutonium in the deposits today.

Modern Nuclear Reactors in the United States

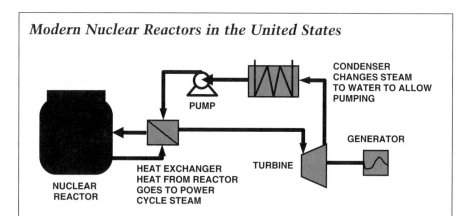

Modern nuclear power plants[24] use a pressurized water reactor to produce the thermal energy to produce the steam to drive a turbine and generate electricity. The fuel is 3% to 4% U-235 enriched uranium oxide pellets sealed in tubes that are held in racks in the reactor pressure vessel. This maintains the geometry of the reactor core. The water that removes the heat from the core leaves the reactor at about 320°C, and it would boil at a pressure of about 70 atmospheres (850 psi). The pressure in the reactor vessel is held at 150 atmospheres (2,250 psi), so it never boils. This hot water is pumped to a heat exchanger, where the steam to drive the turbines is produced. The high-pressure reactor cooling water will always contain small amounts of radioactive chemicals produced by the neutrons in the reactor. This radioactivity never gets to the steam turbine where it would make it difficult to perform maintenance on the turbine and steam-handling equipment.

Large pressurized water reactors produce about 3,900 megawatts of thermal energy to produce about 1,000 megawatts of electric power. The reactor core contains about 100 tons of nuclear fuel. Each of the nuclear fuel racks has places where control rods can be inserted. The control rods are made of an alloy that contains boron. Boron metal absorbs neutrons, so with these rods in position, there will not be enough neutrons to initiate a chain reaction. When all of the fuel bundles are in position and the lid of the pressure vessel sealed, the water containing boric acid fills the pressure vessel. The control rods are withdrawn, and the boron water solution still absorbs the neutrons from U-235 fission. As the water circulates, boric acid is slowly removed from the water and the neutron production rate increases; the water temperature and pressure are closely monitored. When the neutron production rate produces the rated thermal power of the reactor, the boron concentration in the water is held constant. As the fuel ages through its life cycle, the boron in the water is reduced to maintain constant power output.

If there is an emergency that requires a power shutdown, the control rods drop into the reactor core by gravity. The control rods quickly absorb neutrons, and fission power generation stops. The radioactive fission products in the fuel still generate lots of heat, as these isotopes spontaneously decay after fission stops. Water circulation must continue for several hours to remove this radioactive decay heat.

Our use of nuclear fission to make a bomb is based on an uncontrolled chain reaction. A neutron chain reaction results when, for example, two of the neutrons produced by U-235 fission produce two new fission events. This will occur when nearly pure U-235 is formed into a sphere that contains a critical mass: about

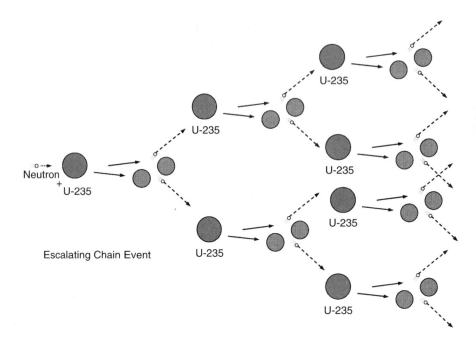

FIGURE 3-4. Escalating chain reaction such as in a nuclear bomb.

60 kilograms of metal. Then in each interval of 10 billionths of a second, the number of fission events grows from 1, 2, 4, ... 64, ... 1,024, 2,048, ... as illustrated by Figure 3-4. The transition from very few fission events to an uncountable number occurs in less than a microsecond. The enormous energy released in this microsecond is the source of the incredible explosive power of a nuclear fission bomb.

This escalating chain reaction is to be distinguished from the controlled steady-state process as depicted by Figure 3-5. In a controlled steady-state process, a nearly constant rate of fission occurs (rather than a rapidly increasing rate) with a resulting constant release of energy.

The first nuclear bomb used in war exploded over Hiroshima, Japan, was a U-235 bomb. Two hemispheres containing half of the critical mass are slammed together with conventional explosive charges. In the resulting nuclear explosion, about 2% of the U-235 mass underwent fission. Everything else in the bomb was instantly vaporized. The fireball and the explosion shock wave incinerated and leveled a vast area of Hiroshima. This is the legacy of nuclear energy that indelibly etched fear into the minds of world citizens. The second explosion at Nagasaki was a plutonium bomb,

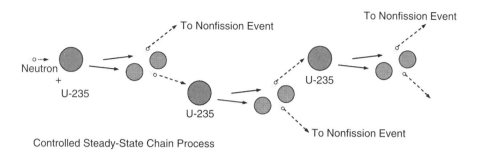

FIGURE 3-5. Controlled steady-state chain nuclear fission such as in a nuclear reactor.

followed by the development and testing of even more powerful and fearsome nuclear weapons during the Cold War period, adding to this legacy of fear.

For a nuclear bomb, the rapid chain reaction depicted by Figure 3-4 is competing with the tendency for the melting/vaporizing uranium to rapidly splat over the surroundings. This "splatting" action tends to stop the chain reaction by separation of small pieces of uranium. Weapon's grade U-235 is typically at least 80% U-235; higher purities give increased release of the nuclear energy (more fission and less splatting).

The enormous energy available from U-235 in a very small space led U.S. naval technologists to consider using nuclear energy to power submarines. The task is to configure the nuclear fuel (U-235 and U-238) so that exactly one of the neutrons produced by a U-235 fission produces one new fission. The shape of the reactor core and control rods (that absorb neutrons) combines to serve as a "throttle" to match the energy release to load. The thermal energy produces steam that propels the vessel and provides electric power. All of this technology development was done with private industrial firms under contract by the military and was classified "top secret."

The industrial firms that built the nuclear reactors for the military also built steam turbines and generators for electric power stations. The first nuclear reactor built to produce electric power for domestic consumption was put into service in Shipingsport, Ohio, in 1957, just 15 years after the "Top Secret" Manhattan Project was assembled to build a nuclear weapon. This represents a remarkable technological achievement. Today, modern nuclear reactors produce electricity based on technology similar to that used in the submarines.

In nuclear reactors and the bomb, neutron sources can be used to supplement the neutrons created by the natural decay of U-235 to create greater control in attaining criticality. The chain reaction is started by inserting some beryllium mixed with polonium, radium, or another alpha-emitter. Alpha particles from the decay cause the release of neutrons from the beryllium as it turns to carbon-12.

Reserves

Uranium reserves are difficult to estimate; however, an estimate can be readily made on the energy in the spent rods from U.S. nuclear power generation. Current nuclear technology uses 3.4% of the uranium in the fuel, leaving 96.6% of the uranium unused. The amount of nuclear fuel in spent nuclear fuel rods in U.S. nuclear facilities has an energy content comparable to the entire recoverable U.S. coal reserve.[iii] The depleted uranium created during the fabrication of the initial nuclear fuel rods has about four times as much energy as that remaining in the spent nuclear fuel rods. Combined, this stockpiled uranium in the United States has the capacity to meet all of the U.S. energy needs with near-zero greenhouse gas emissions for the next 250 years.[iv]

Reprocessing Technology

The 250 years of capacity from uranium that has already been mined will require reprocessing. A typical spent nuclear fuel rod in the United States contains about 3.4% fission products, 0.75%–1% unused U-235, 0.9% Pu-239, 94.5% U-238, and trace amounts of atoms having atomic masses greater than U-235 (referred to as transuranic elements).

Not only would reprocessing tap this 250 years of energy available from stockpiled uranium, the additional nuclear waste would be the fission products produced by the recycled uranium fuel.

[iii] Assuming 30 years of spent fuel at the current rate, this translates to 75 years of capacity to meet all U.S. energy needs at the present rate of consumption with near-zero generation of greenhouse gases.

[iv] 250,000 tons of spent fuel were in storage in 2001 worldwide, with waste inventories increasing about 12,000 tons per year. About 3,000 tons of spent fuel are reprocessed in France.

Reprocessing involves removing the 3.4% that is fission products and enriching the U-235 and/or Pu-239 to meet the "specifications" of nuclear reactor fuel. The "specifications" depend on the nuclear reactor design. Nuclear reactors and fuel-handling procedures can be designed that allow nuclear fuel specifications to be met at lower costs than current reprocessing practice in France. For comparative purposes, the cost of coal, U.S. nuclear fuel from mined uranium, and French reprocessed fuel is about 1.05, 0.68, and 0.90¢ per kWh or electricity produced.

From 33% to 40% of the energy produced in nuclear power plants today originates from U-238 and is released by the Pu-239 formed and subsequently fissions: For every three parts of U-235 entering the reactor, about two parts of U-235 plus Pu-239 leave the reactor and, to date, remain stored at the power plant site. All of the uranium, the two parts U-235 and Pu-239 are the target of reprocessing technology. To tap this 250 year stockpile of fuel, new fast-neutron reactors could be put in place that produce more Pu-239 than the combined U-235 and Pu-239 in the original fuel.

Decades of commercial nuclear power provide stockpiles of spent fuel rods, billions of dollars collected on a 0.1 cent per kWh tax levied, and retained to process the spent fuel rods. A remarkable safety history for U.S. designed reactors is set against a costly history of regulations that limit the ability of the technology to advance. These circumstances provide opportunity or perpetual problems, depending on the decisions made to use (or not) nuclear power.

Figure 3-6 summarizes the accumulation of spent fuel currently being stored on site at the nuclear power plants in the United States.

The United States uses about 98 GW of electrical power generating capacity from nuclear energy. The construction and startup of most of these plants occurred between 1973 and 1989. In 2007 the inventory of spent nuclear fuel will correspond to about 30 years of operation at current generation rates of the nuclear power infrastructure. Figure 3-6 approximates the total spent fuel inventories and cumulative inventories of U-235 and Pu-239 under two different scenarios. Figure 3-6 illustrates that reprocessing is the key to decreasing Pu-239 and U-235 inventories and ending the accumulation of spent fuel nuclear reactor sites.

If reprocessing would have initiated in 2005 to meet all current nuclear power plants, the current inventories, along with the Pu-239 that is generated as part of PWR operation, would provide

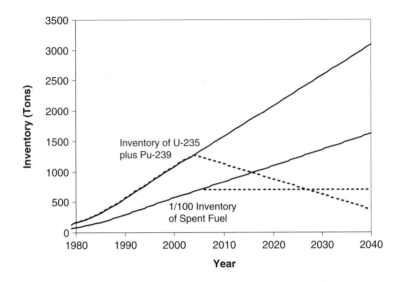

FIGURE 3-6. Approximate inventory of commercial spent nuclear fuel and fissionable isotopes having weapons potential (Pu-239 and U-235). The solid lines are for continued operation without reprocessing, and the dashed lines are for reprocessing (starting in 2005) to meet the needs of current nuclear capacity.

sufficient Pu-239 to operate at existing capacity through 2045. If in 2005 the demand for Pu-239 and U-235 increased threefold (~300 GW capacity), the current inventories would last until 2019. This does not include the use of Pu-239 and U-235 in weapons' inventories, or depend on fast neutron reactor technology to convert the much greater inventories of U-238 into Pu-239 fuel. This partly explains trends of discontinuing breeder reactor research and operation. Breeder reactors will not be needed for some time.

Fast-neutron reactor technology would allow nuclear reactors to meet all energy needs for the next 250 years without generating additional radioactive materials and without mining additional uranium. The potential of this technology should not be ignored.

Geothermal

Geothermal energy is heat that is released from the continuous nuclear decay occurring in uranium that is distributed throughout the Earth. The Earth's core has remained hot for billions of years for two reasons: (1) thousands of feet of the Earth's crust provide a good insulation that hinders the loss of heat to surrounding space; and (2) heavier elements (like uranium) tend to be more concentrated toward Earth's center, where these elements undergo natural radioactive decay releasing heat.

The warmer the geothermal heat source, the more useful the energy. For most locations, higher temperatures are located several thousand feet under the surface, and the cost of accessing them is too great compared to alternatives. At the center of the Earth, some 3,700 miles below the surface, temperatures reach 9,000°F, and metals and rocks are liquid.[25]

At locations such as Yellowstone Park and Iceland, useful geothermal heat is available a few hundred feet under the surface or at the surface (hot springs and geysers). Even at these locations the costs of the underground pipe network necessary to create an electrical power plant is a capital-intensive facility. On a case-by-case basis, geothermal heating has been economical. Much of Iceland's residential and commercial heating is provided by geothermal energy (see box).

> ### *Geothermal Heating in Iceland*
>
> The first trial wells for hot water were sunk by two pioneers of the natural sciences in Iceland, Eggert Olafsson and Bjarni Palsson, at Thvottalaugar in Reykjavik and in Krisuvik on the southwest peninsula in 1755–1756.[26] Additional wells were sunk at Thvottalaugar in 1928 through 1930 in search of hot water for space heating. They yielded 14 liters per second at a temperature of 87°C, which in November 1930 was piped three kilometers to Austurbacjarskoli, a school in Reykjavik that was the first building to be heated by geothermal water. Soon after, more public buildings in that area of the city as well as about 60 private houses were connected to the geothermal pipeline from Thvottalaugar.
>
> The results of this district heating project were so encouraging that other geothermal fields began to be explored in the vicinity of Reykjavik. Wells were sunk at Reykir and Reykjahbd in Mosfellssveit, by Laugavegur (a main street in Reykjavik), and by Ellidaar, the salmon river flowing at that time outside the city but now well within its eastern limits. Results of this exploration were good. A total of 52 wells in these areas are now producing 2,400 liters per second of water at a temperature of 62–132°C.

> Hitaveita Reykjavikur (Reykjavik District Heating) supplies Reykjavik and several neighboring communities with geothermal water. There are about 150,000 inhabitants in that area, living in about 35,000 houses. This is way over half the population of Iceland. Total harnessed power of the utility's geothermal fields, including the Nesjavellir plant, amounts to 660 MWt, and its distribution system carries an annual flow of 55 million cubic meters of water.

Some manufacturers refer to the use of groundwater or pipes buried in the ground used in combination with a heat pump as geothermal furnaces. These furnaces do not use geothermal heat. Rather, the large mass of the Earth simply acts as energy storage to take in heat during the summer and give up heat during the winter.

Recent Solar Energy

Use of Sunlight

Solar energy provides the most cost effective means to reduce heating costs and can be used to directly produce electricity. Both can be cost effective, depending on the local cost of conventional electrical energy alternatives.

Solar heating is the most commonly used and least expensive use of sunlight. Building location, orientation, and window location can be used to displace auxiliary heating such as a natural gas furnace. Windows located on the south side of a northern hemisphere building will bring in sunlight to heat during the winter. A strategically located tree or well-designed roof overhang can block the sunlight during the summer. The number and placement of windows will vary, based on design preference. Aesthetics, solar functionality, and nonsolar functionality (siding on a building) are available for building construction. New building designs are available that provide cost-effective combinations for solar systems.

Solar water heating systems are the next most popular use of solar energy. They use solar heat to reduce the consumption of natural gas or electricity to heat water. Clarence Kemp is known as the father of solar energy in the United States. He patented the first commercial Climax Solar Water Heater.[27] This and competing systems sold about 15,000 units in Florida and California by 1937.

In 1941, between 25,000 and 60,000 were in use in the United States, with 80% of the new homes in Miami having solar hot water heaters. Use outside the United States has developed, especially in regions where electricity costs are high and the climate is warm.

When confronted with the oil boycott and subsequent oil supply incentives, Israel proceeded with a major initiative to use solar water heaters. More than 90% of Israeli households owned solar water heaters at the start of the 21st century.[28] Solar water heaters are also quite popular in Australia. At sunny locations where electricity is expensive and where natural gas is not available, solar water heating is a good option. It is easy to store water so the solar energy collected during the day is available at night.

Considerable research has been conducted using mirrors to focus sunlight and generate the high temperatures required to produce steam for electrical power generation. To date, most of these systems are too costly. Alternatively, the direct conversion of sunlight to electricity is popular in niche markets, and new technology is poised to expand this use.

In the 1950s, Bell Laboratory scientists made the first practical photovoltaic solar cell. Today, photovoltaic technology is widely used on flat screen computer monitors and for producing electrical power for electric devices in remote locations. These remote devices include highway signs that cannot be easily connected to grid electricity and small electrical devices like handheld calculators.

Solar energy is usually not used for power generation in competition with grid electricity. In some locations, photovoltaic cells on roofs provide an alternative for enthusiasts where consumer electrical prices are above $0.10 per kWh. Small solar roof units show a better payback to meet individual electrical needs than commercial units designed to sell power to the electrical grid. While consumers will usually pay more than $0.08 per kWh for electricity, when selling electricity to the grid one typically receives less than $0.04 per kWh.

The south facing walls and roof sections of every building in the United States are potentially useful solar receivers. Materials having both aesthetic and solar function are generally not available today, but such systems will likely be developed. From this perspective, there is great potential for solar energy to replace a portion of grid electrical power. At 0.8 billion kWh in 1999, solar electrical power on the grid provided about 0.02% of the electrical energy production (see Table 3-3).

TABLE 3-3
U.S. electricity power production in 1999 in billions of kilowatt hours.

Coal	1,884.3	50.8%
Nuclear	728.3	19.6%
Natural Gas	556.2	15.0%
Hydroelectric	319.5	8.6%
Petroleum	123.6	3.3%
Wood	37.6	1.0%
MSW	20.2	0.5%
Geothermal	16.8	0.5%
Wind	4.5	0.12%
Solar	0.8	0.02%
Total	3,711.8	

http://www.eia.doe.gov/pub/oil_gas/petroleum/analysis_publications/oil_market_basics/default.htm. "Electrical Energy Consumption."

Hydroelectric

Water stored in high-elevation lakes or held by dams creates high-pressure water at the bottom of the dam. The energy stored in the high-pressure water can be converted to shaft work using hydroelectric turbines to produce electricity. Most of the good dam sites in the United States have been developed, so this is a relatively mature industry. At 319.5 billion kWh in 1999, hydroelectric power on the grid provided 8.6% of the electrical energy production. Environmentalists oppose dam construction in the Northwest, and there is active lobbying to remove dams on the Columbia River.

Wind Energy

There have been over 8 million wind turbines installed since the 1860s in the United States. Wind energy is one of the oldest and most widely used forms of power. Traditionally, the shaft work was used to pump water or mill grain. In the 1930s, an infant electrical power industry was driven out of business by policies favoring the distribution of fossil fuel electricity.[29]

Between 1976 and 1985, over 5,500 small electric utility units (1–25 kW) were installed on homes and remote locations in response to high oil prices. The installation and use dwindled when government subsidies ended in 1985.[30] More recently, wind farms have been installed to provide grid electrical power. Between 1981 and 1990, approximately 17,000 machines were installed with output ranging from 20 to 350 kilowatts with a total capacity of 1,700 megawatts.[31]

The price of electrical power from wind has decreased from more than $1.00 per kWh in 1978 to about $0.05 per kWh in 1998, with costs projected as low as $0.025 per kWh (for large wind farms). At $0.025 per kWh, wind power competes with the fuel costs for most fossil fuel power plants. Projections aside, for most locations in 2002, a low price for wind power was $0.045 per kWh,[32] but some wind farms need $0.06 per kWh to operate at a profit. The primary issue for more widespread use of wind power is not the cost of the wind turbines. Issues are (1) high maintenance costs because of the large number of wind turbines needed to generate as much power as a typical coal-fired power plant, (2) environmental impact (noise pollution and poor aesthetics), and (3) dependable power on demand. (The wind doesn't always blow.)

The dependability issue is the ability of wind power to supply continuous electrical power. The ability of a facility to provide electricity is characterized by its capacity factor. This factor is the actual energy supplied by a wind turbine compared to the theoretical power supplied if it operated continuously at its design capacity. Wind power suffers from low-capacity factors because of the lack of wind at night and the lack of power demand when the wind is blowing. Capacity factors for wind farms range from 0.20 to 0.35 compared to 0.5 for fossil fuel plants, 0.6 for some new gas turbine plants, and 0.85 for nuclear power.[33]

One of the less obvious opportunities for electrical power supply is energy storage. Storing wind energy as it is available for use during peak demand times will increase the value of the wind energy and would increase capacity factors. This could increase the value of wind energy from a wind farm by a factor of three or more. Such an increase in the value of wind energy would change the economic outlook of wind power from marginal to profitable.

Wind power generated 4.5 billion kWh in 1999 or 0.12% of the electrical energy production in the United States.

Biomass

Energy storage limits the utility for both wind power and solar energy. Nature's way for storing solar energy is biomass. Biomass is biological material: vegetation, grass, wood, corn, palm oil, and similar plant material. Time, the absence of oxygen, and compaction (promoted by Earth overburdens) convert biomass to coal, petroleum, or other geological variations of these materials.

Wood has been used through the ages to produce heat. Today, wood supplies heat and is used to generate electrical power, corn is converted to ethanol, and vegetable oils are converted to biodiesel. Unlike fossil fuels, biomass is not available in reserves that have accumulated for years. Rather, biomass grows each year and must be harvested and used quickly to get quality fuel. The supply must be renewed every year.

The availability of biomass is reported as estimated amounts produced per year. A wide range of biomass types, as well as different terrain and climate, control biomass availability. The supply of biomass tends to increase with increasing prices. Table 3-4 summarizes example prices and availability of solid biomass (not including fats and oils).

Solid biomass is used for energy five different ways: (1) burning for heat or electrical power generation; (2) conversion to ethanol; (3) pyrolysis; (4) gasification; and (5) anaerobic (without oxygen) methane production (landfill gas). The high cost of biomass makes conversion to liquid fuels and use as chemical feed stocks the best applications for most of these renewable energy sources. Direct combustion and anaerobic methane are handled on a case-by-case basis, where they are generally profitable if the biomass has already been collected for other reasons. For quality liquid fuel production, two technologies stand out: ethanol production and gasification for Fischer-Tropsch fuel production. When including oil seed crops (e.g., soybeans and rapeseeds) a third option of biodiesel is also becoming quite attractive.

Ethanol and Biodiesel from Agricultural Commodities

Table 3-4 shows the number of gallons of ethanol that can be produced from the most common forms of biomass. The corn data of Table 3-4 are important points of reference. Corn is the largest commodity crop in the United States and provides high yields of dense biomass. While the price per ton of corn is almost twice the price of large-volume switchgrass and wood, the corn prices and volumes are current, while the other biomass prices are based on estimates that may be optimistic.

Dried distiller grains are a by-product sold as a high-protein, high-fat cattle feed when producing ethanol from corn. Over half

TABLE 3-4
U.S. estimate of supplies and production potential of ethanol from biomass.[A] Except for reliable corn numbers, conversions are optimistic.

	Price $/Dry Ton	Quantity Million Dry Tons/Yr	Conversion Gallons Ethanol/Ton	Ethanol Equivalent Millions of Gallons/Yr	Cost of Feedstock/ Gallon Ethanol
$2.40 bu (56 lb) Corn, U.S.[B,C]	85.7	280	89	24,920	$0.96
Refuse Derived Waste	15	80	80	6,400	$0.19
Wood Waste *Cheap*	30	10	110	1,100	$0.27
Wood Waste *Expensive*	45	80	110	8,800	$0.41
Switchgrass *Cheap*	25	5	95	475	$0.26
Switchgrass *Expensive*	45	250	95	23,750	$0.47

Estimated per-ton yields of ethanol from corn, sorghum, refuse-derived fuel, wood waste, and stitchgrass are 89, 86, 80, 110, and 95, respectively. Corn has 56 pounds per bushel with an assumed price of $2.40 per bushel ($85.71/ton) with an annual U.S. production estimate of 10 billion bushels or 280 million tons.

[A] K. Shaine, T. S. Tyson, P. Bergeron, and V. Putsche, "Modeling the Penetration of the Biomass-Ethanol Industry and Its Future Benefits," September 18, 1996, Bioenergy '96, Opryland Hotel, Nashville, TN.
[B] R. M. Tshiteya, *Conversion Technologies: Biomass to Ethanol.* Golden, CO: National Renewable Energy Laboratory, September 1992, pp. 3–5.
[C] http://www.usda.gov/nass/pubs/trackrec/track02a.htm#corn. Corn production U.S.

of the corn costs are recovered by the sale of this by-product.[34] It is because of these by-products that corn is used more than other biomass crops (e.g., sugar cane). Other biomass materials may actually have a higher yield of ethanol per acre, but they do not have the valuable by-products.

Corn is the most commonly used biomass for producing ethanol. The production process consists of adding yeast, enzymes, and nutrients to the ground starchy part of corn to produce a beer. The beer contains from 4% to 18% ethanol, which is concentrated by distillation that is similar to the distillation used to produce whiskey. The final fuel ethanol must contain very little water for use as motor fuel. In gasoline, the water is an insoluble phase that extracts the ethanol and settles to the bottom of the tank. If this water gets to the engine, the engine will stall.

About 90 gallons of ethanol are produced from one ton of corn. The production cost is $1.05 to $1.25 per gallon. Ethanol has about two-thirds the energy content of gasoline,[35] so these prices translate to $1.57 to $1.85 per equivalent gasoline gallon. This is more than gasoline that is produced for about $1.30 per gallon[v] before the $0.42 (average) motor fuel tax is added.

Estimates of gasoline used in U.S. cars, vans, pickup trucks, and SUVs are about 130 billion gallons of gasoline per year.[36] (These numbers agree with the motor gasoline consumption of 3.05 billion barrels reported elsewhere.) About 500 million prime, corn-producing acres would be required for ethanol to replace all of the gasoline. This is about one-quarter of the land in the lower 48 states. The lower 48 states have about 590 million acres of grassland, 650 million acres of forest, and 460 million acres of croplands (most is not prime acreage).[37]

If all of the current corn crop were converted to ethanol, this would replace about 17 billion gallons of gasoline—less than 15% of our current consumption. Estimates of dedicating acreage for ethanol production equivalent (yield-based) to current gasoline consumption would require nine times the acreage used for the current U.S. corn crop. This approach is not realistic. However, if hybrid vehicle technology doubles fuel economy and the electrical power grid further reduces gasoline consumption to about 60 billion gallons, substantial ethanol replacement of gasoline is possible. Use of perennial crops would be a necessary component of large-volume ethanol replacement for gasoline.

Corn is an annual crop and must be planted each year. For this reason, costs for mass production of wood and grasses are potentially less than corn. In the late 20th century, corn-to-ethanol production facilities dominated the biomass-to-ethanol industry. This was due to (1) less expensive conversion technologies for starch-to-ethanol compared to cellulose-to-ethanol required for wood or grasses and (2) generally ambitious farmer-investors who viewed this technology as stabilizing their core farming business and providing a return on the ethanol production plant investment. State governments usually provide tax credits for investment dollars to build the ethanol plants, and there is a federal subsidy for fuel grade ethanol from biomass.

---

[v] $50 per barrel divided by 42 gallons per barrel plus a refining cost.

Because of lower feedstock costs (see Table 3-4), wood-to-ethanol and grass-to-ethanol technologies could provide lower ethanol costs—projections are as low as $0.90 per equivalent gasoline gallon.[38] Research focus has recently been placed on cellulose-to-ethanol. The cost of cellulose-to-ethanol has improved from more costly to about the same as corn-to-ethanol technology. Based on present trends, cellulose-to-ethanol technology could dominate ethanol expansion in the 21st century. It would require large tracts of land dedicated to cellulose production.

The current world production of oils and fats is about 240 billion pounds per year[39] (32.8 billion gallons, 0.78 billion barrels), with production capacity doubling about every 14 years. This compares to a total U.S. consumption of crude oil of 7.1 billion barrels per year[40] of which 1.35 billion barrels is distillate fuel oil (data for the year 2000).[41] With proper quality control, biodiesel can be used in place of fuel oil (including diesel) with little or no equipment modification. Untapped, large regions of Australia, Colombia, and Indonesia could produce more palm oil. This can be converted to biodiesel that has 92% of the energy per gallon as diesel fuel from petroleum.[42] This biodiesel can be used in the diesel engine fleet without costly engine modifications.

In the United States, ethanol is the predominant fuel produced from farm commodities (mostly from corn and sorghum), while in Europe, biodiesel is the predominant fuel produced from farm commodities (mostly from rapeseed). In the United States, biodiesel is produced predominantly from waste grease (mostly from restaurants and rendering facilities) and from soybeans.

In the United States, approximately 30% of crop area is planted to corn, 28% to soybeans, and 23% to wheat.[43] For soybeans this translates to about 73 million acres (29.55 million hectares) or about 2.8 billion bushels (76.2 million metric tons). Soybeans are 18%–20% oil by weight, and if all of the U.S. soybean oil production were converted[vi] to biodiesel, it would yield about 4.25 billion gallons of biodiesel per year. Typical high yields of soybeans are about 40 bushels per acre (2.7 tons per hectare), which translates to about 61 gallons per acre. By comparison, 200 bushels per acre of corn can be converted to 520 gallons of ethanol per acre.

Table 3-5 compares the consumption of gasoline and diesel to the potential to produce ethanol and biodiesel from U.S. corn

---

[vi] 76.2 million metric tons of beans is about 14.5 billion kilograms of soybean oil. This translates to about 16 billion liters, using a density of 0.9 g/cc or about 4.25 billion gallons per year.

TABLE 3-5
Comparison of annual U.S. gasoline and diesel consumption versus ethanol and biodiesel production capabilities.

Gasoline Consumption (billions of gallons per year)	130
Distillate Fuel Oil (including diesel) Consumption	57
Ethanol from Corn (equivalent gasoline gallons)	25 [17]
Biodiesel from Soybeans (equivalent diesel gallons)	4.25 [3.8]

and soybeans. If all the starch in corn and all the oil in soybeans were converted to fuel, it would only displace the energy contained in 21 billion gallons of the 187 billion gallons of gasoline and diesel consumed in the United States. Thus, the combined soybean and corn production consumes 58% of the U.S. crop area planted each year. It is clear that farm commodities alone cannot displace petroleum oil for transportations fuels. At best, ethanol and biodiesel production is only part of the solution. U.S. biodiesel production in 2005 was about 0.03 billion gallons per year compared to distillate fuel oil consumption of 57 billion gallons per year.

Converting corn and soybean oil to fuel is advantageous because the huge fuel market can absorb all excess crops and stabilize the price at a higher level. In addition, in times of crop failure, the corn and soybeans that normally would be used by the fuel market could be diverted to the feed market. The benefits of using soybeans in the fuel market can be further advanced by plant science technology to develop high-oil content soybeans.

Soybeans sell for about $0.125 per pound, while soybean oil typically sells for about twice that ($0.25 per lb). The meal sells for slightly less than the bean at about $0.11 per pound. Genetic engineering that would double the oil content of soybeans (e.g., 36%–40%) would make the bean, on the average, more valuable. In addition, the corresponding 25% reduction in the meal content would reduce the supply of the meal and increase the value of the meal. At a density of $0.879 g/cc$, there are about 7.35 lbs of biodiesel per gallon. A price of $0.25 per lb corresponds to $1.84 per gallon; $0.125 per lb to $0.92 per gallon.

Fuel production from corn and soybean oil would preferably be sustainable without agricultural subsidies of any kind (none for ethanol use, biodiesel use, farming, or not farming). A strategy thus emerges that can increase the value of farm commodities, decrease oil imports, decrease the value of oil imports, and put U.S. agriculture on a path of sustainability without government subsidies. To be successful, this strategy would need the following components:

1. **Develop better oil-producing crops.**
   - Promote genetic engineering of soybeans to double oil content and reduce saturated fat content (saturated fats cause biodiesel to plug fuel filters at moderate temperatures).
   - Promote the establishment of energy crops like the Chinese tallow tree in the South that can produce eight times as much vegetable oil per acre as soybeans.
2. **Pave the future for more widespread use of diesel engines and fuel cells.**
   - Promote plug-in HEV technology[44] that uses electricity and higher fuel efficiency to displace 80% of gasoline consumption. Apply direct-use ethanol fuel cells for much of the remaining automobile transportation energy needs.
   - Continue to improve diesel engines and use of biodiesel and ethanol in diesel engines. Fuel cells will not be able to compete with diesel engines in trucking and farm applications for at least a couple decades.
3. **Pass antitrust laws that are enforced at the border.**
   - If the oil-exporting countries allow the price of petroleum to exceed $70 per barrel ($2.00 per gallon diesel, not including highway taxes), do not allow subsequent price decreases to bankrupt new alternative fuel facilities.
4. **Fix the dysfunctional U.S. tax structure.**
   - Restructure federal and state taxes to substantially eliminate personal and corporate income taxes and replace the tax revenue with consumption taxes (e.g., 50%) on imports and domestic products. This would increase the price of diesel to $2.25 per gallon (red diesel, no highway tax).
   - Treat farm use of ethanol and biodiesel as an internal use of a farm product, and, therefore, no consumption tax would be applied.

Increased use of oil crops would include use of rapeseed in drier northern climates (rapeseed contains about 35% oil) and use of Chinese tallow trees in the South. Chinese tallow trees are capable of producing eight times as much oil per acre as soybeans. If Chinese tallow trees were planted in an acreage half that of soybeans and the oil content of soybeans were doubled, 17–20 billion gallons of diesel could be replaced by biodiesel allowing continued use of soybean oil in food applications. This volume of biodiesel production would cover all agricultural applications and allow the imports to be optional.

Chinese tallow trees are one of the fastest-growing trees. In addition to producing oil crops, clippings and old trees could be used for ethanol production. Chinese tallow tree orchards could readily become the largest agriculture crop in the United States. High-protein animal feed is an additional potential by-product.

The plug-in HEV technology would displace about 104 billion gallons per year of gasoline with electricity and increase efficiency. The electricity could be made available from the reprocessed spent nuclear fuel and adding advanced technology nuclear reactors. About half of the remaining 26 billion gallons of gasoline could be displaced with ethanol and half with continued use of gasoline.

In this strategy, up to 55 billion gallons of annual diesel and gasoline consumption would still need to be met with fossil fuel sources. These could be met with petroleum, coal-derived liquid fuels (like Fischer-Tropsch fuels), and Canadian oil sand fuels. Increase of electric trains for freight could displace much of the 55 billion gallons. It would be a buyer's market for liquid fossil fuels.

These proposed technologies are cost effective and sustainable at $55 per barrel of crude oil and a tax strategy that equally taxes domestic and imported products (the consumption tax). A variety of continued strategies would assure that the United States did not have to import petroleum and that farmers could achieve higher-value nonfood uses for their products.

The consumption tax is emerging as a preferred way to end the stress on U.S. manufacturing with domestic taxes that are not applied to imports. Oil prices are already $55 per barrel level. All the technology to replace petroleum is demonstrated and cost-effective with the possible exception of low-temperature direct-use ethanol fuel cells. Intermediate temperature PEM fuel cells should make direct use of ethanol cost effective in ten years.

Table 3-6 summarizes the current uses, volumes, and prices of ethanol production and compares these to a likely scenario if a consumption tax is implemented and antitrust laws are enforced at the border. Table 3-7 summarizes the impact of the current federal incentive of 5.4 ¢ per gallon tax exemption that goes to blenders placing 10% ethanol in gasoline—this is applied against the 18.4 ¢ federal excise tax on gasoline.[45] The federal tax incentive is paid to the blenders, so if the blender pays $1.25 per gallon for ethanol, the federal government provides a reduction ($0.54 per gallon of ethanol blended to 10% ethanol in gasoline) in the highway taxes that are paid by the blender.

TABLE 3-6
Current and projected costs and production of U.S. ethanol and biodiesel.

	Application	Billions of Gallons per Year	Pricing
2005			
Ethanol[liv]	Octane Enhancer, Oxygenate for CO Nonattainment	3.4	$1.20–$1.50
Biodiesel	Primarily as 2% to 20% Additive in Bus Fleets and Farm Applications	0.03	$1.30–$2.30 per gallon
2015 (2025)			
Ethanol	Direct-Use Ethanol Fuel Cells, Octane Enhancer, Oxygenate for CO Nonattainment Areas	10 (20)	$1.50 per gallon
Biodiesel	Predominant Farm Fuel, 50% Market Share in South, Fleets	5 (15)	$2.10 for farm application (soybean and rapeseed based, no tax); $2.25 for South (beef tallow tree based, including consumption tax)

[liv] 2004 Gasoline Price Increases: An Analysis Summary Prepared by Renewable Fuels Association. (http://www.ethanolrfa.org/, March 2004.)

TABLE 3-7
Example cost of ethanol and impact of tax credit.

Example Ethanol Wholesale Price	123¢/gallon
Alcohol Fuel Tax Incentive	54¢/gallon
Effective Ethanol Price	69¢/gallon
Effective Ethanol Price for Energy in 1 Gallon Gasoline	103¢
Gasoline Wholesale Price ($55/barrel crude oil, $0.14/gallon refining cost)	145¢/gallon

The price of petroleum fuels relative to the price of vegetable oil is an important factor that will impact sustainable biodiesel and ethanol production. Based on $55 per barrel petroleum and a consumption tax strategy that would tax imports the same as domestic production, there is a basis for developing a biodiesel and ethanol industry that can be sustainable and compete with $2.25 per gallon diesel (includes consumption tax, excludes highway tax).

The price of vegetable oil ranges from $0.92 to $1.84, depending on whether the oil is priced at the same value as the soybean or a premium price is received for the oil component of the bean. In principal, higher-oil seed crops could sustainably provide vegetable oil at $1.50 per gallon while maintaining a premium value (more on a per-pound basis than soybeans) for the bean. At a reasonable $0.40 per pound processing cost, a biodiesel cost of $1.90 per gallon would be sustainable. Prices as low as $1.70 per gallon may be attainable with catalyst development. This gives $2.25 per gallon for diesel. With the 50% consumption tax, imported oil would have to be below $45 per barrel for the price of the diesel to be less than $1.90 per gallon.

Current subsidies of $0.54 per gallon for ethanol and $1.00 per gallon for biodiesel approximately compensate for all the U.S. taxes collected on agriculture and processing that go toward the production of these fuels (essentially no U.S. taxes are applied on imported petroleum). These incentives are appropriate in view of current U.S. tax strategies. A continued use of these incentives rather than a consumption tax would have at least three drawbacks: (1) many citizens will perceive the incentive as political favor rather than a mechanism to allow the fuels to compete fairly with imported petroleum, (2) the incentives require periodic renewal and can be eliminated when production becomes high enough, and (3) the incentives currently do not apply to other technologies such as plug-in HEV technology that would help realize the true value of ethanol to replace petroleum.

If large-scale Chinese tallow tree farming were to occur, the farming should be profitable at oil prices as low as $1.30 per gallon ($0.90 per gallon for the oil plus $0.40 for processing). At these prices, the biodiesel would compete in the trucking fuel industry where a 50% consumption tax would take that to $1.95 per gallon—considerably less than petroleum at $2.25 per gallon.

For ethanol use in advanced plug-in HEVs, the technology is demonstrated and cost effective with the possible exception of low-temperature direct-use ethanol fuel cells. Intermediate temperature PEM fuel cells should make direct use of ethanol cost effective in less than ten years.

Consumption taxes will not represent an additional tax burden on U.S. consumers if properly implemented; however, the taxes will be more apparent. The average consumer will undoubtedly welcome the absence of income taxes. However, when the price tag on that new $30,000 pickup truck becomes $45,000, there will be some distress. The implementation should be gradual to allow consumers to become accustomed to paying taxes, at the point of sale rather than on income. An initial phase of eliminating corporate income taxes, about a 5% reduction in all personal income taxes and a 20%–30% consumption tax, would be a good start. The price of a pickup truck should not increase by 20%–30%, because the lack of corporate income taxes, should allow the $30,000 pickup to first have a price decrease to about $28,000 and a total price of about $34,000 when the tax is applied.

The United States is poised to level the playing field between imported and domestic production through use of consumption taxes. This correction to a current tax structure is much needed. If and when this transition happens, U.S. farmers would do well to support the transition and to make sure that ethanol and biodiesel used in agricultural applications would be free of this consumption tax. It should be considered an internal transaction.

Emergence of Nuclear Power

In the pursuit of sustainable energy, nuclear power emerges for four reasons:

1. On a Btu basis, nuclear fuel is the least expensive, and it is economically sustainable. Nuclear fuel has the potential to be ten times less expensive than any alternative (less than $0.10 per MBtu).
2. Nuclear fuel is the most readily available fuel. It is stockpiled at commercial reactors in the form of spent fuel.
3. Nuclear fuel is the most abundant. Enough has already been mined, processed, and stored in the United States to supply all energy needs for centuries.
4. There is no technically available alternative to give sustainable energy supply for current and projected energy demand.

This last point is emphasized in this chapter. It is impractical to try to replace transportation fuels with biomass, let alone nontransportation energy expenditures. The limited availability of petroleum is already inciting military conflict to keep the oil flowing, not to mention the contribution of the trade deficit drag on the U.S. economy. The imports of natural gas are growing rapidly, and at prices greater than $6 per MBtu, it is too expensive for use for electrical power generation.

Coal will be important for decades to produce electrical power and for centuries as a feedstock to the chemical industry. However, coal is already used for about 50% of electric power production (see Table 3-3). Nuclear energy is less expensive on a fuel basis. Chapter 13 provides a more rigorous comparison of electrical power costs for coal versus nuclear.

References

1. http://www.eia.doe.gov/fueloverview.html.
2. http://www.eia.doe.gov/fueloverview.html.
3. http://www.eia.doe.gov/fueloverview.html.
4. Ibid. A–Z.
5. *Kirk-Othmer Encyclopedia of Chemical Technology.* New York: John Wiley & Sons, 1980, Vol. 12, p. 326.
6. From U.S. DOE Public Domain. http://www.eia.doe.gov/pub/oil_gas/petroleum/analysis_publications/oil_market_basics/Supply_Petroleum_Trap.htm, 2003.
7. *Kirk-Othmer Encyclopedia of Chemical Technology.* New York: John Wiley & Sons, 1980, Vol. 12, p. 326.
8. *Kirk-Othmer Encyclopedia of Chemical Technology.* New York: John Wiley & Sons, 1977, Vol. 11, p. 326.
9. S. M. Farouq Ali, Heavy Crude Oil Recovery, in Ender Okandan, *The Hague.* Boston: Martinus Nijhoff Publisher, 1984, p. 50.
10. *Kirk-Othmer Encyclopedia of Chemical Technology.* New York: John Wiley & Sons, 1980, Vol. 11, p. 887.
11. Ibid.
12. http://www.petroleumworld.com/SF021906.htm.
13. Canada's Oil Sands and Heavy Oil. Petroleum Communication Foundation, Canadian Centre for Energy Information, Alberta. See http://www.centreforenergy.com/EE-OS.asp.
14. Evolution of Canada's Oil Gas Industry. Canadian Centre for Energy Information, Alberta, Canada, 2004. See http://www.centreforenergy.com/EE-OS.asp.
15. http://www.agiweb.org/legis105/ch4.html.
16. Mielke, J. E. Methane Hydrates: Energy Prospect or Natural Hazard? CRS Report to Congress. Order Code RS20050, February 14, 2000.
17. *Kirk-Othmer Encyclopedia of Chemical Technology.* New York: John Wiley & Sons, 1980, Vol. 12, p. 326.

18. http://www.eia.doe.gov/fueloverview.html.
19. http://hyperphysics.phy-astr.gsu.edu/hbase/astro/procyc.html, May 15, 2002.
20. http://zebu.uoregon.edu/textbook/energygen.html, May 15, 2002.
21. http://www.lbl.gov/Publications/Currents/Archive/Nov-19-1999.html.
22. Images provided by the Uranium Information Center, Ltd., GPO Box 1649N, Melbourne 3001, Australia.
23. http://www.uic.com.au/uran.htm.
24. Images provided by the Uranium Information Center, Ltd., GPO Box 1649N, Melbourne 3001, Australia.
25. http://www.upei.ca/~physics/p261/projects/geothermal1/geothermal1.htm.
26. http://www.energy.rochester.edu/is/reyk/history.htm.
27. http://www.radford.edu/~wkovarik/envhist/.
28. http://www.californiasolarcenter.org/history_solarthermal.html.
29. http://www.bergey.com/primer.html.
30. Ibid.
31. http://telosnet.com/wind/recent.html.
32. Conversation with Idaho Power's representatives at Bioenergy'02, Boise, Idaho, September 2002.
33. http://telosnet.com/wind/future.html.
34. R. M. Tshiteya, *Conversion Technologies*, pp. 3–21.
35. Lester B. Lave, W. M. Griffin, and H. Maclean, "The Ethanol Answer to Carbon Emissions." *Issues in Science and Technology*, Winter 2001–2002, pp. 73–78.
36. http://www.eia.doe.gov/emeu/aer/, 8.36 million barrels per day consumed.
37. Lester B. Lave, et al., p. 75.
38. K. Shaine Tyson, et al.
39. M. Bockish, *Fats and Oils Handbook*. Champaign, IL: American Oil Chemists Society Press, 1993, p. 11. 20 MMt of soybean oil, SBO at 21% of oil crops, oil crops at 80% of fat and oil.
40. http://www.eia.doe.gov/pub/oil_gas/petroleum/analysis_publications/oil_market_basics/default.htm. 19.48 million barrels per day or 7110.2 million barrels per year.
41. http://www.eia.doe.gov/emeu/aer/. 3.7 million barrels per day or 1.35 billion barrels per year.
42. Peterson, C. L. and D. L. Auld, Technical Overview of Vegetable Oil as a Transportation Fuel. *FACT*–Vol. 12, Solid Fuel Conversion for the Transportation Sector, ASME 1991, page 45. Found at biodiesel.org.
43. Soy Stats™ 2004. American Soybean Association, 12125 Woodcrest Executive Drive, Suite 100, Saint Louis, MO 63141.
44. http://www.missouri.edu/~suppesg/PHEV.htm.
45. 26 U.S.C. 40.

CHAPTER 4

Emerging Fuel Technologies and Policies Impacting These Technologies

Politics of Change in the Energy Industry

Energy conversion and utilization are a multitrillion-dollar business and vital to today's society. The magnitude of these industries presents both mega opportunities and extraordinary challenges. While these industries were once predominantly driven by technology, today politics dominates essentially every aspect. One of today's greatest challenges is to advance an industry in which technology has been displaced from the role of driving the industry to the passive role of yielding to the political policies that drive the industry.

The politics of energy include contributions from both corporations and environmental groups. Mega corporations seek to maintain the status quo because these corporations control billions of dollars of revenue per year. Environmental groups including the U.S. Environmental Protection Agency tend to be single issue and focus on their perception of environmental impacts. Typically, true balances of benefit and risk are much more complicated than the portrayals of environmental groups. A quadrangle of overlapping interests and conflicts is formed by industry, environmentalists, politics, and public welfare.

Technology Emerging to What End?

Several fundamental issues involve defining and discussing emerging energy technologies, including the following:

- When the goal of proposed improvements is defined, the technical challenges can be substantial.
- When the goal of proposed improvements is not defined, the challenge borders on futile.
- Which comes first: the fuel or the engine? Change in energy technology is uncertain and filled with obstacles, but they must match.
- While new technologies may be on the sidelines ready to meet the goals, the momentum of the industry is formidable and will dominate the discussion.

In view of these obstacles in the energy industry, the task of identifying the potential of an emerging technology is difficult. Proposed objectives to be met by new technology should have a reasonable chance of surviving and be substantially free of speculation. To this end, one development goal stands out: *Emerging energy technology must be based on fuel price that includes anticipated environmental costs. The cost of the fuel must include all costs related to collection and delivery to the consumer.*

For the electrical power industry, the energy fuel consumer is the large utility company that generates electrical power. For the vehicular fuel industry the millions of individual owners of cars and trucks are the final consumer. Furthermore, there must be enough of the fuel available to justify the investment in infrastructure, so the fuel supply must be sustainable in the 30-year timeframe. The importance of sustainability after the 30-year timeframe is uncertain in view of potential breakthroughs in science and technology. Sustainability addresses pollution generation based on regulations already in place.

This 30-year timeframe is greater than the 3- to 10-year timeframe that large corporations plan to get a full return on investments but less than the total sustainability pursued by advocates of wind and solar energy.

In addition, historic trends dictate the following:

- Vehicular fuels must be a liquid that is no more hazardous than gasoline. Electrical power must be 60 cycle AC to be compatible with the national grid.

- The final component that will allow future technologies to be evaluated in a systematic way is that both electrical power and engine conversion technologies will be driven by fuel costs within reason.

These objectives assume that the industry is technology driven. In the following discussion the potentials of different energy feedstocks will be reviewed. Corporate and government policies and technology histories will then be presented to explain why certain technologies have not been commercialized to their potential.

Cost of Feedstock Resources

Without detailed process knowledge, as is the case when evaluating the potential of emerging technology, the potential of a process can be estimated by calculating the *gross profit*.[1] In this approach, the cost of the feedstock fuel consumed by a process is subtracted from the value of the final product (a liquid fuel or electricity). The higher the "gross profit," the greater the potential of that process. Technology is developed with the goal of realizing actual profits that approach the gross profits.

Table 4-1 summarizes and ranks feedstock costs by their representative prices at the end of the 20th century. The "Liquid Fuel" column reports the feedstock costs for the energy equivalent to that in a gallon of gasoline. The last column estimates the feedstock cost for producing one kWh of electricity.

The lowest-cost feedstocks provide the best potential for low-cost energy. Municipal solid waste, spent nuclear fuel, uranium, and coal have high gross profits and are sufficiently abundant to evaluate for commercialization. In each case, technology can bridge any gaps between actual and gross profits.

Wastes and By-Products as Feedstocks

The greatest potentials exist with municipal solid waste and spent nuclear fuel. Historically, both vehicular fuel and electrical power technologies were alive and well before there was a municipal solid waste problem of any magnitude and before we were aware of the potential for nuclear energy. Both municipal solid waste and nuclear waste (spent nuclear fuel) provide an opportunity to

78 Sustainable Nuclear Power

TABLE 4-1
Summary of feedstock costs for commonly used and considered fuels.

Fuel	Price ($/MMBtu)	AVG. ($/kWh)	Liquid Fuel* ($/gasoline gallon equ.)	El. Conv. Efficiency** (%)	Electricity Cost ($/kWh)
Municipal Solid Waste (MSW)	$(−2.00) to (−4.00)	−0.0102	−0.36	20–35	−0.037
Spent Nuclear Fuel	$(0.08)	−0.0003	−0.01	25–45	−0.001
Full Uranium	$0.08	0.0003	0.01	25–45	0.001
Remote Natural Gas	$0.50–$1.00	0.0026	0.09	N/A	
U.S. Uranium	$0.62	0.0021	0.072	30–33	0.0068
French Reprocessed Uranium					0.0090
Coal	$1.20–$1.40	0.0044	0.145	35–45	0.0105
Oil sands ($10–$15/barrel)	$2.00–$3.00	0.0086	0.30	28–53	0.021
Natural Gas	$6.00	0.0205	0.68	50–56	0.039
Biomass	$2.10–$4.20–$6.80	0.0149	0.52	20–45	0.044
Petroleum ($45–$75/barrel)	$9.00–$15.00	0.0411	1.44	28–53	0.135

Note: Price does not include $0.42/gal motor vehicle fuel tax.
*Assuming 0.119 MMBtu per gallon of gasoline.
*Electrical conversion efficiency is the thermal efficiency of the cycle. The price of the feedstock fuel is divided by the thermal efficiency to estimate the cost of fuel consumed to generate 1 kWh of electricity.

generate useful energy from a waste that imposes a burden on society.

Today's U.S. nuclear power plants extract about 3.4% of the energy available in the fuel rods before they are set aside as spent fuel. The spent fuel elements are stored at each of the nuclear power plants, and transporting them to a repository in Yucca Mountain is the subject of bitter debate. Reprocessing the fuel reduces the mass of the waste requiring long-term storage and recovers the 96.5% of unspent fuel. This reprocessing is practiced in Europe and Japan, but not in the United States. Excess weapons-grade uranium/plutonium can also be blended with natural uranium to make power plant fuel and eliminate this inventory as a weapons threat.

Similar to nuclear reprocessing, the conversion of municipal solid waste into fuel and electricity can eliminate a waste problem and provide energy. Landfill corporations receive $15 to $60 per ton to dump this waste in a properly prepared hole in the ground and, unfortunately, often see waste-to-energy projects as competition for their revenues. The municipal waste-to-energy must overcome at least four opposing groups: (1) direct political pressure from electrical power providers and landfill corporations, (2) air quality regulations that make it difficult to build the new conversion facilities, (3) the cost and availability of conversion technology, and (4) Mafia-type control on the collection of solid waste. Full use of municipal solid waste for energy is as much a political issue as it is a technology opportunity.

In the absence of American leadership, answers are likely to come from Europe. In Japan and Europe, landfill land is becoming less available. In Germany, it will be illegal to dump waste of high caloric value after 2005.[2] This one German law forces resolution of three of the four issues related to waste-to-energy technology. The opportunity is there, and technology will be developed!

Energy facilities designed to run on waste products typically benefit from the economies of scale made possible by adding other feedstocks to the mix. Municipal solid waste plants will be able to take in biomass produced near the plant locations where hauling distances are short. Coal could also be used in the municipal solid waste plants. Cofiring these facilities with biomass and coal would provide improved economies of scale and feedstock reliability to keep the plant in operation. In the future, the more advanced facilities would be true solid fuel refineries with the option to produce chemicals, liquid fuels, electrical power, and recycled metals.

Liquid Fuels Market

Both nuclear power and municipal solid waste facilities are better suited to produce electrical power rather than liquid fuels. For liquid fuels, the next least expensive feedstock options are remote natural gas, coal, and oil sands (and heavy oil). Here, the liquid fuels market is not synonymous with automobiles. (A case will be presented in Chapter 9 for conversion to electrical power as the primary distributed energy for automobiles.) However, for farm tractors, trucks, aircraft, and remote railways, a viable alternative to liquid fuel dominance is not in sight.

As Table 4-1 indicates, biomass will have a difficult time competing with oil on a cost basis. While it is true that ethanol and biodiesel may actually have periodic cost advantages over gasoline and diesel as these industries expand (2001–2008), cost advantages will disappear when the volumes of ethanol and biodiesel expand using "excess" agricultural commodities competing head on with food applications. A side effect of this expansion will be higher farm commodity prices. This could be a good thing, as it could bring an end to government-subsidized agriculture.

Oil corporations will likely (already are) switch from petroleum raw material to tar sands and heavy oils as the more easily accessed oil reserves are depleted. These fuels will work with incremental modification of existing refining infrastructure. Today, the use of tar sands tends to be a better long-term investment than conventional petroleum. The driver for commercialization will be corporate profitability. This type of industrialization should have little problem providing liquid fuels well past the end of the 21st century—for a price. Lessons of history show that the environment, military conflict, major trade deficits, and huge fluctuations in prices can be the results of industrial expansion driven by corporate profitability.

A problem with the liquid fuel industry is the lack of diversity in feedstocks; petroleum feedstocks and oil corporations dominate the industry. One of the paths forward in this industry is electrical power (including nuclear) and increased automobile efficiency to displace 33%–50% of this 187 million gallon a year industry. If biodiesel and ethanol displace another 10%–20% of this industry, the dominance of petroleum will be diminished. This diversity will tend to stabilize prices, and nations can reduce trade deficits. Diversification is the key.

Fischer-Tropsch fuels are a fourth player in diversification of the liquid fuel industry. Currently, Fischer-Tropsch conversion tends to be an avenue that is present in the oil corporation portfolio but likely to be introduced by other industries.

Fischer-Tropsch Technology

Fischer-Tropsch technology is being used today to produce liquid fuels from both coal and natural gas. The Fischer-Tropsch technology can provide liquid fuels at prices competitive with petroleum today. Estimates indicate that Fischer-Tropsch technology is borderline competitive with crude oil at $20 per barrel (eliminating nontechnical cost barriers). However, the numbers are misleading, since they do not account for the superior quality of Fischer-Tropsch fuels. Unlike crude oil that must be refined, typically resulting in a 28% price increase, vehicle-ready fuel can be produced at a Fischer-Tropsch facility. With this 28% adjustment, the *petroleum oil equivalent* price of Fischer-Tropsch fuels from U.S. coal is $26 to $27 per barrel.

Fischer-Tropsch coal-to-liquids technology is commercial in South Africa and Malaysia and produces vehicle-ready fuels at less than the biomass feedstock costs. Fischer-Tropsch fuels are fully compatible with petroleum pipeline practices, unlike ethanol and biodiesel which are incompatible and will carry premium distribution costs. The Fischer-Tropsch fuels are also near-zero sulfur fuels in compliance with 2007 EPA fuel requirements.[3]

Figure 4-1 summarizes alternative benchmark technologies positioned to complement current electrical power and Fischer-Tropsch liquid fuel infrastructures. The alternative industries of Figure 4-1 have the potential to both reduce energy costs, address waste disposal problems, provide domestic supplies of energy, and provide needed diversity in the liquid fuel market.

Gasification technology was chosen for the Figure 4-1 illustration because it is compatible with municipal solid waste, coal, and biomass feedstocks as well as efficient electrical power generation and liquid fuel production. Gasification technology can increase the efficiency of electrical power production from about 38% to 55% for solid fuels using today's technology. These values are anticipated to increase to maybe 60% by 2015.

As illustrated by Figure 4-1, Fischer-Tropsch can be an important part of advanced solid fuel refinery concepts that include

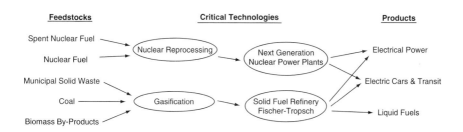

FIGURE 4-1. Alternative benchmark technologies.

production of electrical power and chemicals in addition to liquid fuels. It is an option for the chemical industry, the coal industry, and new players to enter into the liquid fuel market.

The case study of this next section illustrates how policies can have a greater impact on commercialization of a technology than the technology itself. This case study is based on policies that impact corporate decisions on investing into commercializing new technology.

Case Study on Investment Decisions and Policy Impacts

Corporate investment decisions can be understood by performing the same profitability analyses that are used by corporations. These profitability analyses can also be evaluated under assumptions of different government (or corporate) policies to better understand how policies impact investment decisions. A case study was performed on the investment into a Fischer-Tropsch process for converting Wyoming coal into a synthetic oil for replacing imported petroleum. The impact of technology cost was compared to the following four nontechnical barriers:

- **Petroleum Reserves** are the number of years of petroleum crude oil in proven reserves held by the corporation considering an investment into an alternative fuel facility. A base case of 0 years of reserves was assumed. The typical reserves for an oil corporation are from 7 to 14 years, so conservative figures of 5 and 11 years were used in the sensitivity analyses.
- **Intangible Costs** are the costs of the risks associated with investing in a new technology. These costs include the risk of OPEC lowering the price of crude oil to drive the synthetic fuel facility out of business. Intangible costs were incorporated into the sensitivity analysis by either assuming that a higher ROI and shorter payback period (20%, 6 years) would be required to attract investment capital or by assuming that the price of the synthetic fuel would decrease to a very low value ($10 per barrel) shortly after startup (3 years or 5 years).
- **U.S. Tax Structure** is the taxes paid to the United States and local governments, including Social Security and unemployment taxes that must be paid by U.S. employees. A base case of 34% corporate income tax was assumed. The sensitivity analysis included an assumption of 0% corporate income tax and that half of the "threshold" price was due to taxes (corporate income, personal

income, property, FICA, etc.) and that the threshold price would be reduced by 50% if these taxes were not selectively placed on domestic production.
- **Return On Investment** is the % ROI and the time in years (year one is defined as the first year of production) for payback of the investment capital. A base case of 12.5% ROI with a 15-year payback was assumed. For comparison, IRR calculations were performed for a 5% ROI and 30-year payback, since this is reflective of today's municipal bonds for civil infrastructure investment. Also considered was 10% ROI and 20-year payback.

Investment rate of return (IRR) was chosen as the preferred profitability analysis for this case study because the IRR calculation provides the "threshold" price of the synthetic oil product. Petroleum prices above this threshold price would justify investment.[4] The lower this threshold price the more likely the technology will be commercialized and compete with imported petroleum. In the IRR calculation, the "threshold" price for the synthetic oil is adjusted until the net present value of the process is $0 at the end of the process life (e.g., 15 years for the base case). After preparing a base case IRR, the sensitivity of the threshold synthetic fuel price to the four nontechnical barriers was determined by repeating the IRR calculation for the upper or lower limits of each of the nontechnical barriers.

Table 4-2 summarizes the parameters used in preparing the base case from which the sensitivity analysis was performed. Table 4-3 summarizes the sensitivity analysis results showing the impact of the nontechnical barriers.

The base case yields a threshold price of $41 per barrel of synthetic oil; however, this base case assumes no intangible costs and that the investing corporation held zero years of oil reserves.

By relying on oil corporations and including reasonable intangible costs, a price in excess of $150 per barrel would be needed before investment into the needed infrastructure would meet profitability expectations. This threshold price has been known to be elusive, and this calculation confirms that crude oil imports in excess of $600 billion per year are likely to be realized before corporate investments are justified based on current corporate investment criteria. Any plan of relying on oil corporations to take the lead on developing alternatives to petroleum is seriously flawed.

The case study revealed that existing petroleum reserves of a corporation would provide the greatest investment deterrent to that corporation. Remaining deterrents from greatest to least

TABLE 4-2
Conditions for base case Fischer-Tropsch facility used to perform sensitivity analysis.

Property	Value	Justification
Capacity	20 billion gallons per year	This is one-sixth the amount of gasoline consumed in the U.S.
Capital Cost	$3.90/gal/yearly capacity ($78 billion capital investment)	This is the lower cost estimate published by the DOE and some companies commercializing this technology.
Operating Cost	$7.50 per barrel	This is typically published for a coal facility. The cost of coal (essentially the only feedstock for this process) is about $3.75 per barrel.
Construction Time	30% in year 1, 70% in year 2, startup in year 3	Standard for facility of this type.
ROI	12.5%	Standard corporation ROI for low risk ventures.
Investment Payback	15 years	Standard for facility of this type.
Corporate Income Tax	34%	Year 2002 corporate tax rates line out at 35% for taxable income over $20 million per year.
Startup/Working Capital	2 months operating expense	Standard practice.

impact were intangible costs/risks, demands for high returns on investment, unfavorable tax structures, and the cost of the technology.

Hindsight suggests that oil corporations will invest in a U.S. alternative fuel industry only when their petroleum reserves (ΔIRR of >$100/barrel) are depleted to about the time it takes to build the alternative fuel infrastructure, or about two years of reserves. Reduction of reserves to this level is not likely to occur in the near future. The conclusion is that the United States cannot

TABLE 4-3
Summary of changes in the sensitivity factors used in sensitivity analysis.

	Reserves	ROI, P	Taxes	Intangible	COS	Capital ($/gal/yr)	Prices ($/barrel)
Base Case	0	12.5%, 15 yr	34%	N/A	$7.50	$3.90	$41.00
Petroleum Reserves							
Low Res., P-$10	5 yr						$115.00
Typical Res., P-$10	11 yr						infinity
Return On Investment							
Municipal Bond		5%, 30 yr					$20.21
Low ROI		10%, 20 yr					$32.72
U.S. Tax Structure							
No Corporate Tax			34% – -> 0%				$33.31
All Taxes Gone			none				$20.50
Intangible Costs							
High ROI		20%, 6 yr					$73.17
$10 Crude at 3 Years				$10 @ 3			$96.27
$10 Crude at 5 Years				$10 @ 5			$67.50
Technology Costs							
50% Reduction					$7.50	$1.95	$24.24
25% Reduction					$5.62	$2.93	$32.61

rely on oil corporations to make investments that are needed; corporations make investments based on corporate profitability. Oil corporations thrive in the status quo. They would be the last to devalue their petroleum reserve and distribution assets.

Intangible costs (ΔIRR of \$38/barrel) are the second greatest barrier to investment into a domestic alternative fuel industry. Antitrust laws fail to cross international boundaries. Needed investments into a U.S. domestic fuel industry are not made (in part) because OPEC can flood the world oil market with low-price petroleum and drive domestic synthetic production out of business. Possible solutions are international treaties or price-dependent tariffs that effectively extend antitrust laws across international boundaries. These treaties/policies need to be in place at the time investment decisions are made, which means they need to be established now.

Corporate demands for high returns on investment (ΔIRR of \$14.5/barrel) had the third greatest impact on the threshold price. States and communities routinely make infrastructure investments at lower ROI's such as 5% ROI and 30-year payback. The IRR spreadsheets on which corporations base investments include short-term corporate monetary gains in wealth. Long-term, noncorporate, and nonmonetary wealth generation should be included in these calculations.

Domestic taxes (ΔIRR of \$14.1/barrel) have about the same impact as high corporate demands on ROI and have a greater impact on investment decisions than the cost of an otherwise good technology. Presently, about half the price of a barrel of synthetic petroleum produced from Wyoming coal would be taxes (corporate income, personal income, property, FICA, etc.), while essentially nothing (about 10 cents per barrel) is charged to imported petroleum; these import fees represent docking/harbor fees.

It is illogical to burden domestic industry with taxes while foreign producers are allowed to enter the U.S. market without paying a similar tax. While it is true that imported petroleum "can" have similar taxes paid to a foreign country, the key qualifier is "can." If imported oil is produced in a country where taxes are primarily "value-added taxes" (VAT) paid on domestic sales (otherwise known as a *consumption tax*), a foreign producer could export to the United States with little or no tax burden. The current U.S. tax structure is a problem that should be addressed. All four of these nontechnical barriers to commercialization can be readily and sustainably corrected.

True Barriers to Commercialization

Each of the four nontechnical barriers to commercialization of the synthetic oil case study has a greater impact than reasonable advances in technology. If solutions to the nontechnical barriers were implemented, threshold petroleum prices to attract investment into domestic synthetic oil production would be less than $20 per barrel and possibly as low as $13 per barrel. However, this production would be by a nonpetroleum corporation without refineries and a fuel distribution network. This producer would have to sell to existing oil corporations or face tens of billions of dollars of additional investment for separate refineries. In view of possible agreements with foreign producers (e.g., 11 years of reserves), it is not certain that the U.S. major petroleum corporations would displace contracted-petroleum with a synthetic oil alternative.

Intervention or a nonfuel alternative is needed. The plug-in HEV is an example of a nonfuel alternative. The Canadian oil sand industry is an example of how government intervention can be effective. These represent two possible solutions, but other good solutions exist.

The Canadian oil sand (formerly referred to as tar sand) industry is a success story that demonstrates how commercialization barriers can be overcome. At $9–$15 per barrel production costs, oil sand production costs are more than Fischer-Tropsch production costs. But even at these higher costs, large-scale mining of oil sands in Canada began in 1967[5] when oil prices were less than $10 per barrel. It took the better part of a decade to make a profit from the oil sands (when including regional opportunity costs of not allowing imported petroleum to compete with the oil sand fuels).

Oil sand commercialization was made possible by the Canadian National Oil Policy introduced in 1961 that established a protected market for Canadian oil west of the Ottawa Valley and freed the industry from foreign competition. This policy protected companies from the greater intangible costs and provided an environment for smaller companies (other than the major petroleum companies) to develop the technology. In addition, in 1974 the Canadian and provincial governments invested in Syncrude's oil sand project and provided assurances about financial terms.[6] New refineries were built (Shell Canada Limited Complex at Fort Saskatchewan, Alberta). Today, with oil prices in excess of $50 per barrel, the Canadian oil sand industry is profitable beyond most investors' expectations; it provides energy, security, and quality jobs.

The United States is at a disadvantage today because it did not heed the warnings of the oil crises in the 1970s and 1980s. However, it is possible to turn this disadvantage into an advantage. New and better options are available today. One of these is the plug-in HEV approach that can displace the majority of liquid fuel consumption with power from the national electrical grid.

Plug-in HEV vehicles are similar to HEVs on the market today. They use extended battery packs (e.g., 20 miles of range) that charge using grid electricity during the night providing the first 20 (or so) miles out of the garage each day without engine operation. Gasoline is fully displaced with grid electricity for most of each day's transit. Per-mile operating costs for grid electricity are about one-third the cost of gasoline. Rather than going to petroleum producers, the majority of the fuel operating revenues would go to local communities.

If a consumption tax were applied to imported fuels representative of taxes on domestic synthetic fuel production, the higher vehicle cost of a plug-in HEV would be recovered in about three years. Advancing technology would rapidly reduce this time to about two years. Less oil would be imported, domestic jobs would be created, and the new demand for off-peak electricity would allow restructuring of the electrical power grid to include base load generation with increased efficiency for electrical power production and reduced greenhouse gas emissions. Up to 20% of gasoline might be displaced without expansion of the electrical power grid.

Petroleum Reserves and Protecting the Status Quo

The starting point for solving problems related to the deterioration of the U.S. manufacturing base and the inability of the United States to displace its reliance on imported petroleum is to recognize that these are both artifacts of corporations attempting to maintain the "status quo."

In the mid-20th century, the U.S. steel industry decided not to invest in new U.S. steel mills because such mills would displace operational and profitable steel mills they already owned; such investments were not good short-term business decisions. On the other hand, Japanese investors had a greater incentive to invest in a steel manufacturing industry because their infrastructure had been destroyed during the war. Investments were made in more efficient mills that often produced a superior product. Eventually, most of the U.S. steel industry lost out to foreign competition.

From a corporation's perspective, the "upside" potential is clearly greater to invest profits in infrastructure that does not directly compete with existing infrastructure. Whether it be the steel industry, textile industry, or petroleum industry, the path of short-term corporate profitability was and is different than the path of long-term prosperity for countries, states, and communities. From this observation it follows that countries should not let energy corporate giants to decide the long-term national energy strategies.

A solution is to select energy options that do not rely on the refineries or distribution networks of major oil corporations. Increasing the fuel economy of vehicles is such an approach, but it has limited potential. A new technology referred to as "plug-in HEV" technology has the potential to substantially displace oil with domestic electricity and may be a technology that displaces petroleum. Hydrogen gas would be used on fuel cell versions of plug-in HEVs, but no hydrogen distribution infrastructure will be required. Hydrogen would be generated "on board" the vehicle.

Intangible Risks (Costs) and International Antitrust Policies

Investment into new infrastructure brings the risk of losing the monetary investment. Intangible risks are those risks that are difficult to predict and often outside the control of the investors.

When pursuing alternatives to displace petroleum-based fuels, the intangible costs include (1) the risk that oil-producing countries will flood the market with oil to maintain market share and drive competition out of business and (2) the risk that a better alternative will come along in a few years. The latter is an accepted risk for all investments because competition is recognized as being good for the country. The former is recognized as going against the best interest of a country as documented by antitrust cases against corporations like Standard Oil and Microsoft.

Reform in the antitrust laws is needed because John D. Rockefeller demonstrated that even when you are guilty, your competition will be substantially gone, your wealth will be great, and you will merely have to stop certain practices. Reform is needed to provide quicker response to unscrupulous business practices or to even stop such practices from ever eliminating the competition.

To the credit of current antitrust laws, these laws are used to proactively evaluate mergers before they occur to make sure the mergers do not create a monopoly. However, these laws do not prevent corporations from lowering prices to stifle the competition.

90 Sustainable Nuclear Power

In the case of petroleum imports, one solution is to add an "antitrust tax" to imported petroleum if the price is lowered more than, say, 25% from the recent five-year average price (25% and five years are one of many possible combinations). For example, if the price of import oil into the United States was $40 per barrel for five consecutive years and OPEC decided to reduce the price to $20 per barrel, this "antitrust tax" would place a $10 per barrel tax on imported oil. New alternative fuel industries in the United States would then have to compete with an effective price of $30 rather than $20 per barrel. A year later, the new five-year averaged price would be $36 per barrel, and the new minimum effective price would be $27 per barrel. Sustained price decreases could occur, but the reductions would be gradual, and the intangible risk to domestic industry would be considerably less.

An "antitrust tax" on imports would only be applied if significant decreases in prices occurred. It would be proactive, it would not limit how low prices could eventually go, and it would reduce/eliminate the intangible risks that create one of the greatest barriers to the commercialization of alternatives to imported petroleum.

An Obsolete U.S. Tax Structure

U.S. taxes take their toll on the profitability of domestic industries. Figure 4-2 summarizes U.S. taxes on a "hypothetical" domestic synthetic oil (like the case study) for comparison to the taxes on imported petroleum. For these calculations, every dollar of sales going to a domestic synthetic oil producer is interpreted as either

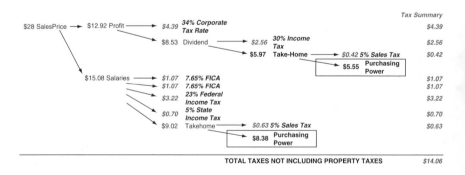

FIGURE 4-2. Summary of tax breakdown on $28 barrel of synthetic crude.

going to the government or becoming "true" purchasing power in the hands of either a worker or investor.

As indicated by Figure 4-2, for a domestic production facility selling a $28 barrel of oil, about $14 in taxes is paid for every $14 in purchasing power that goes to either investors or workers. These numbers will vary depending on specific examples. The inclusion of state portions of employer taxes (unemployment, etc.) as well as property taxes further increases these taxes. Sales taxes should debatably be excluded from this analysis, since it is placed on all products independent of where they are produced. This shows that taxes are about 50% of the domestic sales price of most products produced and sold in the United States—an effective domestic tax rate of 100%!

If $50 is spent on a barrel of crude oil purchased from a foreign government, essentially no U.S. taxes, tariffs, or social costs are paid on that oil. In some cases, little or no taxes would be paid to any government. Foreign taxes may also be excluded from those items exported from foreign countries to improve international competitiveness.

Even though the domestic and imported oils are superficially equivalent purchases to a refinery in the United States (based on current U.S. law), the fundamental question remains whether these options produce the same U.S. societal benefit. This can be stated in a different way: From the perspective of a U.S. citizen, is $1 in U.S. tax money going toward the purchase of a jet aircraft for the U.S. Navy the same as $1 in Iranian tax going toward the purchase of a fighter jet for the Iranian air force? Certainly not!

Why, then, does the U.S. government continue policies that give foreign production competitive advantages and direct cash flow away from the U.S. government and to foreign governments? Who pays the U.S. taxes to make up for the lack of taxes on imported crude oil? U.S. citizens will ultimately pay costs to keep the U.S. government going. Mega energy corporations and several foreign governments benefit greatly by having this portion of the tax bill not show up on their balance sheet. It is clear these are shortcomings of current U.S. tax laws.

Where, exactly, have our classical analyses of this trade gone wrong? The philosophical father of free trade, Adam Smith, identified two legitimate exceptions to tariff-free trade: when industry was necessary to the defense of the country and when tax was imposed on domestic production.[7] Corporate income taxes are an obvious example of a tax imposed on domestic production. Personal income taxes may be a less obvious example, but from the perspective of foreign competition, the personal income tax has the same impact as the corporate income tax—as do property taxes, unemployment taxes, dividend taxes, and FICA. Sales/consumption taxes are fundamentally different, since they are applied without bias to the origin of the item sold.

There are huge profits to be made by taking advantage of the differences in tax structures between countries. Where there are huge profits to be made, corruption follows. In the case of the U.S. tax structure, there is some corruption and much ignorance. There are those who advocate that U.S. politicians strive to keep all U.S. businesses at a tax disadvantage relative to foreign competition so they (the politicians) can selectively provide incentives (political prejudice) to their constituents' businesses.

A typical, poor assumption made in favor of tariff-free trade is that dollars paid to foreign countries are returned in the form of constructive purchases of items produced more efficiently in the United States. Two major flaws with this assumption are (1) a perpetual trade deficit demonstrates that these funds are not returning to purchase U.S. goods, and (2) funds used to purchase oil in unstable political regions, like the Middle East, are known to fund international terrorist activity. Neither increasing trade deficits nor tacit support of terrorism represent constructive use of U.S. economic power.

Classical arguments for tariff-free trade advocate that a comparative (production) advantage is gained through trade that benefits all trading partners. However, in order to quantify the benefit, zero unemployment and zero trade deficit are assumed. Neither of these two assumptions have applied to the United States for decades. Unemployment may be low, but underemployment is not low.

A good alternative exists to the current tax structure: It is the consumption tax that would place a constant tax (e.g., $25 per barrel) on all oil whether produced domestically or abroad. In the implementation of a consumption tax, corporate taxes should be essentially eliminated and income taxes reduced to avoid increasing taxes to U.S. citizens. Under a consumption tax, it is likely that energy prices would go down as the cost of domestic alternatives decreases (since domestic producers would no longer bear all the U.S. tax burden, foreign products would share the tax burden), and there would be "real" competition to imported oil.

While the United States currently selectively taxes domestic production, many countries selectively tax imports (foreign production). Here is a recent statement by the Semiconductor Industry Association to the House Committee on Ways and Means:

> China imposes a value-added tax (VAT) of 17% on sales of all imported and domestically produced semiconductors and integrated circuits. However, current Chinese government policy provides for a rebate of the amount of the VAT burden in excess of 6% for integrated circuits manufactured within China (and the amount of the VAT burden in excess of 3% for integrated circuit designs developed in China).[8]

The U.S. policy of continuing to apply centuries-old tax policies is undermining the entire U.S. industrial base by increasing the incentive to invest abroad.

One approach the U.S. government has used to correct the problems created by selectively taxing domestic production is the application of tax rebates on certain domestic products. The $0.54 per gallon[9] incentive for use of domestic ethanol in gasoline and $1.00 per gallon for use of biodiesel are examples of such rebates. Independent of whether one is in favor of promoting these products, the combination of first taxing all domestic manufacturing and then selectively rebating taxes on certain items replaces capitalism with political prejudice. The former USSR is an example of how politically based investment decisions can lead to economic failure. Two wrongs (a poor tax structure combined with government incentives to promote selected industries) do not make a right.

Corporate Profitability and High Investment Thresholds

Corporations tend to pursue only the most lucrative investments when reinvesting their profits. The spreadsheet analyses of these investments often indicate the potential to recover all capital investment plus a yearly 20% return on investment (ROI) in the first six years (six-year payback) of production. For new endeavors, anticipations of this high return and quick payback are standard. The following represent typical ROI and payback periods for corporate investments:

- 20% ROI, 6-Year Payback, for first time processes with high intangible costs/risks—including risk from foreign competition not covered by U.S. antitrust laws.
- 12.5% ROI, 15-Year Payback, for proven technology with moderate to low risk.

- 10% ROI, 20-Year Payback, for a proven technology with very low risk that fits in well with a corporation's current assets.
- 5% ROI, 30-Year Payback is not acceptable for corporate investment but is used for municipal infrastructure funded through bonds.

The case study results of Table 4-3 demonstrate that if an oil corporation owns years of petroleum reserves in excess of the time it takes to build an alternative fuel facility, the ROI simply cannot be high enough to compensate for the devaluation of those existing assets. Only under the assumption of less than two years reserves can a corporation meet reasonable ROI investment goals.

To place things into perspective, March 2005 petroleum oil prices exceeded $55 per barrel. If imported petroleum was taxed at the same rate as the case study's domestic production of synthetic fuel from coal, the refineries would be paying about $110 per barrel. As summarized by Table 4-3, if a corporation that was not vested in petroleum reserves were to commercialize this technology using municipal bonds, the cost of domestic production would be $20.21 per barrel (including taxes). In this hypothetical environment of "equitable taxes on both foreign and domestic products" and "investment using municipal bonds." The facilities would be built and actual ROIs would be much higher than the minimum expectations of the investors.

The Canadian oil sand industry is an example of a community and government making what was perceived as a low ROI investment in the 1970s. This has resulted in exports of fuel (rather than imports), regional prosperity, and increased national security. With crude oil prices at $55 per barrel and production costs from the oil sand fields at about $12 per barrel, the returns on investments are huge and the industry has self-sustained growth/expansion.

The irony of the Canadian oil sand industry example is that a province made an investment based on anticipation of minimal monetary gains and received great monetary and nonmonetary returns. At the same time, corporation after corporation holding out for high ROI investments have gone out of existence.

Today, some states are attempting to reduce this nontechnical barrier by providing tax credits or land for plant sites. For example, Alabama provided Hyundai $253 million in "economic incentives" to build an automotive plant there. These incentives included sewer lines, highway paving, and tax breaks.[10,11]

Land and tax credits are effective and low-risk approaches for states and municipalities to attract corporate investment. A more direct approach would be for local and state governments to provide capital with bond-based funding. If this direct investment approach was used, communities would be able to match their capabilities and needs with the industry the community is trying to attract. The underlying messages here are that communities realize value from local industry beyond the ROI realized by the corporation and that it is often good policy to provide incentives to attract industry. In this approach there must be assistance to communities to help them make smart incentive decisions.

A particularly intriguing opportunity exists for state and local communities to provide bond funding for local investment with a "balloon" interest rate that pays off well after the 10 or 20 years. Low bond rates for the first decade allow the corporation to receive the quick payback, while high interest rates in the long term make the communities/states winners.

It is important to recognize that profitability (10%, 20%, or higher) can continue for decades after the payback period used to determine the threshold prices that warrant investment. A community can position itself for this long-term upside of a good manufacturing infrastructure investment.

Taxes and Social Cost

The Cost of Driving a Vehicle

Essentially no tax is applied to imported crude oil, but substantial taxes are applied to gasoline at the pump. The taxes at the pump for maintaining the nation's highway system should not be confused with the taxes that increase the cost of producing domestic synthetic oil presented in the case study. Figure 4-3 shows the breakdown for the price of a typical gallon of automotive gasoline.

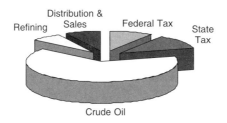

FIGURE 4-3. Summary of price contributions on a gallon of gasoline on $2.01 per gallon of unleaded regular gasoline.

With crude oil selling at about $55 per barrel, about 20% of the price of automotive gasoline is taxes averaging about $0.42 per gallon. The crude oil feedstock is $1.30 ($55 per 42-gallon barrel, March of 2005) of a $2.01 gallon of regular unleaded gasoline. Premium gasolines would have higher refining and sales component costs amounting up to an additional $0.20 per gallon. Table 4-4

TABLE 4-4
Highway taxes on fuels.

Types of Fuels	Cent/gallon*	Types of Fuels	Cent/gallon*
Gasoline	18.4	Special motor fuel, (general)	18.4
Gasohol:		Liquid petroleum gas (LPG)	13.6
10% Gasohol	13.1	Liquid natural gas (LNG)	11.9
7.7% Gasohol	14.3	Aviation fuel (other than gasoline) noncommercial	21.9
5.7% Gasohol	15.4	Aviation fuel (other than gasoline) commercial	4.4
Gasoline Removed or Entered for Production of:		Gasoline used in noncommercial aviation	19.4
10% Gasohol	14.6	Inland waterways fuel use tax	24.4
7.7% Gasohol	15.5	Diesel fuel	24.4
5.7% Gasohol	16.3	Diesel fuel for use in trains	4.4
Kerosene—Highway	24.4	Diesel fuel for use in buses	7.4
Kerosene—Aviation Fuel	21.9	*Including 0.1 LUST tax.	
100% Methanol (natural gas)	4.3	Compressed natural gas (egg)	5.4
100% Methanol (biomass)	12.3	Liquefied natural gas	18.3
100% Ethanol (biomass)	12.9	Propane	18.3

Salvatore Lazzari, *The Tax Treatment of Alternative Transportation Fuels*. CRS Report for Congress, Library of Congress, March 19, 1997.
http://api-ec.api.org/filelibrary/ACF763.doc, May 21, 2002.
*All prices include Lust Tax.

summarizes federal taxes corresponding to the $0.184 federal tax on automotive gasoline for commonly used fuels.

For the gasoline price summarized by Figure 4-3, typical price contributions are: federal tax ($0.184), average state tax ($0.236), crude oil ($1.30), refining ($0.133), and distribution/sales ($0.150).

Federal and state taxes on gasoline go toward the building and maintenance of our nation's highways totaling about $72 billion per year, and most agree that these moneys are well spent. An average of $0.42 per gallon is applied at the pump to gasoline sold for use on highways. For diesel, a red dye is placed in the off-highway diesel (no taxes) to distinguish it from diesel for which highway taxes have been paid. Law enforcement can sample the fuel tank to verify that diesel intended for farm tractors is not used to haul freight on highways.

A Fischer-Tropsch fuel (fuel of case study) would be completely compatible with petroleum-based diesel and could be distributed in the same pipelines. The $0.42 per gallon highway tax is applied to all fuels (imported and domestic, alike) at the pump.

For a hypothetical gallon of Fischer-Tropsch fuel and based on the estimate of Figure 4-2, half of the $1.30 per gallon ($0.65 per gallon) would actually be a compilation of additional taxes collected prior to the fuel reaching the refinery. As the American Petroleum Institute[12] points out, imported crude oil is taxed multiple times prior to the refinery as summarized by the following per barrel taxes:

Import duty	$0.05 to $0.11
Merchandise Processing Fee	Up to $0.06
Harbor Maintenance Fee	Up to $0.025
TOTAL	Up to $0.19, but typically about $0.10 per barrel

These taxes on the imported petroleum total to about 0.25 cents per gallon as compared to the projected 65 cents per gallon for the case study fuel.

In addition to these taxes, approximately half of the refining, sales, and distribution costs (assuming refining occurred in the United States) are also a compilation of taxes. Here, about 14 cents is applied to fuels based on both imported and domestic (e.g., Fischer-Tropsch fuel from Wyoming Coal) oils.

If a $0.79 ($0.14 + $0.65) per gallon consumption tax were applied to all gasoline as an alternative to the Figure 4-2 compilation of taxes, the Fischer-Tropsch gasoline would remain at a price of $2.01 per gallon, while the import-based gasoline would cost $2.66 per gallon. It is possible that under this tax structure the production of Wyoming-based Fischer-Tropsch fuel would increase to totally displace imported oil. It is possible that the price of imported oil would precipitously fall to about $28 per barrel to compete with domestic fuels. It is possible that something between these two extremes might also occur.

How Does the Cost of Fuel Stack Up?

Few people actually sit down and calculate the cost of owning and operating a car. These costs include fuel cost, parking (in some instances including the cost of buying and maintaining a garage parking space), car depreciation, insurance, and maintenance (oil changes, new tires, new brake pads, etc.). Two scenarios are summarized here with different miles traveled each year and different miles per gallon for the vehicle. A gasoline price of $1.79 per gallon is assumed.

Fuel: 12,000 miles/30 miles/gal × $0.42/gallon Highway Tax =	$168
Fuel: 12,000 miles/30 miles/gal × ($1.179 − $0.42)/gallon =	$304
Fuel: 12,000 miles/30 miles/gal × $0.20/gallon for Premium =	$80
Parking at Work and Parking in Home Garage	$500
Car Cost: $30,000 × 12/120 =	$3,000
State/Local Property Insurance ($10,000 vehicle/3 * 6.13%)	$204
Insurance: $700 per year	$700
Maintenance	$150
Total	$5,106
Fuel: 24,000 miles/13.1 miles/gal × $0.42/gallon Highway Tax =	$384
Fuel: 24,000 miles/13.1 miles/gal × ($1.179 − $0.42)/gallon =	$694
Fuel: 24,000 miles/13.1 miles/gal × $0.20/gallon for Premium =	$183
Parking at Work and Parking in Home Garage	$500
Car Cost: $30,000 × 24/150 =	$4,800
State/Local Property Insurance ($12,000 vehicle/3 * 6.13%)	$245

Insurance: $700 per year	$700
Maintenance	$250
Total	$7,756

This analysis shows that the fuel cost is typically less than 10% of the total cost of owning and operating an automobile. If the entire country were using ethanol fuel costing $1.50 per equivalent gasoline gallon to produce and distribute, the additional burden would be equivalent to a 5%–10% increase in the per-year operating expense of an automobile.

Alternative fuels like Fischer-Tropsch fuels would not increase the cost of owning and operating an automobile because Fischer-Tropsch fuels can be produced for the same price as petroleum at $20 per barrel. Domestic Fischer-Tropsch fuels would actually save consumers about $375 if the increased taxed revenues from domestic production were returned to consumers in the form of reduced income taxes.

Corporate Lobbying Retrospect

The impact of corporate lobbying on energy policy is far greater than the impact of Washington's oil lobbyists on today's pending legislation. Years of legislation that have been created under their influence is a problem.

Legislation providing essentially tax-free crude oil production, especially in foreign countries, gives foreign governments and oil companies competitive advantages of more than $25 per barrel of crude oil (about 50% of the price of imported petroleum) over alternative fuels. Legislation inhibiting the development of reprocessing technology has created nuclear waste disposal problems and increases the price of electricity.

Corporations recruit and hire many of the brightest graduates of our universities. The regulations are so complex that only the mega corporations have both the legal and technical teams to identify and understand them. Fortunately, the rightness of practices can be accurately available based on fundamental principals that do not require an MBA.

The industrial economic analysis methods show that the federal tax structure impacts the viability of an indigenous synthetic fuel program and is unfair to U.S. fuel providers. In order to address the energy problems of the United States or other countries, the first step is to acknowledge the problem. Alternative fuels appear to be uneconomical because they cost too much to produce and use. Actually, the federal tax structure places an unfair burden on indigenous production. A burden that is not placed on imported crude.

Certainly, it is better to have the $28 per barrel stay in the United States as taxes and buying power rather than to go to foreign countries with unknown agendas. If $28 of a domestic barrel includes $14 in U.S. federal/state/local taxes, then of the $28 spent on imported petroleum, $14 should also go to the U.S. taxes. The impact of policies on past and future development of an alternative fuel infrastructure is summarized in Chapter 8. The most likely scenario of a policy promoting commercialization of competitive alternative fuels in the U.S. is lower pretax gasoline prices. This translates to lower costs to consumers under the assumption that total taxes collected by the government remain constant.

The gross profit analysis of Table 4-1 indicates several options with feedstock prices less than $0.40 per equivalent gasoline gallon. For coal and oil sands, proven technology is ready to implement. For hydrogen from nuclear power, smart approaches could be implemented cost effectively (see section on Energy Wildcards later in this chapter).

The mega corporations, by definition, are on top of their industry. For them, change is interpreted as status reduction. They do and will resist alternatives to petroleum. They do and will resist nuclear power that will decrease the value of natural gas and coal reserves and facilities.

In response to corporations that serve their shareholders, government leaders must recognize that these corporations will not be objective in their advice. Government leaders must understand energy options and represent the people in their actions. Commercialization of alternatives to petroleum would be much easier with one or more of the corporative giants on board. The one thing that should bring them on board is recognition that a substantial part of the more than $200 billion (in 2005) per year of oil moneys going to foreign producers could be corporate revenue.

The electrical power industry takes a more modular approach to systems than the liquid fuel industry, so greater versatility exists leading to healthy competition. For example, the paper industry produces a by-product called black liquor. Factories are able to custom design boilers to burn the black liquor. The steam produced by these boilers can be used to drive steam turbines and produce electrical power. Similar sustainable and economical liquid vehicular fuel applications are rare.

Because of the versatility, viability of localized operations, and domestic nature of the electrical power industry, problems similar to the inequitable tax structure of crude oil are less common. Foreign competition is not a big issue. Federal regulations that inhibit development and commercialization of nuclear waste reprocessing are another exception.

Reprocessing technology that taps into 20%, potentially 100%, of the energy contained in uranium is potentially the low-cost option for sustainable electrical power production. Uranium reprocessing technology could take markets away from the coal and natural gas industries. The role of reprocessing in the U.S. electrical power infrastructure needs to be reconsidered in the absence of the influence of special interest groups.

Nuclear power appears to be gaining favor with increasing global warming concerns. The major objection to nuclear power is what to do with spent fuel. Reprocessing technology holds the potential to solve problems related to waste disposal by recovering fuel values from the spent fuel while reducing the total mass of radioactive waste. Valuable uses are already known for some of the fission products (ruthenium for example, rare in nature) found in nuclear wastes. Valuable uses may be found for sources of the fission products that are at present only recognized as waste. Alternatively, these wastes can be converted to benign materials through nuclear processes (currently considered too costly). When considering the hundreds of years of energy available in the stored spent fuel, uses for available fission product metals may be found long before current stockpiles of spent fuel are reprocessed.

One hundred years ago, nuclear energy and most of the chemical products we use every day were unknown. Advances of science and technology will continue to bring treasures that can include converting spent nuclear fuel to energy along with uses for the valuable by-products. Reprocessing spent nuclear fuel is now practiced in Europe. Reprocessing produces about 750 kg of nuclear waste from a 1,000 MW power plant as compared to 37,500 kg of waste produced by a current U.S. nuclear power plant without reprocessing. Current U.S. policy is to bury the 37,500 kg per GW-year of spent fuel, even though 36,200 kg of this is unused nuclear fuel.

Pressure is mounting (by those who do not understand and those vested in other energy sources) to bury the spent nuclear fuel and excess weapons grade material. This would be an expensive disposal process and make this energy resource impossible to retrieve. If this plan is successful, it will be a strategic victory for those wishing to eliminate nuclear energy as an option for electrical power production.

Diversity as a Means to Produce Market Stability

There is no substitute for competition to create and maintain low prices and increase consumer buying power. Even in the era of strict government regulation, U.S. consumers have benefited from healthy competition in the electrical power industry. The most important competition has been among energy sources. Figure 4-4 summarizes the distribution of electrical power by energy source[13] in the United States. By example, in the 1970s when petroleum became more expensive, coal replaced petroleum (with a slight delay to build new power plants capable of using coal), and electrical power prices were relatively stable.

It's evident from Figure 4-4 that two factors have contributed to relatively constant electricity prices in the United States for the past two decades. The primary factor is diversity: If coal prices go up, natural gas gains a larger market share and vice versa. No single source dominates even half the market. The second factor is the strong foundation provided by an abundant supply of coal.

Significant price fluctuations have not occurred in consumer prices for electricity in recent history. Furthermore, prices have been cheap by most standards. An exception is the failed attempt to convert California to a "free market" region (see the box "The 2001 California Electrical Power Debacle").

The 2001 problems in California were caused by strict government regulation on prices and emissions. California purchased

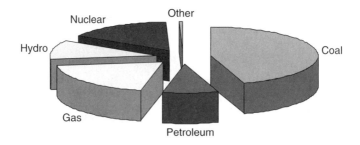

FIGURE 4-4. Summary of electrical power-generating capacity by fuel sources for electrical power generation in the U.S. (1999).

electrical power because some environmental groups promoted policies that eliminated building of new nuclear or fossil fuel power plants. The mistake was the free-market bidding for power by distributors required to sell electrical power at a fixed price. The distributors were forced to sell electricity at a loss. Even though this situation was only temporary, the financial burden on the distributors brought long-term debt and increased the cost of electricity to the customers.

The 2001 California Electrical Power Debacle

In the past, the Federal Trade Commission controlled the price of electricity to customers. Private power producers got approval to use a "free-market" model in California with a promise that the price to consumers would be reasonable. In this model, a power distribution utility would request bids for power from the producer. This would assure low cost power to the customers.

Historically, the large population centers of Southern California placed strict environmental restrictions on fossil fuel power plants. Few fossil fuel power plants were built. They also refused to allow nuclear power plants to operate near the geological faults in the interest of safety. The region soon became dependent on imported power, electricity produced in surrounding states and sold to California.

The new California free-market model contained the seeds of disaster. The distribution utilities were required to set low, fixed rates they could charge the thousands of customers. They purchased power from public and privately owned electric utilities on a "low bid" basis. The bid price was adjusted daily (later, hourly). When a power production shortage developed, the distribution utility had to bid more to meet customer demand, but could not recover the cost because the sale price was fixed low. The bid price for electricity suddenly spiked to nearly one dollar per kilowatt-hour, a factor of ten above the selling price to the customer. In a matter of days, a multibillion-dollar distribution utility went bankrupt and a second was salvaged by the California legislature. The huge energy conglomerate, ENRON Corporation, reportedly made huge profits in this electricity bidding war.

The governor of California recommended to the legislature that they underwrite long-term contracts to purchase electricity. These contract prices were well above the cost of electricity before the failed free-market experiment. It will take several years to complete the analysis of this free-market failure. Reliable, affordable sources of electricity are as important to a local economy as water is to life. Construction permits for natural gas fired power plants have been approved in Southern California and with this plan, public confidence should return with local ownership.

To the extent that diversity is a strength of the electrical power industry in the United States, the lack of diversity in the vehicular fuel industry is a weakness. The majority of liquid fuels are derived from petroleum, and over half of the consumed petroleum is imported. As a result, petroleum prices fluctuate and the economy can be driven into recession in a matter of weeks. The oil producers know how vulnerable the U.S. economy is to oil prices, and oil is the hostage they can use to influence U.S. energy policy.

The development of at least one large-volume and sustainable vehicular fuel alternative is needed to counter this threat to security. Such an alternative could stabilize fuel prices, stabilize the national economy, and provide an incentive for otherwise hostile states to be good neighbors. Neither ethanol nor biodiesel have the low cost or capacity to fit this task. Fischer-Tropsch conversion of coal to liquid fuels does. Collaboration with Canada on oil sands could also provide an alternative solution. Use of rechargeable hydrogen (using electricity from nuclear or wind power) in hybrid vehicles for reducing engine use during commuting could provide reductions in cost, increases in national security, and reductions in greenhouse gas emissions.

Fischer-Tropsch fuels have a built-in none-sole-source capability, since they can be produced from coal, natural gas, biomass, and municipal solid waste. This can provide a fuel alternative to petroleum.

The Details Are Important

The combination of nuclear reprocessing and advanced solid fuel technology (synthesis gas pipeline or solid fuel refineries) provides valuable additions to the arsenal of energy utilities. Solid fuel technologies that produce both liquid fuels and electricity at efficiencies exceeding 50% provide improvements in electrical power generation efficiency and indigenous vehicular fuel production. Combined, these technologies conserve energy resources and lower greenhouse gas emissions.

The operation of these two energy production facilities is most efficient when they operate continuously at constant load. On the other hand, consumption tends to be dominated by on-again/off-again machines. These include devices such as air conditioners, electric stoves, and clothes dryers that tend to run between 9:00 A.M. and 9:00 P.M. The "peak demands" caused by these uses are both a problem and opportunity.

Meeting peak demand for electrical energy by the most economical means is the responsibility of electrical providers. The typical approach involves the use of special power plants as "peak demand facilities." These peak demand facilities are powered by natural gas or petroleum, operating at 25% to 30% efficiency rather than the 35% to 53% of baseline power systems. They are cheap to build but are expensive to operate and produce more greenhouse gases per kWh of electrical energy.

Nuclear and solid fuel facilities have problems meeting peak power demand because it can take hours for them to go from 0% load to high load service. Solar and wind power options supply electricity only when the sun shines and the wind blows and do not operate at our convenience. Production facilities for meeting peak demand are needed in addition to new and alternative technologies for baseline loads and come at a high premium.

An alternative to converting fuel to electricity upon demand is to store electrical energy during low demand and then return it to the grid during high demand. Battery storage is too costly. Other options are used to a limited extent. Pumped water storage is the alternative that has shown commercial impact. Not included in Figure 4-4 on fuel sources is *pumped water storage*. Hydroelectric power contributes about 10.8% of the electrical power generated in the United States. An additional 2.7% of hydroelectric power is provided by water that is pumped to higher elevations by excess baseline power during off-peak demand periods. One of the disadvantages of water storage is that energy is lost during both the storage and recovery processes. For example, energy that is initially generated at 50% thermal efficiency may contribute at a thermal efficiency of 36.1% (30% may be more realistic) if it is used to drive a pump and later converted back to electricity by a turbine. The overall efficiency is better than a peak demand facility, but the pump storage system is capital intensive because it requires real estate with favorable elevation change.

> ### Thermal Energy Storage
>
> During the summer a refrigeration system can be used to freeze ice at night using off-peak power rates. During the day, ventilation air passed over the ice (high cost, peak power) provides cooling as an alternative to running the air conditioner during the day. It has been decades since such a primitive system has been used; however, modern versions are in use. At Curtin University in Western Australia, air conditioners are run at night to chill water with the cool water stored in a large tank. This chilled water is circulated to buildings throughout campus to provide cooling during the day. Such systems use less-expensive nighttime electricity (typically available to commercial customers but not to residential customers) and require smaller air conditioners since the air conditioners can be run at a constant load during the 24-hour cycle rather than during just the day.
>
> Thermal energy storage can be used to make both heating and air conditioning systems operate in better harmony with electrical power supply. For solar heating, energy storage is particularly important. The need for supplemental heating is at a minimum during the day when solar devices are operating at their greatest efficiency. Heating is needed when the sun is down and the solar device is providing no heat. Thermal energy storage allows the heat to be stored during the day for use during the cool nights.

A successful and common method for reducing peak demand is to provide lower electrical rates to industrial customers for off-peak hours. Especially in industrial settings, some energy-intensive operations can be scheduled for off-peak operation. In these cases, the equipment is turned off during peak demand hours to avoid the higher penalty rates of peak-demand periods. In one form or another, price incentives promote a number of solutions to peak demand energy needs. Price incentives are successful, but alone, they cannot bring electrical demand to constant levels throughout the day.

Better options are needed for electrical energy storage. A disadvantage of converting electrical energy into either chemical or potential energy and then back to electrical energy is the lost efficiency at each conversion. In general, each conversion is less than 90% (often less than 85%) efficient and the conversions to storage and from storage have a combined efficiency of less than 75% (often less than 70%). This reduction in efficiency translates to 25% more air pollution, 25% more fuel consumption, and 25% more greenhouse gas emissions for the peak power than would be possible if the electricity demand was level.

A promising technology for shaving peak demand is thermal energy storage. Thermal energy (stored heat or stored refrigeration) is a lower form of energy than electricity or stored hydraulic (potential) energy. In principle, conversion to thermal energy is irreversible; however, for heating and air conditioning applications, thermal energy is the desired form of energy.

Thermal energy storage systems can approach 100% efficiencies and can be used by all customers using electricity for heating or air conditioning. Since heating and air conditioning represent a major component of peak demand loads, this technology can have a major impact.

Modern thermal energy storage options include ice storage, chilled water storage, and use of phase change materials. Phase change materials are chemicals that freeze near room temperature. For example, a material that releases energy at 74°F as it freezes and takes in energy at 76°F as it thaws might be used to keep a house at 78°F in the summer and at 72°F in the winter. A material with a large heat of fusion is ideal for this application.

The utility of thermal energy storage goes beyond converting inefficient peak demand electricity to more efficient baseline load. Solar energy storage is an example of how the need for auxiliary heating can be partially eliminated. During parts of the air conditioning season, nighttime temperatures are sufficiently cool to cool a medium (water, solid grid, etc.) to offset air conditioning demand during the heat of the day. This is especially true for commercial buildings that often require air conditioning even during spring and fall days.

Another promising technology is the plug-in hybrid vehicle[14] that charges batteries and produces fuel cell hydrogen during the night for use during the day (the engine on the vehicle provides backup power). Night hours are the best time to recharge the vehicle since it is typically not in use. Also, electrical power is used to displace imported petroleum. The diversity of the electrical power infrastructure can be used to stabilize and decrease the prices of petroleum. Major investments are not necessary in this approach, since hybrid vehicles are on the market and advances can be made incrementally. It is projected that up to 20% of gasoline could be replaced without building additional power plants; however, when baseload electrical power demand increases, history shows there will be investors stepping forward to make investments. These investments will be for new generation, more efficient electrical power plants. If coal were used as a fuel source (see Table 4-1; $0.011 per kWh), the coal costs would be less than $0.02 per mile compared to $0.05–$0.10 per mile for gasoline ($2.00 per gallon).

Environmental Retrospect

Benchmark Alternatives

Two technologies stand out as offering competitive alternatives to benchmark petroleum and coal energy industries. These technologies are (1) reprocessing of nuclear fuel for electrical power production and (2) solid fuel refineries for producing both liquid fuels and electrical power. These two technologies use the lowest-cost feedstocks and allow feedstocks such as biomass to be used when locally available. The quantity of municipal solid waste can supply a fraction of the energy supplied by nuclear and coal.

Sustainability

It is certain that science and technology will continue to produce yield discoveries as in the past. The inability to predict future breakthroughs is the reason a 30-year sustainability goal is a better planning criterion than perpetual sustainability. While the wind will blow and sun will shine for thousands of years, any particular wind turbine or solar receiver may only be functional for 20 years. Which is more sustainable—a wind turbine that must be replaced in 20 years or a nuclear power plant that must be replaced in 40 years? Which is more sustainable—a solar receiver that will take five years of operation just to produce as much energy as was consumed to manufacture it or a nuclear power plant that produces as much energy in four months as it took to manufacture the facility?

No technology should be developed simply because it appears to offer perpetual sustainability. Few scientists are presumptuous enough to say they know what energy options will be used in 100 years, so why should we assign a premium value to an energy source having perpetual sustainability over an energy source with a 100+ year energy reserve? On the other hand, an energy source having 100 years of reserve should be considered premium relative to an energy source having a 20 year reserve.

Basing energy decisions on single-issue agendas like sustainability is not productive. Rather, hidden costs associated with nonsustainable technologies should be evaluated and included in the economic analysis of a technology. An economic analysis that includes hidden costs and analysis based on cost alone has broader implications. For example, what medical breakthrough did not happen because the resources were spent on very costly solar receivers? If there is a hidden cost associated with potential greenhouse warming due to carbon dioxide emissions, should an appropriate "CO_2 tax" be placed on a worldwide basis? At the same time, no reasonable technology (e.g., nuclear reprocessing) should be barred by federal law. Such restrictions are subject to abuse by special interests such as corporations vested in alternative energy technologies.

At any point in U.S. history, total sustainability could have been put in place. For example, if in 1900 environmentalists were successful in persuading the U.S. government only to use sustainable energy sources, today's world would be substantially different. There is no reason to believe that single issue environmentalism is any more appropriate today than it would have been in 1900. The environment must be protected, but other factors must be part of the future.

Efficiency and Breakthrough Technology

Breakthroughs in grid-based electrical power production are not likely to occur, but opportunities do exist for breakthrough discoveries in nongrid electrical power. Considering that (1) about 10% of electricity is lost during distribution, (2) much of the cost of electricity is associated with grid maintenance and accounting, and (3) the many "remote" locations in need of electricity, the opportunities for nongrid power are immense. Can a photoelectric roofing material be made that costs slightly more than normal roofing material? Can cost-effective storage units be manufactured for home use? Could cogeneration (using the waste heat of power generation) be more widely used?

New Solar Devices

The greatest opportunities in enhancing residential solar devices is the use of solar panels in place of other structures during new building construction. The Solar Wall[15] is an example of such a device. The solar collecting surface replaces the building siding and structural elements and collects thermal energy.

A solar device that costs $5.00 per square foot to build may cost an additional $1.00 per square foot to install. If this device were added to an existing structure, it would cost $6.00 per square foot in addition to the cost of the wall on which it is placed. If during new construction, an 8 by 8 foot section of the Solar Wall was placed on the south side of the building in place of $1.00 per square foot siding and $1.00 per square foot of wall construction, the net cost for the solar addition would be $4.00 per square foot.

The cost of solar devices could approach the cost of the siding and wall structures they replace. This net cost of $4.00 could reduce to less than $2.00 with improved production efficiency. The device would pay for itself in a year or two. Solar walls on the south sides of buildings collect thermal energy and reduce heating costs, while solar shingles on the southern exposures of roofs could produce electricity. Ideally, the wall and shingles could be designed to be aesthetically pleasing. Solar shingles are an example of technology that could generate electrical power cheaper than conventional power plant production technologies.

Electric-Powered Automobiles and the Hydrogen Economy

The marriage of the electric car, fuel cell technology, and the hybrid car (plug-in hybrid vehicle) provides a promising market opportunity. Electric cars have inherent problems associated with the weight and cost of batteries that limit the vehicle range. 1,100 pounds of batteries are required to provide 100 miles of range for an automobile.[16] Rechargeable fuel cell power systems could use power grid electricity to convert water to hydrogen and oxygen. Since they weigh a fraction of equivalent battery systems, they can overcome the limitations of rechargeable battery automobiles. A combination of batteries and regenerative fuel cells on these vehicles may overcome today's problems of each.

When large numbers of rechargeable, hybrid fuel cell vehicles are in use, there will also be a demand for quickly refueling hydrogen for those consumers who have an extra 20 miles to go but do not have the time for electric grid recharging. This is the most likely method for the development of a hydrogen infrastructure: one station at a time where they are needed to extend the miles of rechargeable fuel cell vehicles. More likely, advances in fuel cell technology will allow liquid fuels (like methanol or ethanol) to be used in the fuel cell after the "recharge" hydrogen has been consumed.

Rechargeable fuel cells are an excellent alternative to batteries. On the other hand, fuel cell development has a long way to go before they are a good alternative to a light duty diesel engine.

Plug-in hybrid fuel cell vehicles may use liquid fuel storage (diesel or gasoline) to achieve greater vehicle ranges (200–400 miles) and hydrogen storage to provide the "hybrid" advantages. Depending on the capacity of the hydrogen storage, commuters may be able to meet most of their commuting needs by recharging hydrogen with grid electricity. This will provide a much-needed diversity in the vehicular fuel market. Price and economic stabilities should follow. Furthermore, recharging can be timed to stabilize electricity demand and reduce the cost of electricity. When the electricity is supplied by wind or nuclear energy, near-zero carbon dioxide will be generated. This transformation is underway.

In the United States, a great emphasis has been placed on developing fuel cell vehicles powered by distributed hydrogen. While this vision has secured years of funding for some government laboratories and creates hope to those seeking an alternative to petroleum, such a path to a hydrogen economy is unlikely to happen. It would cost hundreds of billions of dollars to build a hydrogen distribution/refueling infrastructure to parallel the gasoline refueling infrastructure. History shows that no entity (commercial or government) is likely to pay for this infrastructure if there is not a fleet of vehicles to use the fuel. Consumers certainly will not buy a hydrogen vehicle that relies on refueling from a nonexistent or distant refueling station. For this reason, this path to the hydrogen economy is not discussed in depth in this text. The PHEV path is more realistic. The future belongs to those who have the vision to overcome obstacles and help a technology reach its potential rather than those who can only see the obstacles.

Diesel and HEV Technology

In much of Europe, about half of the new automobiles are diesel powered. Diesel vehicles get 30%–40% more miles per gallon of fuel due to the higher thermal efficiency of diesel engines.

Unfortunately, this advantage is lost in the United States because no automaker produces a light-duty diesel engine using high-production runs that allow economy of scale comparable to gasoline engines.

Americans still remember the noisy and dirty diesel engines of the 1970s. These engines are gone, but the mindset remains. This has caused American automaker and engine manufacturers to limit development of light-duty diesel engines, and it will make it difficult for new light-duty diesels to be produced in large quantities anytime soon. Recently, quieter and cleaner diesel pickup trucks are being introduced, and these may change the attitude toward diesel-powered cars.

Farm Commodities and Land Utilization

At the end of the 20th century, the conversion of farm commodities to fuel was recognized and began as a means of reducing dependency on foreign crude oil. It is creating new markets for farm commodities. In the fuels arena, ethanol was the primary contender for gasoline engines and biodiesel for diesel engines. In both instances, the cost of these agriculturally derived fuels was more than twice the before-tax cost of gasoline and diesel. To overcome these price obstacles, federal and state governments passed laws to subsidize the use of agriculturally derived fuels.

Table 4-5 summarizes several of these laws. Given the task of promoting a fuel costing twice as much to produce, laws provide cash subsidies and fuel restrictions designed to improve the market environment. Among the most important of these regulations is a tax credit of 5 cents per gallon of gasoline that contains 10% ethanol. This translates to about $0.50 per gallon subsidy for fuel ethanol. This subsidy combined with ethanol to enhance gasoline's octane rating made 10% blends of ethanol in gasoline widely available as a competitive midgrade gasoline

In 2001, biodiesel had a market of about 20 million gallons per year, considerably less than one-tenth of a percent of the diesel market. Depending on the feedstock used to produce biodiesel, the before-tax cost of biodiesel ranged from $1.30 to over $2.00 per gallon. This compares to less than $0.75 for the pretax price of petroleum diesel (year 2002). The lack of biodiesel market penetration is understandable. Only consumers that really want to use a renewable and ecologically friendly fuel are willing to spend

TABLE 4-5
Example legislation promoting alternative fuels.

Law	Description
Energy Policy Act	Mandated use of alternative fuel vehicles by vehicle fleet owners.
1990 Clean Air Act Amendments	Mandated use of oxygenates (like ethanol) in nearly all carbon monoxide nonattainment areas.
Energy Tax Act of 1978	$0.54 per gallon ethanol price advantage to ethanol when used in 10% blends with gasoline.
New Energy Policy Act	Promotes the use of biodiesel.
Minnesota BD Use	Requires use of 2% biodiesel in all diesel. The 2% level will be gradually increased over time to over 5%.

the additional money for biodiesel. More favorable market conditions are possible as petroleum prices increase and biodiesel prices decrease (through coproduction of other chemicals with biodiesel).

Promoting such inherently expensive fuels appears questionable except for one important factor. The United States has a history of providing at least some subsidy and stability to farm commodity prices. When confronted with options to increase the market for corn or soybean oil, providing incentives to allow these products to be sold for economic fuels has distinct advantages over other more costly options. For example, paying farmers $1.00 per bushel not to plant corn and providing an incentive of $0.50 per bushel to convert the corn to ethanol, both achieve the result of removing a bushel of corn from the market. Incentives to convert the corn to fuel simply cost less than paying the farmer not to farm. Tax subsidy dollars are leveraged by the fuel value of the corn or soybeans.

In the 1990s a hot and dry summer led to decreased corn supply and sharply higher prices. During this period, many ethanol plants decided to sell corn futures and close down the ethanol production facilities. This shows another advantage of price incentives. The incentives indirectly increase corn production above what the market could otherwise bear. During years of crop failures, the production that would have gone toward ethanol production can be used to feed livestock.

To some extent, corn-to-ethanol fuel market promotes grain commodity overproduction that can be valuable when crops fail. This is a good policy for agriculture and the country, but it will not eliminate the need to import crude oil.

At the end of the 20th century, soybean oil was the largest-volume vegetable oil produced. The 21st century brings on palm oil as an even larger volume commodity. Unlike soybeans that must be planted every year, palm trees are perennials. In addition, oil yields of new palm tree varieties are expected to be three to five times the per-acre oil yield of soybeans. These yields and low-cost production may provide an oil that can directly compete with imported crude oil. Biodiesel will be an increasingly good option for those countries with suitable climates for palm oil production. Genetic engineering could create a similar energy crop for the Florida Everglades or other areas of the United States with high rainfall and a long growing season.

The cobenefits of a healthy corn-to-ethanol and soybean-oil-to-biodiesel justify federal and state subsidies. These subsidies actually level the playing field, since taxes are paid by the fertilizer supplier to the employee of the ethanol plant in the production of these fuels (and no U.S. taxes are paid on imports). When used in combination with plug-in hybrid technology (e.g., with more than 80% of miles covered by the plug-in), biofuels like ethanol might some day displace liquid forms of fossil fuels.

Global Warming

The scientific community is in agreement that carbon dioxide concentrations have increased in the atmosphere as a result of combustion of fossil fuels. Increases have been measured, and the reasons for the increases are generally understood. There is less agreement about the consequences of these increased carbon dioxide levels in the atmosphere, with cries of "catastrophe" and "forget it."

Since the impact of higher carbon dioxide levels in the atmosphere is unknown, the benefits of slowing or reversing trends in carbon dioxide are also unknown. The corresponding risk-benefit analysis of using technology to reduce greenhouse gas emissions has even greater uncertainty, since the cost/risk associated with reducing emissions on individual national economies is also unknown.

The consistent arguments put forward by American politicians opposed to reducing greenhouse gas emissions are the high costs and adverse national economic consequences of any new technologies to address the problem. The facts suggest that these arguments have no basis, since there are a range of options that include electrical energy storage, hybrid cars, nuclear energy, wind power, and use of rechargeable car fuel cells. Unfortunately, good solutions to greenhouse gas emissions could compromise competitive advantages of mega corporations that have major influence in national politics.

Figure 4-5 summarizes carbon dioxide emissions by sector sources.[17] At 34%, the electricity generation produces the most carbon dioxide. Upon recognition that the predominant source of greenhouse gas emissions from the residential and commercial sectors are HVAC heating and hot water heating, this 34% increases to about 45% of the greenhouse gases due to electricity and heating. Another 27% comes from transportation.

It is important to note that Figure 4-5 shows that both the industrial and agricultural sectors have had essentially constant or even decreasing carbon dioxide emissions since 1990. Concerns about the impact of reducing greenhouse gas emission on industry are put to rest by simply recognizing that neither industry nor agriculture are the source of increasing carbon dioxide emissions. If the goal is to maintain 1990 carbon dioxide emission levels, industry and agriculture have achieved the goal and should not be burdened with further reducing emissions.

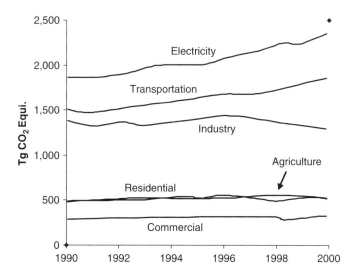

FIGURE 4-5. Carbon dioxide emissions by sector.

Reprocessing spent nuclear fuel could cost effectively increase the share of wind and nuclear power generation from about 17% to 42%. This alone would reduce carbon dioxide emissions by about 10% (34% reduced to 24% from electrical power generation). Increases in efficiency of coal-fired facilities could provide another 5% reduction. Replacing fuel-fired heating of residences and commercial buildings with heat pumps could provide another 5% reduction. Diesel engines could provide another 5% reduction in the transportation sector. Use of rechargeable fuel cells in hybrid cars could provide another 3% reduction through use of electrical grid recharging and another 5% from increased miles per gallon. All of these technologies promise to be low-cost alternatives. They represent a 33% reduction in total carbon dioxide emissions without much improvement in the technology of these alternatives.

Contrary to the propaganda about how costly it would be to reduce greenhouse gas emissions in the United States, these reductions could be made while saving consumers money, reducing oil imports, and putting the country on an energy track that is sustainable for the next century.

Diversity and the Role of Nuclear Power

This chapter argues that policies are every bit as important as technology to attain sustainable energy. While many discussions of policies focus on new policies that require the use of alternative technology, such new policies are not likely to succeed because they bypass the free market that has made the United States the superpower it is today. Policy reform rather than new policies is required. The solution is identifying and promoting those technologies that can become economically sustainable through success in the free market. Policies that need reforming are those that lock in advantages to petroleum with tax structures that actually inhibit the commercialization of domestic alternative fuels.

Examples of technologies that should be advanced because they are economically sustainable are new generation nuclear power facilities that are compatible with reprocessed spent nuclear fuel, and increased use of electrical power for transportation.

Major benefits from the use of PHEV technology include meaningful diversification of transportation fuels and the ability to use nuclear power to provide the energy for a substantial portion of the transportation sector.

References

1. W. D. Seider, J. D. Seader, and D. R. Lewin, *Process Design Principles*. New York: Wiley, 1998, p. 47.
2. Debora Hairston, "Newsfront: Pyrolysis Gets All Fired Up." *Chemical Engineering*, March 2002.
3. G. J. Suppes, M. L. Burkhart, J. C. Cordova, R. M. Sorem, and B. Russell, "Performance of Fischer-Tropsch Liquids with Oxygenates in a VW 1.9L TDI." SAE Paper 2001-01-3521.
4. W. D. Seider, J. D. Seader, and D. R. Lewin, *Process Design Principals*. New York: Wiley, 1998, Ch. 10.
5. D. Deming, "Oil: Are We Running Out?" Second Wallace E. Pratt Memorial Conference, "Petroleum Provinces of the 21st Century," January 12–15, 2000; see http://geology.ou.edu/library/aapg_oil.pdf.
6. "Canada's Oil Sands and Heavy Oil." Petroleum Communication Foundation. Available at http://www.centreforenergy.com/EEOS.asp, 2001.
7. Adam Smith, *Wealth of Nations* (facsimile of 1776 original). Amherst, NY: Prometheus Books, 1991, p. 361.
8. http://waysandmeans.house.gov/hearings.asp?formmode =printfriendly&id=1059.
9. Brent D. Yacobucci and Jasper Wornoch, *RL30369: Fuel Ethanol: Background and Public Policy Issues*. A CRS Report for Congress. See http://www.ncseonline.org/NLE/CRSreports/energy/eng-59.cfm? &CFID=19293910&CFTOKEN=12099740#_1_1, March 22, 2000.
10. http://www.conway.com/ssinsider/bbdeal/bd020401.htm.
11. http://www.sb-d.com/issues/winter2002/currents/index.asp.
12. http://api-ec.api.org/filelibrary/ACF763.doc.
13. http://www.eia.doe.gov/cneaf/electricity/ipp/html1/t17p01.html.
14. http://www.iags.org/pih.htm.
15. http://www.solarwall.com/.
16. Lester B. Lave, W. M. Griffin, and H. Maclean, "The Ethanol Answer to Carbon Emissions." *Issues in Science and Technology*, Winter 2001–2002, pp. 73–78.
17. *Emissions by Economic Sector*. Excerpt from the Inventory of U.S. Greenhouse Emissions and Sinks: 1990–2000. U.S. Greenhouse Gas Inventory Program, Office of Atmospheric Programs, U.S. Environmental Protection Agency, April 2002. http://www.epa.gov/globalwarming/publications/emissions/index.html.

CHAPTER 5

History of Conversion of Thermal Energy to Work

The story of energy conversion is the story of life. It is the story of people coming out of the caves to harness the environment. It is the story of conquest. And it is the story of the rise and fall of empires. The many forms of energy surround us every day, including the energy that fuels life itself.

As illustrated by Figure 5-1, the human body and the steam engine are both parts of the marvelous conversion process that propagates life and drives machines. Starting with the nuclear fusion of hydrogen atoms in the sun, energy is transferred and converted, ultimately being dispersed to a useless form of waste heat. The industrial revolution came of age when we began to understand how all these forms of energy are related.

The use of fire to cook food, to win metal from ore, and to work metal into useful tools is described in ancient historical documents. In the late 18th century, methods began to appear that converted the heat of a fire (thermal energy) to work that could replace humans or animals to perform tiresome daily tasks. The industrial revolution of the 19th century was fueled by fossil fuels feeding steam engines. The 20th century brought the nuclear age, where the burning of coal could be replaced by the fission of uranium and nuclear bombs elevated nations to superpowers. The opportunities of the 21st century are limited by what we choose to do and not by what can be done.

This is a fascinating story of trial and error with the successful inventions providing the many devices we use every day.

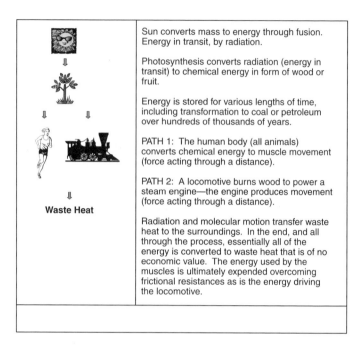

FIGURE 5-1. On Earth, most energy comes from the sun and ultimately becomes heat. This is a fascinating story of trial and error with the successful inventions providing the many devices we use every day.

Use of Thermal Energy

It was the middle of the 19th century when experiments were performed that convinced some scientists that the energy associated with heat was exactly equivalent to the energy that produces work. It took a new generation of scientists to finally end the arguments of those who could not accept this principle of energy conservation. The concept is now stated as the postulate *energy is conserved*. It is a cornerstone of classical physics that has been elevated in status and is the first law of thermodynamics.

Classical thermodynamics is an exact mathematical structure that describes the behavior of gases, liquids, and solids as the temperature, pressure, and composition change. The simple form of these mathematical relationships was established using experiments performed to improve the design of engines that convert thermal energy to work. The theory applies equally to energy conversions involved in heating and cooling a house, conversion of energy from coal combustion to electricity, combustion of gasoline to power an automobile, and the food you eat converted to the energy you use to walk and talk.

Our narrative will proceed with some examples of tasks we understand as work. These examples will give us a quantitative definition of work, and we can then assign numerical units to energy. We will show how any gas can be used to do work. We then show how we can use water as a liquid and vapor (steam) as the fluid to accept thermal energy, do work, and then reject some thermal energy to complete the work-production cycle. Today's machines that convert energy to work all use this basic principle.

The Concept of Work

Everyone understands that work is the use of energy to perform a task. It is work to lift a box from the floor to a table. Shoveling sand or snow is work. When we carry a bag of groceries from the store to the parking lot, it is work. All tasks we do that have us moving an object from one place to another we understand to be work. When we repeat the task several times, we need rest, and we need food to restore the energy we used performing those tasks. This qualitative description forms the basis for the scientific definition of work.

The scientific definition of work used in physics must be mathematically exact. This is necessary because the numerical value of the work we assign to a task must be the same for every scientist or technologist who does the analysis. This allows scientists to communicate efficiently. Physics has its own definition of work: *Work is the numerical value of a force multiplied by the distance over which that force is applied.* We can use this definition to compute the work required to lift a box from the floor to a table.

The planet Earth exerts a force on all objects near it: the force of gravity. We measure this force in pounds. Where a 10-pound box is lifted the 2 feet from the floor to the top of a table, 20 foot-pounds of work are done against the force of gravity. If a 10-pound box is lifted 2 feet, 20 foot-pounds of work are performed.

The analysis of the work one does shoveling snow or sand involves replications. The weight of each shovel of snow times the distance the shovel travels upward against the force of gravity is the work per full shovel. Multiply the work per shovel by the number of times you must load the shovel and this is the total amount of work required to clear the sidewalk. It is more work to shovel snow into the back of a truck than to shovel snow to the side of the driveway because the truck is higher.

If we threw 100 pounds of sand instead of 100 pounds of snow into the truck, the same amount of work would be required to load snow or sand. This physics definition of work allows us to calculate the numerical value of the required work, and the work is independent of the material we move. We can replace the sand with corn, coal, or cucumbers, but it is the weight in pounds times the height of the truck bed that determines the work required to load the truck.

When we move groceries from the supermarket to our car in the parking lot, the definition of work appears to fail. In particular, if the counter, the shopping cart, and car are at the same level, the groceries have not moved upward against the force of gravity, yet we know it took work to push the grocery cart from the store to the car. To make this definition of work more useful, early scientists identified that gravity was only one of many forces that is routinely overcome to perform work.

Where Did the Work Go?

You have certainly noticed that pushing the grocery cart up an inclined parking lot to your car is more difficult than on a flat parking lot. In this case, you must provide the work it took to push the cart on a flat parking lot, plus you are "lifting" the groceries from the store counter to the height of your car. Where does the work we do pushing the grocery cart go?

Let's consider a box sliding across a table. If you push on the box to slide it across the table, you apply force to move the box times the sliding distance. The box did not move up or down, so there was no work raising or lowering the box. You must push with enough force to overcome this friction between the box and the table as the box slides. Where did that work energy go?

Rubbing your hands together to keep them warm on a cold winter day is an example of frictional force. When you press them together harder, or rub faster, or rub longer, your hands get even warmer. Your muscles do work to overcome the friction between your hands, and this work is converted to heat through the frictional contact. This shows that you can convert work to thermal energy (heat), since rubbing your hands back and forth requires a force through/times the total distance you rub them, and this is our definition of work.

For the grocery cart, the energy expended in performing the work ended up in two forms. The energy overcoming gravity was stored by its location—specifically by its height. The potential to recover this energy remains and this is called potential energy. This potential energy can be converted to kinetic energy by letting the cart speed up as it rolls down the hill.

The energy expended to overcome friction of the cart became thermal energy. Likewise, energy expended to overcome friction from sliding the box became thermal energy. The box actually became warmer. The work done sliding the box is converted to heat. Our grocery cart is designed to reduce the friction, wheels with bearings turn easily, and this reduces the work required to move groceries to the car. You could have put the groceries in a box, attached a rope, and dragged the box to the car. This would take much more work than using the cart, even if one of the cart's wheels doesn't turn very easy.

What would happen if the cart were flung off a cliff or loading dock? Literally, the cart would go faster and faster as it falls. When it hits the ground, the speed of the cart would be converted to thermal energy, as it stopped at the end of the fall.

Notice, the work required to push the cart to the car can be computed if we measure the force, in pounds, required to push the cart times the 200 feet to the car. The resisting force of the shopping cart is an example of a frictional force. In this case, the work required depends on how easily the wheels on the cart turn. This is the way one can compute the work required to move a load on a level road using a cart, truck, or a train.

If work can be converted to heat, is it possible to convert heat into work? The first steam engine demonstrated this could be done. It is this puzzle that scientists solved during the middle of the 19th century. They were able to show that work could all be converted to heat, but there is no machine we can design that will convert all of the thermal energy (heat) into work. A fraction of the thermal energy (heat) must be transferred from the "work producing" machine or engine.

Converting Thermal Energy to Work

We are all familiar with fire. When wood, coal, oil, or natural gas burns, it produces thermal energy. The energy is released when the molecules that make up the fuel break apart and chemically combine with oxygen in the air (combustion) to form new molecules. This oxidation reaction destroys the fuel chemicals to produce mostly carbon dioxide and water vapor, releasing heat (thermal energy) in the process. The ash remains of a campfire are the oxides of the minerals that were in the wood. The thermal energy that is released is diffuse; it moves in all directions with equal ease. The development of heat engines to convert this diffuse thermal energy to useful work set us on the path of the industrial revolution.

Early experiments were done with air trapped in containers to develop the science that describes how to build heat engines. Consider the following modification of their experiments to show how to produce work from thermal energy.

An open can is equipped with a free-sliding piston that seals at the walls so that heating the can from 30°C to 60°C causes the piston to move upward. Any weights on the piston would move up or down with the piston. Movement is caused by the pressure force of the trapped air on the bottom of the piston. This force in pounds times the displacement of the piston in feet is the work done lifting the weights in units of foot-pounds. For all practical purposes, the work performed by this primitive heat engine is the same as work performed by a man lifting the same weight over the same distance.

In addition to the ability to perform work, fundamental observations were noted for this primitive machine comprised of a free-sliding piston in a can. First, the volume inside the can at 60°C is always 1.1 times the volume at 30°C. Also, the pressure of the contained gases remains constant during heating. The pressure can be increased by placing more weights on the piston.

If superglue is put on the sides of the piston and the piston is locked into place, our constant pressure experiment is converted to a constant volume experiment. When heated from 30°C to 60°C at constant volume, the pressure increases. In fact, the pressure at 60°C is 1.1 times the pressure in the can at 30°C—the same 1.1 multiple that described the volume change at constant pressure for the same temperature change.

You can use a bicycle tire pump as a "modern" laboratory equipment to show that the temperature of air increases when it is compressed. Increase the bicycle tire pressure from about 30 psi (pounds per square inch) to 60 psi. When you have completed a few quick strokes of the tire pump, touch the pump cylinder near the bottom. It will be warm. Had you continued to operate this pump to fill a large tank from 55 to 60 psi, completing many quick strokes over several minutes, the cylinder of the pump would be too hot to grasp and hold.

The tire pump experiment is a qualitative demonstration that work used to compress a gas raises its temperature as the pressure increases. The pump requires work. Is it possible to reverse the process starting with a hot gas at some high pressure and allowing the gas to expand moving a piston against a force to do work? The answer is, "Yes."

Start with the piston near the bottom of the cylinder. Fill the space with a hot gas at an elevated pressure. Reducing the force on the piston a little allows the hot gas to expand causing the temperature and pressure of the gas to decrease. Continue reducing the force and the expansion will continue until the piston gets to the top of the cylinder. The work done will be the force on the piston times the piston displacement. This work-producing stroke that expands hot gas to produce work is exactly the reverse of the compression stroke that required work. Both the temperature and pressure of the gas will decrease as the gas expands to produce work.

Early investigators cleverly designed experiments to demonstrate that the compression/expansion (work in/work out) cycle is reversible under ideal conditions. Is the total work required to perform the compression stroke recovered in the expansion stroke? In the most ideal of circumstances, the answer is, "Yes." However, total work recovery can not be attained in a real machine. There will always be friction between the piston and the cylinder in a real engine. Work wasted overcoming friction can never be recovered.

Early Engine Designs

Like the free-sliding piston in a can, the "ideal engine" is important in visualizing how heat is converted to work in a practical engine. The ideal engine operates under the following rules:

- The piston moves in the cylinder without friction.
- There can be no heat transfer to (from) the gas from (to) the piston/cylinder during the expansion (compression) stroke.

While both of these rules are not realistic, real engine performances can approach these "ideal engine" specifications. For example, lubricating oil and Teflon O-rings can reduce friction, or piston compression can be made to occur so fast that the heat simply does not have time to escape through the cylinder walls.

The free-sliding piston in a can (cylinder) suggests that a gas such as air may be used in an engine to produce work. The gas is a fluid for doing work and is referred to as a *working fluid*. This work-producing machine should operate in cycles, with each cycle producing a small increment of work. An important theoretical result was obtained from early analysis of this "ideal engine cycle." Such an ideal cycle is traced schematically in Figure 5-2. The line drawing represents the position of a piston in a cylinder as the engine performs the work-producing cycle. Focus your attention on the gas. Everything else is mechanical equipment required to contain the gas and transfer the work from the gas to a task we assign. It is the gas that does all of the work.

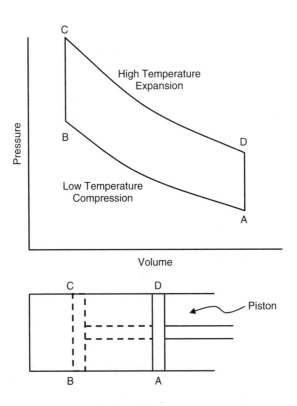

FIGURE 5-2. Illustration of how pistons perform work.

Suppose we start with a piston fully extended so the volume is maximum. Assume this corresponds to a volume of 3 quarts (point A; see Figure 5-2). Fill this volume with a gas at room temperature and pressure. This will fix the quantity of gas in the air-tight cylinder. Here are the four steps for our engine:

Step 1. Do work on the gas by pushing the piston from 3 quarts to 1 quart volume (from point A to point B). Like the tire pump, work is put into the process, and the working fluid (gas) heats in response to taking in this work.

Step 2. Heat the gas at constant volume—the 1-quart volume position. Similar to when the free-sliding piston was superglued to the can wall, the pressure increases and no work is done.

Step 3. Expand the gas to its initial volume (position D). Work is done by the gas moving the piston against a force.

Step 4. Cool the gas at constant volume to the initial temperature and pressure—the same condition as at the start of the cycle.

The gas is exactly the same temperature and pressure as the start of Step 1, so the cycle can be repeated again and again. More work is done in Step 3 than is required in Step 1, since the expansion Step 3 has greater force (greater pressure) than the compression Step 1, so net work is produced each cycle. Modern engines convert the back-and-forth motion of the piston to rotating shaft motion using a crank shaft.

There is a problem operating this simple hot gas expansion engine. A heat engine must take in thermal energy to produce the gas at high temperature and pressure. This hot gas then expands to produce work as the temperature and pressure decrease. At the end of this power stroke, the gas must be cooled before it is compressed for the next cycle. If we do the gas heating and cooling through the walls of the cylinder, it takes a long time to complete one cycle. This piston/cylinder combination is an obvious choice for an engine, but we must design the engine so we reduce the time required to complete each four-step work-producing cycle.

Two approaches are available that are better than heating the working fluid through the walls of the cylinder. For steam engines, valves can be used to control the flow of high temperature and high pressure steam into an engine as the piston moves to produce work. Alternatively, heat can be generated in the engine by actually burning a mixture of air and fuel in the engine. This latter, "internal combustion engine" requires fresh air and fuel in the engine at the start of each cycle.

The Science of Heat Engines

The details of the engine cycle of Figure 5-2 will provide an improved understanding of this cycle. During Step 1, the temperature and pressure of the gas will increase, and the volume will decrease to the minimum volume defined by the piston stroke-length. The work required for this step will be the force on the piston times the distance the piston travels from the initial to final position. No heat is transferred to or from the gas during this step according to the ideal engine assumption.

In Step 2, the temperature and pressure of the gas increase during this constant volume step to the pressure shown at position C. Since there is no change in volume, there is no work associated with this step. During Step 3, the temperature and pressure of the gas will decrease. The work produced will be the force on the piston times the distance the piston travels from C to D. No heat is transferred to or from the gas. During Step 4, no work is performed, since the volume is constant.

What can we learn from this ideal engine cycle? If we did not transfer heat to the gas in Step 2, the work required to compress the gas from A to B is exactly equal to the work produced in Step 3, expanding the gas from C to D. With no heat transfer, the points B and C would be identical to the points A and D. This engine produces no net work because the work produced is equal to the work required to compress the gas.

Both the input of heat to the engine and release of heat by the engine are necessary for the engine to produce work. The heat input could come from the burning wood. Outside air could be used to cool the engine during step four—this heat is rejected or lost to the environment.

When we transfer thermal energy (as heat) to the gas in Step 2, the path from A to B is separated from the path C to D. The work required to compress the gas in Step 1 is now less than the work produced in Step 3. This is true because at each position of the piston (volume of the gas), the force on the piston along the expansion path from C to D (producing work) is always greater than it is along the compression path from A to B (requiring work). Since energy is conserved (i.e., thermal energy and work are equivalent), the net work done by this ideal engine will be exactly equal to the difference between the heat added in Step 2 and the heat extracted in Step 4. We have assumed that there is no thermal energy transfer to the gas during the compression stroke, Step 1 and the expansion stroke,

Step 3. The gas started at the temperature, pressure, and volume of point A and when Step 4 is completed, the temperature, pressure, and volume of the gas are exactly the same as the starting values.

The Steam Engine Operating Cycle

The steam engine was developed over several decades. Working engines using steam were in operation before the theory of the work producing cycle just described was developed in the century. Water was the working fluid of choice for heat (steam) engine design starting in the mid-17th century.

One of the first devices that used steam was designed to lift water from mines. In 1698, Thomas Savery was issued a British patent for a mine water pump based on a simple device. It was known that when steam condenses in a closed container, a vacuum forms. The mine pump consisted of a metal tank and three valves. The only moving parts were the valves.

Ideal Engine Work and a Newly Defined Temperature Scale

The "ideal engine" cycle can be repeated any number of times, each cycle producing work from the thermal energy we add to the gas. The fraction of the thermal energy we put in at Step 2 that is converted to work by this ideal engine is represented by the simple ratio:

$$FracConverted = \frac{HeatIn - HeatOut}{HeatIn} = \frac{Work}{HeatIn}$$

The thermal energy removed in Step 4 is discarded and this ratio (Work/Heat In) is the fraction of the thermal energy converted to work.

The remarkable conclusion of these experiments is that the maximum work produced by any heat engine is independent of the working fluid. One can calculate the maximum fraction of thermal energy in a fluid that can be converted to work, and it depends only on the difference between the high temperature at which thermal energy passes into the engine and the low temperature at which thermal energy is taken out.

$$FracConverted = \frac{Thot - Tcold}{Thot}$$

The temperatures in this expression must be modified to correspond to the absolute temperature scale of thermal physics by the addition of 273.15 + T°C (or 459.7 + T°F) to complete the calculation. This simple expression is the upper limit of the work that an engine can produce per unit of thermal energy transferred to that engine at the high temperature. The materials used to build the engine cylinder and piston will set this maximum working temperature. The cold temperature will be close to the temperature of the atmosphere since the low temperature must be above the ambient temperature (temperature of a nearby river or the air) for thermal energy to pass from the cylinder.

With the vent valve open, a second valve was opened to admit steam to the tank from a steam generator (boiler). The steam and vent valves were closed when the tank was flushed out and water sprayed on the outside of the tank to condense the steam producing the vacuum. A valve in the pipe that extended from the tank down to the water in the mine was opened and the vacuum drew water from the mine into the evacuated tank. When the steam was condensed, the water flow stopped and this valve was closed. The vent valve was opened to drain the mine water and the condensed steam. The cycle was then repeated. This pump is not very efficient, but feeding wood and water to the steam generator and operating the valves requires much less physical effort than manning the mine water pumps to lift the water from the mine.

The Savery mine water pump could only raise water about 8 or 9 feet. Pumps had been used for many years to raise water from deep wells by placing the piston/cylinder of the water pump close to the water surface in a well. When the piston is pulled up from the bottom of the cylinder, water flows into the cylinder through a flapper valve in the bottom of the cylinder. At the end of the stroke, the valve at the bottom of the cylinder closes and a valve in the piston opens so water can flow through the piston as it moves down to complete one pump stroke. A rod connected to the piston is placed inside the pipe that brings the water to the mine or well surface. A pump handle moves the connecting rod and piston up and down to lift the water from the well. This pump can lift water many feet. The diameter of the piston, the length of the stroke, and the number of strokes per minute determine the volume of water removed from the mine. The first pumps were hand operated and it was tedious work.

The design of the water pump suggested the next design that used steam to operate the mine water pump. In 1712, Thomas Newcomen developed a pump that replaced the steam chamber of the Savery pump with a piston/cylinder. This engine is shown in Figure 5-3.

When the piston was at the top of the cylinder, the space was filled with steam and the steam valve closed. Cold water was sprayed into the cylinder and the vacuum pulled the piston down. This pulled down the rocker beam that raised the water pump piston lifting water out of the mine. Water from the condensed

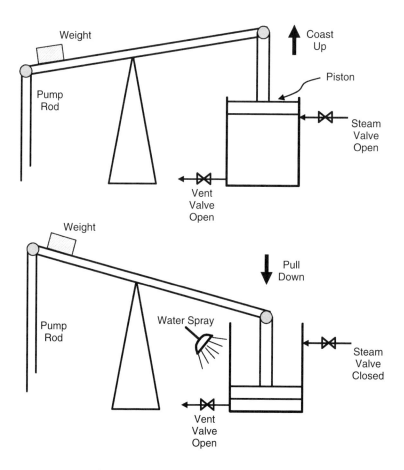

FIGURE 5-3. Condensing steam used to move a piston.

steam was drained from the steam cylinder as fresh steam was blown in. A weight on the rocker arm pulled the piston to the top of the steam cylinder completing one pump cycle. This engine cycle was slow, but by making the diameter of the steam piston and the pump piston large, the engine could move lots of water from the mine in a day. Operating the valves and putting wood in the boiler was easier than using a pump handle and operating the pump by hand.

Figure 5-4 illustrates the next significant improvement. By this time, designers had learned to produce steam from water in a closed vessel. This boiler produced steam at pressure—a hot gas that can "push" a piston on the power stroke of a heat engine rather than using a vacuum produced by condensing steam to "pull" the piston. Using steam pressure to push the piston, the connecting rod on the engine is attached to the rocker arm from the top and pushes the pump handle down rather than pulling it down in the vacuum pump design. At the end of the power stroke, the steam valve is closed and a vent valve opened to allow the steam to

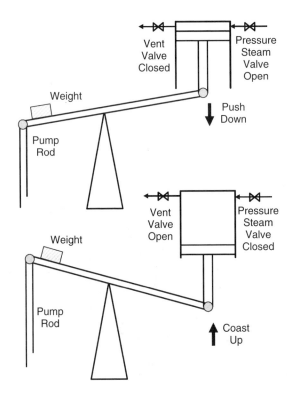

FIGURE 5-4. Use of high- and low-pressure steam to power a piston.

escape from the cylinder. The counterweight on the rocker arm returns the steam piston and the water pump piston to the starting position.

Closing the vent valve and opening the steam valve begin the next cycle. This operating cycle repeats as quickly as the valves can be opened and closed. It was no longer necessary to do the slow step of condensing the steam in the cylinder. This new design increased the number of strokes per hour and greatly increased the volume of water pumped per day.

The next major development was the steam engine design described in a British patent issued to James Watt in 1769.[1] Watt's design included three features that led to the modern steam engine: (1) He used a boiler to produce steam at an elevated temperature and pressure to drive the steam engine. (2) Steam was alternately introduced on opposite sides of the piston so work was produced on both the push and the pull stroke of piston motion. This "double action" piston was the standard in most steam engines. (3) His engine converted the oscillating motion of the piston-cylinder into continuous rotational motion using a connecting rod that turned a crank attached to a flywheel. This engine produced continuous rotation that could be used to pump water but, even more important, to turn shafts on machines.

Watt's engine design used a cylinder that was closed at both ends and fitted with a piston attached to a drive rod. Each end of the cylinder had a steam valve and an exhaust valve. To operate the engine, the steam valve on one side of the piston is opened and the exhaust valve is closed. On the other side of the piston, the exhaust valve is opened and the steam valve closed. The piston moves toward the low pressure (toward the open exhaust valve) with a force equal to the difference in the pressure on the two sides of the piston multiplied by the area of the piston. When the piston reaches the end of the cylinder, the position of the steam and exhaust valves has reversed and the piston moves in the opposite direction. The work produced is the product of the net force on the piston times the stroke the connecting rod moves.

The drive rod on the piston is coupled to a connecting rod that turns a crank that turns a shaft connected to the load. The piston applies a pushing force on the crank as it moves to the right and a pulling force as it moves to the left. A flywheel on this shaft keeps the crank turning at nearly constant speed at the end of each piston stroke when no work is produced as the piston changes direction. The steam and exhaust valves are connected to the engine crankshaft by a cam so they are "timed" to open and close in proper sequence with the position of the piston. A pressure control valve in the steam supply line adjusts the steam pressure so the rotational speed of the engine can be held nearly constant as the load on the engine changes.

The basic design of the Watt steam engine was the workhorse of the industrial revolution. This was the design used in machine shops, on ships, for train locomotives, for farm tractors, to generate electricity—pretty much any application that required a rotating shaft. The mechanical parts of the engine were customized to satisfy the requirements of the many applications. Improved materials of construction and the design of the steam boilers improved the efficiency and safety of the conversion of thermal energy to steam using the locally available fuels. The art and technology of the steam engine design could never overcome the problem of vibration from the rapid oscillation of the piston and the enormous weight required for large engines with high-power output. The steam turbine gradually replaced the steam engine when it was an advantage to increase the rotational speed.

For these early engines, boiler design improved with engine design. The very early engines used steam at low pressure. The boilers that generated low-pressure steam were relatively safe. When the steam engines began to use pressurized steam well above the normal boiling point of water, the boiler became an explosion hazard. Boiler failures were common with injury and death often the result. The shape of the boilers, new metal alloys used for construction, and replacing riveted joints with welded seams improved boiler safety. Boiler construction experience written into steam generator design codes and pressure testing procedures for new and "in-service" boilers have nearly eliminated steam boiler explosions in modern steam power plants.

Turbine-Based Engines

The Steam Turbine

In its simplest form, a turbine is a windmill enclosed in a tube that directs the air flow across the windmill blades. Wind has been used as an energy source to produce work for centuries. Ancient art shows sails were used to assist or replace men rowing ships. Much later, "sails" were set on wooden spokes that turned windmills centuries before the invention of the steam engine. Those who developed the steam engine knew that steam vented from a boiler through a nozzle produced a "wind." When the pressure in the boiler increased, the exit steam velocity also increased. A logical question is why not use this "steam wind" to drive a windmill? Certainly, one real advantage is this wind direction and velocity will be constant and controlled by the temperature, pressure, and flow rate of the steam to this "steam windmill."

The first practical steam turbines were built about 100 years after Watt's steam engine. An early turbine design by Charles Parsons was patented in 1884. Parsons' turbine used a jet of steam to turn several turbine wheels mounted on a single shaft. The turbine blades were placed on the outside edge of a wheel, and stationary vanes were set between each of the rotating wheels, which redirected the steam flow onto the blades of the next turbine wheel. Steam flow is parallel to the turbine shaft (perpendicular to the turbine wheels), just like a classic windmill. This makes it possible to gradually extract energy from the steam as the temperature and pressure of the steam drop as it passes through each turbine wheel.

Parsons showed that the most efficient conversion of the thermal energy in steam to work occurred when the pressure decrease across each turbine wheel was the same. Furthermore, at each turbine wheel, the pressure drop should be equally divided between the stationary vanes and the turbine blades.

Sufficient progress had been made with the design and operation of the Parsons turbine by 1894 that a syndicate was formed to test it in a small ship. This ship, named the *Turbinia*, ultimately attained the then-spectacular speed of 34.5 knots. This test established the steam turbine as the power plant of choice for marine applications.

The Parsons steam turbine used the flow of steam as a high-velocity wind to turn the turbine blades much like a windmill. A second practical design changed the shape of the turbine blades and had the steam pass over them much like water passing over a water wheel. This turbine was designed by Carl G. deLaval in 1887 and also was very successful. The turbine wheel has "bucket-shaped" blades that are "pushed" by the steam flowing from one or several nozzles directed onto the turbine buckets mounted on the turbine wheel. The steam nozzles were designed to increase the velocity of the steam as it approaches the turbine wheel and add to the force of the steam as it passes to the turbine wheel increasing the power generated. This turbine was developed to drive a cream separator, which requires very high rotational speed. One of these turbines that developed about 15 horsepower was actually used for marine propulsion before Parsons' turbine.

In 1895, George Westinghouse obtained the American rights to the Parsons turbine. He built a turbine to generate electricity for his Westinghouse factory in New York City. Electric motors were used on each machine in the shop. This was the first fully electric powered manufacturing plant in the United States. The steam engines that turned all of the drive shafts with their belts and pulleys were all replaced by electric wires and switches to electric motors mounted on each machine. This is the model that dominates the design and layout of all modern machine shops, assembly lines, offices, and homes today. A steam turbine drives a generator to produce electricity that is easily distributed to the point of use.

The first turbines did not efficiently convert the thermal energy in steam to power on a rotating shaft. Much of the steam "blew" past the turbine blades without producing work. Modern turbine design uses very close spacing between the multiple rows of turbine blades and the stationary blades that redirect the flow of steam as it passes through the turbine. The shape of each blade is as carefully designed as an airplane wing to reduce the energy loss to friction and maximize the power generated as the steam passes through the turbine.

The power output of a turbine increases as the temperature and pressure of the inlet steam increase. Higher temperatures require special metal alloys for the turbine blades and for the turbine casing that must contain the steam as it flows through the turbine. The high steam pressure and rotational speed of the turbine rotor place forces on the turbine blades that must be designed to withstand the stress without bending or breaking. The rotating blades must stay perfectly aligned so there is no contact with the stationary blades. The blades must be flexible enough to avoid brittle fracture, but they must also be designed so they do not flutter (like a window blind on an open window on a windy day) because even a small vibration could destroy the turbine.

Material science and technology research funded by the U.S. military developed and demonstrated that new high-performance metals could meet the demands on Navy ships. The military turbine technology was soon available to private industry because better performance reduces the cost of operating the turbine. The firms that produced the military turbines also produced turbines for industry, and it was natural to transfer the technology from the military to the private sector.

> *Process Steam and Its Use*
>
> Steam turbines are very versatile. They have been designed to operate with very hot, dry, high-pressure steam or with lower-temperature, low-pressure, wet steam. Large central heating plants produce steam that is used in petroleum refineries and chemical plants to heat process streams. Controlling the pressure of condensing steam controls the temperature that the thermal energy passes to the process. We get a high processing temperature when the condensing steam pressure is high and low temperature at low pressure. It is simplest to build a steam boiler that operates at constant temperature and pressure so the steam plants are usually designed to produce all of the steam required at a pressure greater than required for any process in the plant.
>
> Low-pressure steam is obtained by passing all of the steam at high pressure through a turbine and drawing off part of the steam at lower pressure a few turbine wheels into the turbine. Additional side streams can be withdrawn as the steam pressure decreases through the turbine. The turbine drives a dynamo to produce electricity as it provides steam at the desired pressure for the process. Central heating plants for large buildings or building complexes—a university is one example—always produce electricity with the steam before it is distributed to heat rooms or buildings.
>
> Steam turbines that deliver 50 to 10,000 horsepower have been designed to drive pumps and blowers and are distributed throughout chemical plants and petroleum refineries. These turbines are designed to operate with "available steam." The steam discharged from these turbines is then used in the plant as process steam (a heat source for processing chemicals). This combination of producing work for rotating machines and providing process steam is an efficient way to use more of the thermal energy from the fuel, which produced the steam.

The steam turbine is the primary source of power to drive dynamos in electric power stations. There are many power stations with electricity as the only product. In these power plants, the steam from the turbine is condensed, and the low-temperature thermal energy is discarded. The power plants that use fossil fuel and the nuclear power plants are different. Fossil fuel plants use higher-pressure steam and higher temperatures, and nuclear plants

138   *Sustainable Nuclear Power*

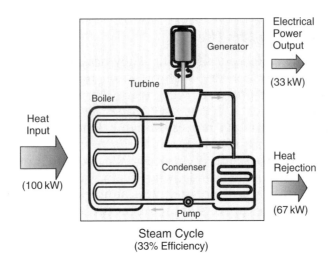

FIGURE 5-5. A basic steam turbine power cycle.

are limited to lower temperature. Figure 5-5 illustrates the steam cycle and the key components of that cycle.

The plants that use fossil fuels—coal, oil, or natural gas—generate high-temperature (typically 1,000°F or 538°C), high-pressure (3,550 psi or higher) steam. The steam actually passes through three or four turbine stages, all mounted on one shaft. As the temperature and pressure of the steam decrease, the turbine wheels get larger and the metal alloy in the turbine blades changes to match conditions of the steam and the rotational stress on the blades. The initial hot, dry steam expands through the turbine until the pressure is very low (actually a vacuum) and the temperature is about 120°F. After expanding to about 120°F, the steam passes to a condenser, where it condenses to liquid water. The liquid water is pumped back to the steam boiler at the pressure of the boiler.

After a century of design improvements, most of the modern fossil fuel power plant turbines turn at 3,600 revolutions per minute, and the largest units produce more than 2 million horsepower to generate 1,000 megawatts (1 billion watts) of electricity. Steam flows through the turbine at the rate of about 90,000 pounds per minute, and it takes 12,000 tons of coal per day to generate the steam to run the turbine. These are huge plants!

The thermal efficiency of a modern coal-fired power plant is about 40%—about 40% of the fuel's chemical energy is converted to electrical power. Figure 5-5 illustrates the energy flow for a cycle at 33% thermal efficiency. The thermal efficiency is defined as the work produced divided by the heat put into the cycle. It is a dimensionless number indicating that the work and heat must be in the same units to compute the thermal efficiency.

Nuclear Power

The first nuclear reactors designed to produce steam to drive a turbine were developed to replace diesel engines on submarines. The diesel engines charged batteries that powered the submarine underwater, but air was required to power these diesel engines. Surfacing to run the diesel engines, or even using "breathing tubes" to recharge the batteries while submerged, revealed the location of the submarine. Modern nuclear powered subs remain submerged for 90 days or more, thereby reducing the risk of detection as they cruise deep underwater. These nuclear power plants are now standard on many military surface ships. This increases their range and eliminates the difficult (dangerous) task of refueling at sea. The nuclear reactor technology to produce steam to drive a turbine soon passed to domestic power plants. The U.S. government with the "atoms for peace" program of the 1960s provided additional financial incentives to speed this domestic use of nuclear energy.

Water-moderated nuclear reactors are the most common for commercial electric power production. Liquid water is required in the nuclear reactor to "slow down" the neutrons produced by fission, so the reactor continuously produces the thermal energy to make steam. The water also serves to extract the thermal energy produced by the nuclear fission process.

There are two types of water moderated reactor designs used commercially. The boiling water reactor (BWR) has the reactor core and water contained in a pressure vessel. The water is allowed to boil, and the steam goes directly to the turbine. Figure 5-6 shows a BWR.

The pressurized water reactor (PWR) has the reactor core and water contained in a pressure vessel, but the pressure is high enough that the water never boils. The hot, high-pressure water is pumped through a heat exchanger that produces the steam that goes to the turbine. The PWR design keeps the water that comes in contact with the reactor core isolated from the steam that goes to the turbine. Figure 5-7 shows a pressurized water reactor.

The steam turbines in modern nuclear power plants operate at lower steam temperature (about 560°F; about 290°C) and pressure (about 1,000 psi) than the fossil fuel plants. The steam produced is

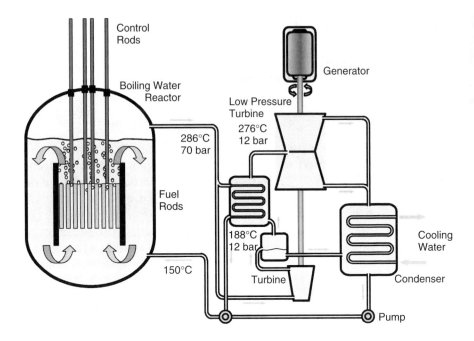

FIGURE 5-6. A boiling water reactor (BWR) and steam power cycle.

"wet," formed in contact with boiling liquid water, and it passes to the high-pressure stage of the turbine, where the pressure drops to about 150 psi and the temperature to 350°F (177°C).

As the steam temperature and pressure drop in the high-pressure turbine, about 14% of the steam condenses. This water must be removed from the remaining steam before it goes to the low-pressure turbine. The steam continues to condense as the temperature and pressure of the steam drop in the low-pressure stage of the turbine. The small water droplets that form move at high velocity with the steam through the turbine. They collide with the turbine blades that are also moving at high velocity. The turbine blades must be made of a special alloy to keep the water droplets from "sand blasting" the turbine blades away. Liquid water must be continuously removed from the low-pressure turbine casing to avoid damage to the turbine blades.

The turbines in nuclear power plants turn at 1,800 revolutions per minute and also develop over 2 million horsepower to produce 1,000 megawatts of electric power. They run slower because the diameter of the turbine wheels must be larger to allow much more steam to flow through them. The lower energy content of low-pressure steam requires about 160,000 pounds per minute of

History of Conversion of Thermal Energy to Work 141

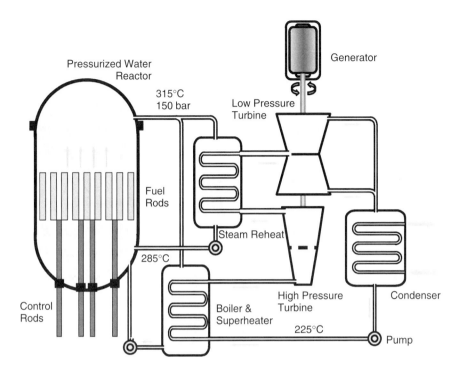

FIGURE 5-7. A pressurized water reactor (PWR) and steam power cycle.

steam flow. This is a huge machine that weighs about 5,000 tons with a rotating shaft that is approximately 73 meters long.

These huge turbines are remarkably reliable. It is necessary to shut down the power plant to refuel and perform routine maintenance on all of the mechanical equipment. Nuclear power plants usually run on an 18-month refueling schedule, and this becomes the interval for turbine maintenance. The greatest efficiency is obtained when the turbine operates continuously at full load between reactor refueling. The turbines in coal-fired power plants operate on about the same schedule. These steam turbines are a remarkable technological achievement matching special materials of construction with precision mechanical design.

Thermal Efficiency

The primary measure of performance for steam engines and turbines is the thermal efficiency. The thermal efficiency is defined as the amount of shaft work produced divided by the amount of heat taken from the boiler. The best of our modern natural gas power plants are able to convert about 55% of the heat provided by the natural gas into work. The remaining 45% is lost to the surroundings in the form of hot exhaust gases leaving the power plant stacks or steam leaving the cooling tower. The coal-fired power plants have thermal efficiencies up to 45%. The thermal efficiency of the nuclear plant is 30%–33% because the steam temperature and pressure are lower. The first steam engines were doing well to convert 10% of the fuel energy into shaft work.

Gas Turbines

The gas turbine has a long history. The theory of gas turbines and how they should work was known long before the first one was built. It appears the first gas turbine patent was issued to John Barber in 1791. There is no record that this gas turbine was ever built, but it did establish a basis for future development.

The first U.S. patent that described a complete gas turbine was issued to Charles G. Curtis in 1895. The first turbine that ran was built in France in 1900, but the efficiency of this unit was about 3%—not very encouraging. The early gas turbine builders tried to use the steam turbine wheel because it was available and it worked well with steam. It is a simple matter to mix fuel and air and burn it in a chamber, but how do you increase the pressure of the hot gas to make it flow through the turbine to produce work the same as a steam turbine?

On paper, the gas turbine promised to deliver considerably more power than a gasoline or steam engine of similar size. Early workers observed that higher initial steam temperatures and pressures provided more work per gallon of fuel oil—efficiency increased with increasing temperatures and pressures. The large quantity of thick metal alloy pipes necessary to take steam to higher temperatures and pressures was costly, but a small, inexpensive combustion chamber could provide vast amounts of hot combustion gas for a gas turbine. This created new challenges. First, the pressure of the air and fuel had to be increased before combustion. Next, the steam turbine metals would soften and deform at these high temperatures. Metallurgists began the search for new, high-temperature alloys and ceramics to make the turbine blades durable at higher temperatures, a search that continues today.

In principal, the same pistons and cylinders used to compress air in the gasoline engine could be used to compress air for the gas turbine. However, in practice much of the pressure gained was lost in the tortuous flow path through the many valves, turns, and manifolds needed to direct the air to a combustor chamber. The gas turbine engine needed a simple compression process that did not rely on valves, pipes, and manifolds. While piston-in-cylinder (reciprocating) compressors work well when compressing to pressures greater than 150 psig, reciprocating compressors did not work well for the lower pressures of about 60 psig that were needed in gas turbines.

The axial flow gas compressor (the flow of air is parallel to the compressor shaft) provided this simpler compressor that was effective at lower compression ratios. (Pressure ratio is defined as the pressure after compression divided by the pressure before compression. A pressure ratio of 5 corresponds to increasing pressure from 15 psia to about 75 psia.) It was the mid-1930s when the aerodynamic theory developed to design aircraft wings was applied to the shape of the axial gas compressor blades, ensuring the commercial future for gas turbines.

An axial flow compressor operates on the same principle as the steam turbine. The design engineers changed the shape of the blades in the turbine so that the gas pressure increases as the wheel turns. You supply the power to spin the compressor shaft and the pressure increases at each set of rotating and stationary blades, and you have an axial flow compressor. This axial compressor avoids the use of valves and gas flow through tortuous manifolds by placing the compressor on the same shaft as the gas turbine. With fuel delivery and a well-designed combustion chamber located between the compressor and turbine, the major engine components are in place. Some of the compressed air bypasses the combustion chamber to reduce the temperature of the gas to the turbine. Good design adjusts the size of the compressor and the turbine to deliver maximum power at the turbine driveshaft.

The turbine blade is the critical mechanical part in a gas turbine. The aerodynamic shape of the outside surface of each row of blades must be matched to the conditions of the hot gas flowing past them. The first row is exposed to the highest temperature because the gas cools as it expands through the turbine; each turbine wheel after the first is exposed to cooler exhaust gases. In some modern designs, air passageways are machined inside each turbine blade to allow cool, compressed air to blow through the blade to cool it as the hot gas flows by on the outside. Ceramic coatings can be applied to the outside surfaces of the blade to give additional heat protection to the turbine blade. These are technical fixes to "beat the heat" and keep the turbine running at higher gas inlet temperatures. Manufacturing these complex blades increases the cost of the turbine, but running at slightly higher temperatures increases the fuel efficiency. For example, increasing the gas inlet temperature from 900°C to 1,000°C increases the thermal efficiency of the turbine about 10%.

The greatest boost to gas turbine development came from the development of jet engines for aircraft propulsion. The British and the Germans both tested jet aircraft engines from the mid-1930s. The Germans had jet-powered fighter planes in service during the final stages of World War II. The development of the military jet engines was classified "Secret," and advanced technology remained classified following World War II.

The gas turbine as an aircraft engine was successfully demonstrated during World War II. Immediately following the war there was a period when redesign and testing produced the first commercial turbo-prop airliner. It was Vickers Viscount turboprop that was first flown in 1949 with conventional twin propellers powered by gas turbines. It entered commercial service in 1953. This was the beginning of the end for the reciprocating engine on all but small aircraft. Commercial aviation was powered by gas turbines spinning the propellers (turbo-props) and gas turbines providing vast amounts of hot exhaust gases that were the start of more advanced jet engines.

Gas turbines exceed in two performance criteria:

1. A high power output per pound of engine weight makes the gas turbine ideal for aircraft service.
2. A gas turbine can be started and run at full power in a few minutes.

Gas turbines also work well with a variety of fuels: natural gas, fuel oil, waste gas in an oil refinery, and so forth. As long as the fuel combustion does not produce solid particles that "sand blast" the turbine blades, it can be fuel for a gas turbine. Given a fuel supply, the gas turbine combustion chamber can be designed to optimize performance for that fuel.

The period from 1945 to 1970 represents consolidation of the gas turbine as a power source for many applications. The first commercial gas turbine train locomotive was put in service in 1950. The low thermal efficiency of the gas turbine was a disadvantage relative to diesel units. The simple gas turbine is less efficient than either the gasoline engine or diesel engine with maximum efficiencies of about 30%, 37%, and 50%, respectively. Interest in developing turbine technology for locomotives disappeared as freight traffic moved from railroads to the highways.

The first trials placing a gas turbine in a personal boat came in 1950. Again, the gas turbine offered no real advantage over the conventional diesel-powered ocean fleet. It has been only recently that gas turbines have replaced steam turbines on Navy ships.

Pumping natural gas through pipelines from the gas wells to customers all over the country requires lots of power. A gas turbine fueled with natural gas driving a centrifugal compressor powered by an 1,800-horsepower gas turbine was installed on a 22-inch diameter pipeline in 1949. Gas turbines have found a place on oil platforms located at sea to provide electricity and to turn pumps and compressors. They use the oil or gas produced on the platform as fuel.

It was in 1949 that the first gas turbine electric generating unit was put into service in the United States. A unit rated at 10,000 kW represents an important achievement in the development of electric power generation, pointing toward modern electric power plants.

The first gas turbine installed in an automobile traveled across the United States in 1956. Chrysler Corporation built 50 gas turbine–powered cars in 1963 to 1965 to be used by typical drivers on daily trips. The advanced technology of the internal combustion piston engines at that time offered better fuel efficiency and reliability. The market for the gas turbine car did not develop, and the project died. There was no reason to develop this application of the gas turbine.

Gas turbines are now widely used to supply electrical power because they are the least-expensive engine for a given power output in the large engine market. However, the simple gas turbine's disadvantage of low fuel efficiency (i.e., high fuel costs) can more than offset the money saved in purchasing the engine when placed in continuous service. Obviously, the gas turbine is not best for all electrical power applications.

Electric power demand changes in predictable ways based on season and time of day. The long-term use cycle corresponds roughly to the seasons, and a short-term 24-hour cycle corresponds to variations in human activity during the day as compared to night. The vast quantity of electricity cannot be stored, so it must be produced at the rate it is used. There are three levels of electrical power production designed to meet this cyclic demand.

Baseload power must be continuously supplied every day of the year. The baseload power generators must show efficient conversion of fuel to electricity, must be very reliable, and must provide the lowest cost per kilowatt-hour of electricity produced. These are usually efficient coal-fired power plants or nuclear plants. The low fuel costs for these plants more than compensate for their high capital cost for continuous power production. They run at constant power output for months without shutting down or throttling back.

Intermediate load generation units operate about 75% of the year, and they take care of the swings in seasonal load. They also assume the load when a baseload unit is shut down for routine maintenance or for an emergency shutdown of any generating unit on the power grid. These plants should be capable of operation at partial load. The energy conversion efficiency of any power plant is reduced when it operates at partial load because the steam boiler and turbine are most efficient at full load.

Peak load generation units are used in the summer for a few hours after noon when all the air conditioners are turned on. Peak power production will be required about 5% of the year (certainly less than 10%) and usually only part of the 24 hours each day. The thermal efficiency of these units is not as important as the ability to start them quickly, run them for a few hours, and shut them down. The gas turbine is best suited for this peak load assignment. A wide range of horsepower ratings are available, and it is often an economic advantage to use several small gas turbine units for peak demand service. This allows the use of just enough units operating at full power to cover the peak demand for that day. Gas turbines can be started and brought to full power in minutes. This is in contrast to coal-fired steam plants that take many hours to start or shut down.

The recent strategy for intermediate electric power production is to use a *combined cycle* power plant. The exhaust gas from a gas turbine is still very hot. This hot gas can be passed through a boiler to produce steam to turn a turbine and generate electricity. Combining the power output of a gas turbine with that of the steam turbine increases the total thermal efficiency of the combined cycle plant to over 50%. This compares to efficiencies of about 30% for simple gas turbines. Improved thermal efficiency becomes very important when fuel costs are high, but the flexibility of "quick" start and shutdown is lost because boilers are slow to heat up and produce steam.

Natural gas is often the fuel of choice for gas turbine plants. There is the added advantage that the carbon dioxide released to the atmosphere per kilowatt is less than for a coal-fired power plant. Combined cycle plants are about half the price of coal-fired plants due to elimination of costly coal-handling facilities, use of smaller boilers (only part of the electrical power is provided by steam), elimination of precipitators to remove the particulates in coal exhaust gases, and the elimination of scrubbers to remove sulfur oxides from coal exhaust gases. Natural gas is considered a premium fuel because it does not require the costly solids-handling and exhaust-treatment facilities. Control of the supply of natural gas by a few pipeline distributors has created wide price fluctuations, which makes it difficult to estimate the future production cost of electricity from natural gas. This makes investment planning for the natural gas plants difficult. Government loan protection is given to utility investors in some regions of the United States where electricity is in short supply and demand is increasing.

There will always be demand for specialized uses for gas turbines. Today, electric power generation and aircraft propulsion are featured. Any device designed to power a commercial aircraft must satisfy many criteria. It must meet the performance demands of the aircraft and be fuel efficient, low weight, reliable (never fail in flight), and easy to maintain with a maximum number of operating hours before the engines must be replaced. These are the problems gas turbine design engineers have worked on during the past 20 years.

Government regulations regarding safety must be satisfied by the design. The exhaust emissions are federally controlled. The amount of carbon monoxide, unburned hydrocarbons, and nitrogen oxides in the exhaust gases must meet EPA standards. Noise standards have also been established that require special tests to find the source of the noise and the mechanical design changes to reduce the noise. Each of these requirements places demands on the design of the combustion chambers and mechanical components of the gas turbines on an aircraft.

Fuel Cells

Fuel Cells

Practical fuel cells were designed during the 1960s and were first used during the NASA Apollo program. The fuel cells were fueled with stored hydrogen and oxygen. This was a particularly valuable technology for space travel, since the process that produced electricity also produced water for the astronauts. Transporting drinking water into space could be substantially eliminated by matching the minimum fuel cell hydrogen consumption to the drinking water needs.

Similar to steam turbines, rockets, and jets, the military (specifically NASA) paid for fuel cells during the early and costly development stage. It is this technology adapted for use in automobiles that is being developed for domestic use with financial support from the U.S. Department of Energy.

Fuel cells are like batteries, but instead of storing energy using the chemistry of lead and acid, they are powered by a fuel and oxygen. Both fuel cells and batteries convert the energy of chemical bonds into electrical power. When connected to electric motors, fuel cells can be used to power vehicles much like gasoline or diesel engines.

Fuel cells weigh less than batteries and they produce electricity as long as the fuel (hydrogen and oxygen) is supplied. The chemical reaction that produces electricity in a battery is reversible. The quantity of chemical is limited by the size of the battery. When the electrical current slows or stops, an electrical current is passed through the battery to recharge it (reverse the chemical reaction). This "remakes" the chemicals that produce electricity. The discharge/charge cycle can be repeated many times, but there must be a source of electricity to recharge the battery. This is a clear advantage for the fuel cell when a reliable source of low-cost fuel is available.

Fuel cells are also considerably lighter than batteries. For a car with a 100-mile range, it takes about 1,100 pounds of lead/acid storage batteries to power the car.[2] Depending on design, fuel cell systems complete with on-board fuel to achieve the same range would weigh less than 100 pounds. This is why fuel cells worked so well for space travel.

At the beginning of the 21st century, much attention is focused on fuel cells. The excitement comes from the potential of fuel cells to achieve high fuel efficiency and low emissions. Table 5-1 lists several performance categories to compare the advantages of fuel cells to gasoline and diesel engines, gas turbines, and batteries.

Work and Efficiency in Fuel Cells

In an ideal fuel cell, 100% of the energy that would otherwise go to heat during combustion goes to electrical power (e.g., to a

TABLE 5-1
Performance strengths of different mobile power sources. E = Excellent, G = Good, F = Fair, P = Poor, I = Insufficient Data.

Property	Fuel Cell	Gasoline Engine	Diesel Engine	Battery	Gas Turbine
Fuel efficiency potential	E	G	G	E*	F
Present and near-term fuel efficiency	F	F	G	E*	F
Ability to deliver good fuel efficiency with varying loads	E	F	F	E*	P
Will work with variety of fuels	F	E	E	P	E
Power output per system weight	G	G	G	P	E
Engine cost per power output	P	E	G	F	G
Durability and expected performance life	I	F	G	F	E
Emissions at vehicle	E	G	F	E	G
Emissions not at vehicle	F	E	E	G	E
Feasibility of practical rechargeable systems	E	P	P	E	P
Performance synergy with hybrid vehicles	E	G	E	E	G
Lack of hidden performance problems	P	E	E	E	E

*While batteries are very efficient for converting their stored chemical energy to electrical power, the efficiency of converting an available energy source (coal or nuclear) to the chemical energy in the battery is poor. The reported efficiency is based on electricity to charge the battery and not the fuel to produce electricity.

motor). The ideal fuel cell converts 100% of the chemical energy into electricity.

In the worst case, essentially all the chemical energy would go to heat, which could burn up the fuel cell. The heat must be transferred from the fuel cell to keep it from being damaged.

How Does a Fuel Cell Work?

Understanding fuel cell operation is much like understanding combustion.

Hydrogen (H_2) and oxygen (O_2) will burn to form water (H_2O). This is a chemical reaction depicted by the following chemical equation:

$$H_2 + O_2 \rightarrow H_2O + \text{Energy Release}$$

Combustion, as with all energy-releasing reactions, occurs because the final product is more stable than the reactants. The progression of compounds going from less-stable higher-energy states to more-stable, lower-energy states is the natural route for all energy technology.

We want the reaction to occur when and where the energy can be used. For hydrogen combustion, heat or a spark applied to a mixture of hydrogen and oxygen will cause the reaction to proceed quickly and release lots of heat. Are there other ways to control the rate of this reaction to produce electricity rather than heat?

Consider the following series of steps that occur when platinum dust is coated on a membrane surface:

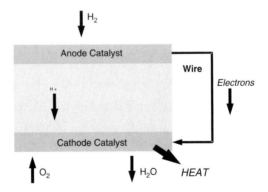

The hydrogen molecule, H_2 is pulled apart to form two H+ (hydrogen cations) and two electrons. It is the platinum powder that makes this happen.

The H+'s are able to travel to an oxygen molecule by passing through the membrane that does not pass electrons.

The electrons travel in a wire contacted to the platinum anode through an external circuit where it can turn a motor ending up at the cathode platinum dust on the other side of the membrane where there is oxygen.

On the oxygen side of the membrane, two of the electrons and two hydrogen cations combine with an oxygen atom to form a very stable water molecule. The driving force for forming water is so strong that this reaction literally pulls the electrons through the wire to the oxygen/cathode side.

As improbable as this process sounds, it works for proton exchange membrane fuel cells. This process does occur at room temperature because platinum is such a good catalyst.

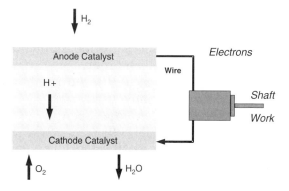

If this occurred for the system shown by the diagram, the same amount of heat would be released as when you burn hydrogen. If the wire is connected to the armature wires of an electric motor, the oxygen side of the membrane acts as though it is "pulling" the electrons through the wire/motor due to the electromotive force produced by the reaction of hydrogen and oxygen to form water. The current passing through the electric motor produces work.

The best practical fuel cells operate at about 55% efficiency. This means that 55 Btus of work are generated for every 100 Btus of combustion energy released by "burning" hydrogen in the fuel cell. The other 45 Btus of heat are released in the fuel cell and must be removed to prevent the fuel cell from overheating.

Issues that have pushed fuel cells into the spotlight are fuel efficiency, low emissions, and compatibility with the hybrid motor vehicle. Fuel cells avoid the thermodynamic limitations of combustion by directly converting chemical energy into electricity at about ambient temperatures. Practical fuel cells do operate at 80°C or higher to allow the waste heat to be easily removed.

Fuel cells appear to be the "natural choice" as a power source for hybrid vehicles, since they store energy as fuel and use electricity to turn the wheels. There is little doubt that the potential of fuel cells will eventually outperform conventional engines in essentially every category. Keep in mind that the people who describe fuel cell technology in the technical literature have vested interests in fuel cells. It is easy (almost natural) for them to be optimistic about the progress that can be made in the next one or two decades.

There is no corporate incentive to change from internal combustion engines to fuel cells as long as petroleum is plentiful and cheap. It took legislation that set pollution standards (eliminate lead and sulfur from motor fuels) and increased the miles per gallon (fuel efficiency standards) to move the auto and fuel industries to reach performance levels available today.

For a state or nation, the economic incentive to change from internal combustion engines to fuel cells rests on the simultaneous replacement of an imported fuel with a locally produced fuel. By replacing imported gasoline with electrical power produced from any low-cost fuel (uranium?), a state could keep billions of dollars spent on imported fuel. This could create tens of thousands of jobs. The last chapter of this text discusses this economic picture.

Fuel Efficiency

An ideal fuel cell running on pure oxygen and hydrogen produces 1.229 volts of electromotive force to push electrons through the electrical devices when there is no load on the circuit (the switch is open). Since the chemical reactions of a fuel cell provide the same flow of electrons independent of the fuel cell efficiency, this voltage is used to compute the efficiency. For example, when the fuel cell is connected to an external load, the "open circuit" voltage of the fuel cell is split between the resistance to anion flow in the cell and the resistance to electron flow in the external load. If the fuel cell produces 0.615 volts when connected to a lightbulb, the efficiency of the fuel cell with that load is 50% ($\frac{0.615}{1.229} \times 100$). The conversion efficiency improves when the external load is decreased. It takes careful design of the fuel cell to handle the variable power requirements of commuter car travel and maintain high efficiency.

At the start of the 21st century, the efficiency of a good hydrogen fuel cell running with atmospheric air rather than pure oxygen varied from about 45% at maximum design load to about 65% when operating at low power output corresponding to the "most efficient" load.[3,4] If electrical power is used to make the hydrogen, and this electric power is generated with the efficiency varying from 30% to 55%, this gives an overall fuel cell efficiency of 13.5% to 36%. If the fuel cell operates on hydrogen produced by converting gasoline to hydrogen on the vehicle, the conversion of energy from gasoline to hydrogen is about 70% to 80% which gives an overall efficiency of 31.5% to 52%. The technology to use gasoline to power a fuel cell car requires the technology development to reform the gasoline to hydrogen. The gasoline reformer and fuel cell cars must be road-tested before they will come to the dealer's showroom.

The reality for fuel cells in vehicles today is that they are less efficient than the best diesel engine. The current goal of an overall fuel cell efficiency of 70% is not realistic neglecting the energy loss during hydrogen production. Operation at maximum efficiency requires a large fuel cell to keep the electrical current produced per unit area of the cell low. A fuel cell does not respond to the quick starts and acceleration required for urban driving. The fuel cell power output can be "leveled out" by using a battery pack that provides surge power and is recharged during cruise and stop phases of the trip. The gasoline-powered hybrid cars on the market today provide an ideal test for an electric-powered car. Design a fuel cell to replace the gasoline engine generator and battery in the hybrid, and you have a fuel cell car.

Battery and Fuel Cell Options

Whether we use a fuel cell or battery to power a car, the following components are necessary:

- A wire that conducts electrons (an electrical current) to the external load. The load can be the electric motor that turns the wheels, the lights, the CD player, and so on.
- A membrane or "salt bridge," a solution that conducts cations (cations are the positively charged particles that remain when the electrons are stripped off a molecule during the chemical reaction) or any other medium that conducts cations but not electrons.
- A chemical reaction that will proceed by producing cations.

- A reaction surface/catalyst that promotes the reaction and that frees the electrons to flow to the wire and the cations to flow to the membrane (salt bridge solution or membrane) to complete the formation of products of the "burned" fuel. For the hydrogen fuel cell, this is hydrogen and oxygen combining to form water.

Different chemical reactions will produce different fuel cell voltages. The total current a fuel cell produces depends on the total catalyst surface area and the number of electrons produced when a fuel molecule goes to its reaction products.

We have the technology to make a fuel cell work with hydrogen, methanol, and natural gas. However, the technology available today requires that gasoline, diesel fuel, or coal be converted to one of the fuels used in fuel cells.

The active research associated with each of the preceding four steps for an improved fuel cell includes (1) develop better and cheaper membranes to conduct cations in the fuel cell; (2) catalysts that are less expensive than platinum; and (3) improved design to remove the heat that is generated when the fuel cell is operating. Fuel cells can operate at theoretical efficiencies in excess of 90%, but the typical practical efficiencies are 30% to 55%. This means that 35% to 70% of the energy available from the hydrogen fuel must be removed as heat. Other practical issues include preventing the fuel cell membranes from drying out, how to keep the water in the fuel cells from freezing in cold weather, and other process-specific problems associated with hydrogen generation and the use of hydrogen that contains impurities.

Methanol fuel cells rather than hydrogen fuel cells have great potential because they use a liquid fuel. It is much easier to carry liquid fuel on a vehicle. The methanol fuel cells are not as efficient as hydrogen fuel cells, and they are also more expensive to build.

One of the advantages of fuel cell–powered vehicles is that they produce a low level of emissions. When methanol or hydrogen is used, problem emissions associated with gasoline engines are essentially eliminated, since water is the primary emission with carbon dioxide added from the methanol cell. When gasoline is autoreformed to produce hydrogen for fuel, it is a high-temperature process. It will produce carbon dioxide, some nitrogen oxides, and particulate matter, much like a gasoline engine.

It is easy to exaggerate the true emission benefits of fuel cells. The history of the gasoline and diesel engines shows that when high pollution levels were tolerated, cities did prosper. Claims of premature deaths due to auto emissions probably are true, but the case for these claims was not strong enough to lead to drastic resolutions such as banning cars from the city. Individual vehicle emissions in the late 1960s were less than in the 1950s, and since then, the 1960s emissions have been reduced by more than a factor of ten. Based on current regulations, by 2010, emission standards for gasoline and diesel engines, the emissions will be reduced to 1% of late-1960s levels. For diesel vehicles, implementation of tier 2 standards in 2007 will reduce emissions for new vehicles to 10% of the engines manufactured in 2002.

It is reasonable to ask the value of reducing vehicular emissions to 1% of previous levels. In September 2002, the EPA reported[5] that "the exposure-response data are considered too uncertain" to produce a confident quantitative estimate of cancer risk to an individual. The EPA reported, the "totality of evidence from human, animal, and other supporting studies" suggests that diesel exhaust is likely to be carcinogenic to humans by inhalation and that this hazard applies to environmental exposure. While these data on cancer risk were less than definitive, the EPA reports state that long-term exposure has been shown to be a "chronic respiratory hazard to humans." A 100-fold reduction from levels that produced less than definitive cancer risks and require long-term exposure to cause chronic respiratory problems should be sufficient to protect the public. What is the value of going to an impossible zero emissions from a level very near zero that is in the emission standards? The answer is not obvious. If emission reductions require a change to fuel cell systems that cost more to purchase and operate, expect to meet consumer resistance.

The perceptions and priorities given to fuel cells today speak to the current status of energy politics in the United States. While there is potential in fuel cell technology, the general public and politicians responsible for setting energy policy have been subjected to selective distribution of information. The federal government has approved financial support to auto manufacturers to develop a fuel cell car by 2015. This new legislation relaxes the increased miles-per-gallon standards for new cars and continues to exempt trucks from the standards. This strategy assures that petroleum will remain the only source for transportation fuel. Great expectations for fuel cell–powered cars in 15 years are used to replace any real efforts to implement new technology to address the critical issue of imported oil.

While the benefits of fuel cell technology have been substantially exaggerated, one thing is certain: Fuel cell technology has great potential to bring societal and economic benefits. If fuel cells are improved with the use of grid electricity to power automobiles, the impact could be great. Fuel cell research and technology development should be pursued.

Recommended Reading

1. Bathie, William W., *Fundamentals of Gas Turbines*, 2nd ed. New York: John Wiley & Sons, 1996.
2. Constant II, E. W., *The Origins of the Turbojet Revolution*. Baltimore, MD: Johns Hopkins University Press, 1980.
3. Cumpsty, Nicholas, *Jet Propulsion*. Cambridge, UK: Cambridge University Press, 1997.
4. Garvin, R. V., *Starting Something Big: The Commercial Emergence of GE Aircraft Engines*. Reston, VA: American Institute of Aeronautics and Astronautics, Inc., 1998.
5. Von Braun, Wehrner, and F. I. Ordway, III, *History of Rocketry and Space Travel*, 3rd ed. New York: Thomas Y. Crowell Co., 1975.
6. Wilson, D. G., and Theodosios Korakianitis, *The Design of High-Efficiency Turbomachinery and Gas Turbines*, 2nd ed. Upper Saddle River, NJ: Prentice Hall, 1988.

General References

1. *Academic American Encyclopedia*, Grolier Inc., Danbury, CT, 1996, Vol. 17, p. 375.
2. *Encyclopedia Americana*, Grolier, Inc., Danbury, CT, 1997, Vol. 27, p. 241.
3. *McGraw-Hill Encyclopedia of Science and Technology*, 8th ed., McGraw-Hill, New York, 1997, Vol. 17, p. 375.
4. *The New Encyclopedia Britannica*, Encyclopedia Britannica, Inc., 1997, Vol. 18, p. 343.
5. *The World Book Encyclopedia*, World Book, Inc., Chicago, IL, 2000, Vol. 18, p. 413.
6. *A History of the Automotive Internal Combustion Engine*, SP-490, Society of Automotive Engineers, Inc., July 1976.
7. Cummins, C. Lyle, Jr., *Internal Fire*, Carnot Press, Lake Oswego, OR, 1976, Ch. 14.
8. Lilly, L. R., Ed., *Diesel Engine Reference Book*, Butterworths, London, 1984.

References

1. See R. H. Thurston, Cornell University Press, Ithica, NY, 1939. This book was first published in 1878 and provides a detailed history of the inventions and technology developments that made the steam engine the key player in the industrial revolution.
2. Lester B. Lave, W. M. Griffin, and H. Maclean, *The Ethanol Answer to Carbon Emissions*. Issues in *Science and Technology*, Winter 2001–2002, pp. 73–78.
3. http://www.e-sources.com/fuelcell/fcexpln.html.
4. 2000 National Design Contest Problem. American Institute of Chemical Engineers. New York, 1980.
5. EPA: Long-Term Diesel Exposure Can Cause Cancer, September 4, 2002. Posted 8:28 A.M. EDT (1228 GMT) at cnn.com.

CHAPTER 6

Transportation

In the competition during the industrial revolution, steam turbines and internal combustion engines emerged as they provided the best combination of usefulness, engine cost, and fuel cost. Each met market demands. The less-expensive engines tended to require more expensive fuels, while the more expensive engines could use less-costly fuels.

Burning solid trash to produce electricity is an example of an application with a high engine/system cost and a low fuel cost. A gasoline auto engine uses a relatively expensive fuel. The automobile engines may seem expensive, but building a coal-fired power plant for an automobile with the same horsepower makes the gasoline engine look like a real bargain.

There are novel transportation power systems like the nuclear power plants in navy ships and submarines. Civilian transportation power is dominated by petroleum derived fuels: gasoline engines, diesel engines, jet engines, and electric-powered trains/streetcars. Petroleum is the primary source for modern transportation fuels.

Transportation Before Petroleum Fuels

We entered the 19th century on horseback, in wagons, and on sailing ships. We traveled through the 19th century with horses and oxen and began to use steam engines. We emerged from the 19th century trying to use steam engines, batteries, and the new gasoline engine technology to drive the developing horseless carriage. This was the beginning of building the infrastructure that would be necessary to supply fuel for the internal combustion engines that power transportation vehicles.

As cities approached populations in the millions, the use of horses as the sole source of transit was not an option. Feeding and caring for millions of horses plus disposal of manure and animal carcasses occupied much of the population. The gruesome toll taken by the steep hills of San Francisco on the horses that pulled streetcars motivated Andrew Hallidie to develop the San Francisco cable car in 1873 (see the box "History of San Francisco Cable Car").[1] In New York City, street congestion led to groundbreaking for the New York subway at Borough Hall in Manhattan in March 1900.

Streetcars and subways met the challenge of city transit and were used in metropolitan United States. In many cities, effective public transit ended suddenly when automakers and oil companies purchased public transit systems and then shut them down and replaced them with buses.[2] Through the 20th century the automobile achieved domination of the transit industry with increasing infrastructure (highways and fueling stations) and a regulatory base that now makes any significant change difficult.

As we begin the 21st century the transportation infrastructure is dominated by petroleum fuels used in spark-ignited engines (gasoline), compression-ignited engines (diesel), and turbine engines (jet aircraft engines). Whether we got here by providing the "best" transportation alternative or by destruction of the competition, this is our transportation system.

History of San Francisco Cable Car

The driving force behind the San Francisco cable car system came from a man who witnessed a horrible accident on a typically damp summer day in 1869. Andrew Smith Hallidie saw the toll that slippery grades could extract when a horse-drawn streetcar slid backward under its heavy load. The steep slope with wet cobblestones and a heavily weighted vehicle combined to drag five horses to their deaths. Although such a sight would stun anyone, Hallidie and his partners had the knowhow to do something about the problem.

The next step that brought Hallidie closer to his mission was moving his wire-rope manufacturing business to San Francisco. Just witnessing that accident was enough to spawn

the idea of a cable car railway system to deal with San Francisco's fearsome hills.
Cable Car Chronology:
1852—Andrew Hallidie arrives from Great Britain.
1869—Hallidie witnessed horse-car accident and had inspiration for a cable railway.
1873, August 2—Andrew Hallidie tested the first cable car system near the top of Nob Hill at Clay and Jones Streets.
1873, September 1—Clay Street line starts public service at an estimated construction cost of $85,150.
1877—Sutter Street Railroad converts from animal power to cable with no break in service.

Source: From http://www.sfcablecar.com/history.html.

Petroleum Fuels: Their Evolution, Specification, and Processing

Petroleum, vegetable oils, animal fats, and alcohol-turpentine mixtures sold for premium fuel prices in the 19th century. Two properties made these fuels more valuable than alternatives like wood or coal. They released considerable energy when burned (high heating values), and they were liquid. A reasonably high heating value is needed for any fuel to work in combustion applications, and this higher heating value translates to traveling farther on a tank of fuel. Liquid fuels were much easier to handle when refueling and metering the fuel to the internal combustion engines.

Drilling for petroleum (1859; Colonel Drake at Titusville, PA)[3] became common, and liquid fuels so abundant in the United States the price of petroleum dropped from $20 to $0.10 per barrel in one year. This abundant liquid fuel was absolutely the best source of energy to power the developing internal combustion engines.

Liquid fuels are used in vehicles because they can be handled with inexpensive pumps, carburetors, and injectors. Interestingly, diesel engines were developed to run on coal dust. These modified diesel engines were expensive to build, and the pistons wore out quickly from coal ash that acted like a grinding compound when it mixed with the lubricating oil. The coal-fueled diesel engine simply was not economically competitive.

Gaseous fuels can also be used to power internal combustion engines. Natural gas is commonly used to fuel gas turbines for stationary power plants. For mobile applications, the heating value of a gas fuel per gallon (per unit volume) is considerably less than petroleum fuels. This generally makes gaseous fuels inferior for transportation applications. There are special applications where natural gas and hydrogen can be used in trucks and buses.[4]

In the early 20th century, engines were custom-built or modified to operate with locally available liquid fuels. As certain engine designs gained favor, refiners began to provide liquid fuels specifically prepared for the engines. During the military conflicts of World War I, the need to standardize fuel specifications became apparent. During the technology-rich World War II, fuels and engines developed together. The designs of the fuels and engines have been closely coupled since World War II.

Liquid fuels can be broadly classified by application: spark ignition (gasoline) fuels, compression ignition (diesel) fuels, and gas turbine fuels (both stationary and jet engines). Turbines use combustion chambers where the fuel burns continuously. The power output is adjusted by the rate of fuel consumption. This is much easier to achieve than the spark- or compression-ignition engines where the combustion must start, burn at a uniform rate, and end in a very short time interval (usually milliseconds).

The first petroleum fuels were fractions of the crude oil produced by distillation. The five fractions, from the most to the least volatile, included (1) liquefied petroleum gas, (2) naphtha spark-ignition fuel, (3) light distillate (jet fuel), (4) middle distillate (diesel fuel), and (5) residual fuel (thick gummy oil) and road tar. There is a sixth fraction—natural gas—that cannot be liquefied at room temperature. This fraction is typically separated at the crude oil wellhead or in a gas plant located near the oil field.

Figure 6-1 summarizes the natural breakdown of petroleum using the slightly different European nomenclature. These fractions are liquefied petroleum gas (not shown), gasoline, kerosene and gas oil, and fuel oil. By European notation, jet fuel is kerosene, and diesel fuel is called gas oil. Figure 6-1 has kerosene and gas oil lumped as one fraction.

During the early history of petroleum use, refineries sold the useful fractions and dumped everything else into any available market. This natural fraction process often did not match the demand for a specific product.

According to Owen and Coley,[5] in the years immediately following the exploitation of petroleum resources by drilling (initiated by Colonel Drake at Titusville, Pennsylvania, in 1859), kerosene lamp oil was the most valuable petroleum fraction. Surplus gasoline was disposed of by burning, surplus heavy residue

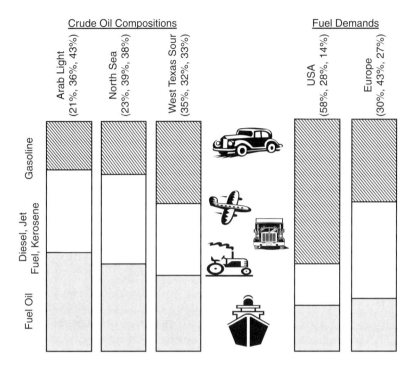

FIGURE 6-1. Crude oil fractions and market demands.

was dumped into pits, and the "middle distillate" was used to enrich "town gas," which explains why it is often still referred to as gas oil. Only with the invention of the diesel engine was a specific role found for the middle distillate fraction.

Practical spark-ignition engines (and the use of spark-ignition fuels) date back to 1857, with three in service in 1860.[6] The gasoline fraction was the middle distillate fraction. Dr. Rudolph Diesel developed the compression-ignition engine (the diesel engine and diesel fuel are named after the inventor) that was both more energy efficient than the gasoline engine and the fuel was this middle distillate fraction of petroleum.

Following World War II, there was a demand for a fuel for jet aircraft. Diesel fuels would not ignite easily, and they became gummy or froze at the low temperatures of high altitude. Gasoline would boil off (vaporize) at the low pressure at high altitude. A narrow-boiling fraction with properties between that of diesel fuel and gasoline was developed to meet these jet fuel requirements.

Today, petroleum is too valuable to dump unused fractions to low value markets. The modern refinery meets the challenge by converting the available crude oil (crude oil from different oil fields have different compositions and flow properties) to the exact market demands for the wide array of products. They use a complex of chemical reactors to change the chemical composition and distillation columns to separate these into the desired products. Petroleum refining technology research has produced many of the 20th-century developments in chemical science and engineering.

When entering the 20th century, petroleum marketers had little concern for the composition of the fuel. The focus was to make a kerosene lamp oil that was volatile enough to ignite easily but not so volatile that it would burn out of control. By the end of the 20th century, the modern refinery takes in the crude oil and then tears apart and reassembles the molecules so they are in one of the classifications of a desirable product. Engineering at this molecular level makes the desired products; it uses all of the feed petroleum and improves the performance of these new fuels over the old fuels made from the "natural fraction processes."

At the beginning of the 20th century an engine was designed to operate on the local fuel supply. Today, the fuels are produced to perform in an engine designed for that fuel. This matching of the engine and the fuel comes from the 100 years of experience building motor vehicles. A description of each of the fuel categories will help us understand how alternative fuels can replace those petroleum fuels.

Spark-Ignition Fuels

Spark-ignition engines use gasoline but can be fueled with propane, ethanol, methanol, or natural gas. Each of these fuels has four important performance characteristics: (1) a high-energy release during combustion; (2) vaporizes (or is a vapor or gas) at about the boiling point of water; (3) when mixed with air and compressed to about one-sixth the initial volume, the heat produced by compression will not ignite the fuel/air mixture; and (4) a spark ignites the compressed mixture, and it burns without detonating (sudden, complete combustion or explosion). The fuel is vaporized and mixed with air in the proper ratio before it is drawn into the cylinder of a spark-ignition engine.

When the spark ignites the fuel/air mixture in the engine, the heat released increases the pressure in the cylinder, and it is this force that turns the crankshaft of the engine. Best engine performance is obtained when the fuel continues to burn as the piston moves down, keeping the gas hot as it expands. Combustion is complete at about half-stroke, and the piston continues to move down, producing work as the gas cools until the piston reaches the end of the power stroke. The burned fuel/air mixture is then expelled through an exhaust valve as the piston moves up and a new charge of fuel/air is drawn in as the piston moves down to complete the cycle. Some of the work produced is used to compress the fuel/air mixture and to turn all of the moving parts in the engine. It is the rotating crankshaft attached to the load that produces the useful work.

The important performance characteristic of this engine is to avoid pre-ignition of the fuel during the compression stroke. The octane number, posted on the fuel pump at the gas station, is the index of resistance to pre-ignition. A high octane number corresponds to a fuel that pre-ignites at a high temperature (produced by a higher compression pressure). Since pre-ignition occurs before the piston reaches the end of the compression stroke, the cylinder pressure increases too soon, and it takes more work to complete the compression stroke. Part of the fuel is burned before the spark ignites it for the power stroke. Pre-ignition produces "knock" or "ping," and it increases the mechanical stresses on the engine.

It is essential to match the octane number of the fuel (careful preparation of the fuel), a proper mix of fuel and air, and control of the compression ratio of the engine (proper design of the engine). The mechanical design fixes the engine compression ratio.

The power output of the engine is controlled by the quantity of fuel/air mixture that is drawn into the engine. At low power (during idle or cruise) the pressure of the fuel/air mixture to the engine is kept low by throttling the flow. Then the amount of fuel and the pressure of the mixture in the cylinder are low as the compression begins, so the pressure during the power stroke is also lower. This lowers the power produced by the engine. At full power (during acceleration) the engine operates at more revolutions per minute and pulls in more fuel/air. These adjustments are made on modern cars by a computer that monitors all of these functions as you drive in stop-and-go traffic or on the highway.

A typical gasoline engine in an automobile requires an octane number of about 87. This number was selected because it can be obtained by modern refining methods, converting about one-half of the petroleum into gasoline. Automobile engines are now designed with a compression ratio of about 6.5:1, low enough to avoid pre-ignition using this fuel. During World War II, the gasoline-powered aircraft were designed with a high compression ratio to increase the power with the low-weight, air-cooled engines. Tetraethyl lead was added to the gasoline to give an octane number of 100 (adding more lead will increase the octane number to a maximum 115). High-compression, high-performance automobile engines were common following World War II, until the lead additive was banned to eliminate toxic lead from the engine exhaust gas. The federal requirement to eliminate lead in fuel resulted in the industrial response, lower engine compression ratio that works just fine on 87-octane gasoline.

The American Society for Testing Materials (ASTM) is an independent group that prepares and publishes standardized tests and specifications for a wide range of industrial materials and consumer products. ASTM Standards D2699 and D2700 are used to certify that gasoline has an 87 octane number. These standards are cited and usually required in federal and state laws controlling the quality of gasoline. Customers have assurance that the product will work independently of the supplier. Different suppliers often use the same pipeline to move the fuel from the refinery to local distribution terminals. The difference between gasoline purchased from Texaco and Shell, for example, is the additives they blend into their gasoline before it goes to the filling station.

Compression-Ignition Fuels

A compression-ignition engine is usually powered by diesel fuel and, recently, by biodiesel. Some desirable performance characteristics of diesel fuel include (1) a high heat release during combustion, (2) a volatility that keeps it liquid until the temperature is well above the boiling point of water, (3) rapid compression-ignition (without a spark) when the compression ratio is about 15 to one or higher, and (4) formation of a fine, uniform mist when pumping the fuel through the fuel injectors in each cylinder.

Diesel fuel specifications are almost the opposite of those for gasoline. Gasoline is designed to readily evaporate into air and not to ignite during compression in the engine cylinder. Air is compressed in the diesel engine cylinder before the fuel is injected so there can be no pre-ignition. Diesel fuel evaporates as the fine mist particles from the fuel injectors ignite in the hot, compressed air. The fuel also lubricates the fuel injector pump. The cetane rating of a diesel fuel characterizes the tendency of the fuel to ignite. U.S. standards for diesel fuel require a minimum cetane number of 40. The mechanical difference between a diesel engine and a gasoline engine is that the spark plugs are replaced with fuel injectors.

It is not a good idea to put gasoline into a diesel fuel tank, or vice versa. Many gas stations sell both fuels. The nozzle on the gasoline pump is larger than on the diesel fuel pump. The hole below the fuel cap on the diesel fuel tank is smaller than the gasoline fuel nozzle, so you can't fill a diesel tank with gasoline. The diesel fuel nozzle, however, *will* fill a gasoline tank, so buyer beware!

When designing alternative fuels, the fuel scientist/engineer first translates physical properties such as volatility and ease of ignition into molecular properties such as the size and shape of molecules. Designing a fuel becomes a manageable task, since the molecules primarily contain carbon, hydrogen, and oxygen atoms, with few exceptions.

Small molecules containing ten or fewer carbons are more volatile and make gasoline, spark-ignition fuels. The word octane in the "octane scale" is the chemical name for an eight-carbon molecule that is found in gasoline. It is a good representative molecule for gasoline. Pure isooctane is assigned an octane number of 100, and it was used to establish the empirical octane scale in 1930.

Diesel fuels contain molecules with eight or more carbons and are less volatile than gasoline. They have cetane numbers that characterize good compression-ignition fuels. The word cetane in "cetane scale" is the name for a 16-carbon molecule that is representative of a "good" diesel fuel. Molecules with the carbons arranged in straight chains have high cetane numbers and are better fuels for compression-ignition engines. Molecules where the carbon atoms form rings (benzene or toluene) or branched chains (isooctane is an example) tend to be better spark- ignition fuels.

Refineries today use molecular rearrangement (catalytic reforming) to produce six to eight carbon atoms with branched configurations. This increases the fraction of gasoline produced per barrel of crude oil, and the gasoline has a higher octane number than can be obtained by simple distillation. Diesel fuel specifications are easier to reach by simple refinery processes, so little molecular design is necessary to produce diesel fuel. Diesel fuel is a hodgepodge of different refinery streams sent to a mixing tank and blended to give the right volatility and cetane number to be a "good" diesel fuel.

Turbine Fuels

Gas turbines have burners where combustion takes place continuously—very different from the spark-ignition or compression-ignition engine. Ignition characteristics of this fuel are much less important. The fuel does not lubricate the fuel injection system. This means a wide range of fuels can be used to power a turbine. Turbines require a fuel that burns cleanly without producing an ash that can erode the turbine blades. There are two properties absolutely required for aircraft jet engines: (1) Jet fuel must not freeze at the low temperatures we find at high altitudes, often −40°F. (2) The fuel should have a high heating value per pound of reduce the ratio of fuel weight to total load on the aircraft.

Diesel fuels can and do power turbine engines, but the high freezing points make them unsuitable for jet aircraft. If the fuel freezes, the jet loses power when the fuel pumps cannot deliver the jelly-like fuel to the engine.

Gasoline fuels would work in a jet engine, but these fuels are too volatile, and the vapors mixed with air lead to explosions under certain conditions in the fuel tank. The low atmospheric pressure of high altitude causes this volatile fuel to "boil off" and be lost. TWA flight 800 exploded and crashed near Long Island on July 17, 1996, as a result of fuel volatility problems. The aircraft was filled with the proper fuel, but it was heated above design temperature on this summer day due to a takeoff delay on the runway. The extended time on the runway allowed heat from air conditioners and other equipment on the aircraft to warm the fuel tank. The explosion occurred when a spark, perhaps from a short in a fuel pump, ignited the fuel-air vapor mixture above the liquid in the fuel tank. This is the reason that low-volatility fuel is preferred for aircraft. For military aircraft subject to enemy arms fire, low fuel volatility is certainly important.

Low volatility and low freezing points are conflicting specifications for a fuel. Jet fuel is designed to provide a reasonable compromise between these two important properties.

Alternative Fuels

Modern fuels are designed at the molecular level to meet the demands of 21st-century engines. There are processes developed to convert available feedstocks that include coal, natural gas, and vegetable matter into modern fuels. Electricity can be used to produce hydrogen that can be used in fuel cells or spark-ignition engines. Hydrogen air burns very fast—not a very good fuel for spark-ignition engines.

Liquid fuels contain almost exclusively carbon, hydrogen, and oxygen atoms. The ratio of hydrogen to carbon atoms should be greater than one, since ratios less than one tend to be solids (like coal). All potentially competitive alternative transportation fuels fall within these guidelines, with the exception of hydrogen. This means we are limited to feedstocks that contain these atoms to make the alternative fuels. The useful feedstocks include natural gas, coal, municipal solid waste, and biomass. Biomass is another term for materials produced by plants and animals and includes wood, vegetable oils, animal fat, and grass.

Liquid Fuels from Coal

Coal resembles a soft rock that contains mostly carbon and hydrogen, the two components needed to form a liquid fuel. These two components provide the high heating value of coal. To convert coal to a liquid fuel, the coal molecules with between 20 and 1,000 carbon atoms must be broken down and rearranged to form molecules with between 4 and 24 carbon atoms. These new molecules must have at least as many hydrogen atoms as carbon atoms. At least half of the energy in the coal should end up in the final liquid product, or the process will be too expensive. Such transformations are possible with two approaches that have been used.

Liquefaction provides a low-tech conversion where the larger molecules of coal are torn apart by heating the coal. We can visualize that the atoms in the large coal molecules are held together by rubber bands (actually, these are interatomic forces that bind the atoms to form the coal molecule). As the temperature increases, the atoms "jiggle" more and more rapidly, and the weakest bands start to break, forming smaller pieces (small molecules). As the temperature continues to increase, the small molecules become even smaller. Liquids are formed at low temperature (large molecules) and vapor is formed at high temperature. If some air (or oxygen) is present, some of the vapors burn and supply energy to increase the temperature. If insufficient oxygen is present, the broken-off molecular segments rearrange into smaller particles that are mostly carbon and form a black smoke. You have probably seen this black smoke from a smoldering wood fire or a diesel engine that was not running properly.

How is coal liquefied? Heating coal in a vessel with little air (no air) forms vapors. The vapor is made up of the smaller molecular segments of the coal. As the vapor is cooled, it will condense to form a liquid or paste. The properties of the liquefied coal will vary depending on (1) whether the coal is in large chunks or a fine powder, (2) how quickly the coal was heated, (3) how much air (oxygen) is present, and (4) how much water vapor was present. This process is called *pyrolysis* chemical change caused by heat.

Many processes have been developed for the different steps of preparing the coal, heating the coal, processing the vapors, and condensing the vapors. Processes have also been developed that improve the quality of the liquid or paste formed by pyrolysis. Pyrolysis is one of the simplest coal conversion processes, but it has shortcomings. The quality of the liquid formed is substantially inferior to petroleum crude oil. A major fraction of the energy in the coal remains in the char, the part that does not form a liquid. The char is much like commercial coke and contains the coal ash and a major fraction of the carbon. This char can be used as fuel. Technologies to improve the fraction of coal converted to liquid using pyrolysis are not competitive with alternative processes.

Coal gasification is the complete destruction of the coal molecules followed by chemical recombination of the gas mixture to form liquid fuels. This is a better alternative to pyrolysis liquefaction. This process has been established as a sustainable industry. The coal is fed to the reactor as a fine powder together with steam (to supply hydrogen to the product gas) and some air (oxygen enriched air or pure oxygen) to supply oxygen. The oxygen quickly burns some of the coal to supply the energy to keep the temperature high to break up the coal molecules. At this high temperature, the carbon reacts with the water to form carbon monoxide, carbon dioxide, and hydrogen; high temperatures break apart the coal molecules, similar to liquefaction. The fraction that is carbon monoxide increases as the reactor temperature increases. It is carbon monoxide and hydrogen that will be used to make the liquid fuels, and this mixture is called *synthesis gas*. Coal passes through these reactors in milliseconds, so the equipment to gasify lots of coal is relatively small.

Gasification has two important advantages over pyrolysis: (1) essentially all of the coal is consumed during gasification, while there is only partial conversion during pyrolysis liquefaction, and (2) the well-defined product composition from the gasification process simplifies the process to obtain the molecules that will become the liquid fuel. In the past, liquefaction products rich in benzene, toluene, and xylene provided potential competitive advantages for coal liquefaction. Recent trends away from these compounds in fuels (they are toxic) have reduced the value of liquefaction processing. Most experts see gasification as the future way to process coal into liquid fuels.

Carbon monoxide and hydrogen are the desired products of gasification. This mixture is referred to as a synthesis gas because it can be used to synthesize (make) a wide variety of chemicals.

Two things are required to make useful chemicals from synthesis gas: (1) lower temperatures (less than 600°C), so the larger molecules are "thermodynamically" favored over the carbon monoxide and hydrogen building blocks, and (2) a catalyst to increase the rate of formation and the fraction that is converted to the desired liquid chemical products. Good chemical process engineering seeks the optimum combination of temperature, pressure, reactor design, and the ratio of hydrogen to carbon monoxide that will efficiently produce the desired liquid fuel product.

A variety of chemicals can be made from synthesis gas, including alcohols (for example, methanol and ethanol), ethers, and hydrocarbons. Synthesis gas produced from coal or natural gas is the primary source of commercial methanol, which can be used directly as a spark-ignition engine fuel. The octane number of methanol is considerably higher than 93. When it is blended with gasoline, it increases the octane of the fuel, since methanol has an octane rating of 116.5.[7]

The early 1990s represented a turning point for methanol. The demand was close to the production capacity that depended on a few large natural-gas-to-methanol facilities. When one of these facilities was shut down for maintenance, the price of fuel methanol went from about $0.50 to $1.50 per gallon.[8] The cost to keep a bus running on $1.50 methanol was then more than twice the cost of using gasoline. The methanol fuel industry still suffers from the poor impression created by price fluctuations combined with a lack of political support for this alternative fuel.

Ethers can be used in compression-ignition engines and were commonly used to assist during "cold start" for diesel engines prior the 1990s. Diesel engines require substantial modification to use ether fuels, since the common ethers must be stored in pressurized tanks to reduce losses due to evaporation.

Hydrocarbons are made commercially using Fischer-Tropsch synthesis. German scientists Frans Fischer (1877–1947) and Hans Tropsch (1889–1935) initiated their hydrocarbon synthesis work in

1923.[9] The process starts with synthesis gas and produces hydrocarbons by adding CH_2 groups to make larger molecules. Like building blocks, these groups are added to a molecule until the Fischer-Tropsch fuel mixture is similar to the most useful fuel fractions of petroleum. During World War II, the Germans used this technology to produce quality fuels for aircraft and tanks. There are commercial plants using this process in South Africa today. A commercial plant in Malaysia starts with natural gas to make synthesis gas followed by Fischer-Tropsch synthesis to make liquid fuels.

Many in the petroleum industry consider Fischer-Tropsch synthesis to be the alternative fuel process that has been commercialized without subsidy. (This view would consider Canadian oil sand fuels as petroleum of low volatility and therefore not a synthetic fuel.) Based on the number of patents filed by major oil corporations related to Fischer-Tropsch synthesis, this process is considered the best option for replacing petroleum. As long as there is crude oil, energy corporations will not invest in this process.

Natural Gas

Natural gas can be used as a spark-ignition fuel, or it can be used to produce synthesis gas. Natural gas as a transportation fuel is gaining momentum for bus fleets, and it is used in U.S. Postal Service vehicles. The excellent natural gas pipeline infrastructure across the United States makes natural gas a candidate to replace gasoline as a transportation fuel. Disadvantages of natural gas as a vehicle fuel include (1) refilling at room temperature will require pressurized gas storage, (2) it costs more to store natural gas on vehicles, (3) the energy content of natural gas per unit volume is less than a liquid fuel, so it will be necessary to refill tanks frequently, and (4) there will be concern about safety with pressurized gas on a vehicle, especially when involved in an accident.

Natural gas is the primary feedstock for commercial production of synthesis gas with then converted to a variety of liquid products. Gasification and conversion to liquid fuels are commercial method to produce "remote" natural gas fields not connected to a gas pipeline system.

Natural gas has a pipeline distribution infrastructure across the United States that makes it readily available as an alternative transportation fuel. For fleet owners, such as municipal buses or the U.S. Postal Service, the installation of refueling stations can be justified. The vehicles are more expensive, since they must be modified to store and handle high-pressure natural gas, but the combination of EPACT law requirements and a desire to use alternative fuels make natural gas a candidate alternative fuel.

Solid Biomass Utilization

Biomass is a term used to identify any product produced "naturally" in the past few decades (there are annual crops and trees that take decades to mature). Fossil fuels are biomass materials that have been transformed and stored in geological formations over thousands of years. Examples of biomass fuels include wood, grass, corn, and paper. Biomass differs from coal in that it is less dense, has a higher oxygen content, and its molecular structure is recognized as food by animals and microscopic organisms.

Biomass materials can be used as feed for liquefaction and gasification processes to produce pyrolysis oils and synthesis gas. Because microscopic organisms recognize biomass as food, their enzymes can convert biomass to other chemicals. Yeast converting sugars to alcohol is one of the oldest commercial biochemical processes. Microorganisms that produce antibiotics represent another established technology. Beverage and fuel alcohol represents a large-volume, low-cost product (if you subtract the beverage tax). Antibiotics are high priced; they are low-volume products commonly using microorganisms to "do the chemistry."

Ethanol is the most common vehicular fuel produced from biomass. It can be used directly in a spark-ignition engine. The ethanol octane rating[10] is about 110. When blended with gasoline, ethanol behaves as if it had an octane rating of 115.[11] The starchy parts of corn and sorghum are the most common feedstocks for the ethanol process that is similar for beverage alcohol. Enzymes and heat are used to convert starches to sugars. Yeast or bacteria are used to convert the sugars to ethanol. Distillation removes the water from the fuel ethanol, and about 2 billion gallons of fuel ethanol were produced during 2001 in the United States.[12]

Ethanol is also produced from petroleum or from nonstarchy plant components. The U.S. government promotes the use of ethanol from biomass, with a 5.4 cents per gallon subsidy on gasoline blended to contain 10% ethanol. The ethanol increases the gasoline octane number and meets the oxygenated fuel requirements imposed in several cities during winter months. The use of ethanol is also growing to replace part of the MTBE additive that has recently been banned. MTBE was used to increase the octane number and to meet the oxygenate requirement prior to the policy to phase it out.

Cellulose is the nonstarch plant material that can also be converted to ethanol. It is the primary component of grass and wood. A cow can digest cellulose, but a human cannot. Certain microorganisms can convert cellulose to sugars and then to ethanol. These cellulose feedstocks are less expensive than corn or sorghum. However, the process is more expensive, and the technology does not have the support of farmers, since they usually have excess production of corn and sorghum. The industry is at a crossroads where economics should favor cellulose-to-ethanol production. However, there is little political support to develop cellulose-to-ethanol technology.

Table 6-1 summarizes the prospects of common alternative liquid fuels.

TABLE 6-1
Comparison of alternative fuels and their ability to displace petroleum.

Fuel	Advantage	Disadvantage
Hydrogen	Works efficiently in fuel cells. Can be produced from electricity using electrolysis.	Hydrogen stores energy and requires an energy source to make it. It is a gas and it is difficult to store more than about 50 miles' worth of hydrogen on an automobile.
Ethanol	Can be produced from renewable biomass without gasification step. Oxygen content makes ethanol clean-burning.	The most optimistic estimates for production from biomass give bulk, pretax > $1.10 per equivalent gasoline by 2025. Current prices > $1.50 per equivalent gallon.
Methanol	One of least-expensive liquid fuels. Considered most compatible for spark-ignition engines and fuel cells. Can be produced from natural gas, coal, municipal solid waste, and biomass.	Unstable prices in recent years created poor perceptions with bus fleet managers. (Several managers made attempts to use methanol in the early 1990s.) Not good fit for distribution in current petroleum pipeline network. Transportation adds a few cents per gallon to the cost of methanol.

TABLE 6-1
(*continued*)

Fuel	Advantage	Disadvantage
Natural Gas	Relatively inexpensive, since it can be used with minimum processing.	Fuel sold as a gas may not be accepted by motorist. Current supplies are insufficient, and remote production sites must be developed to meet demands.
Propane	Alternative with good infrastructure in place and in common use.	Supply is directly related to petroleum supply. Prices go up with increased consumption. Not truly an alternative fuel.
Biodiesel	High conversion efficiency when produced from vegetable oils and animal fats.	Optimistic estimates place bulk pretax costs greater than $1.50 per equivalent gallon with production from soybean and rapeseed oils. Feedstock oils are insufficient to give a noticeable market impact. Supplies only impact diesel fuel.
Fischer-Tropsch	Least-expensive alternative fuel (as low as $0.60 per equivalent gasoline gallon) compatible with current diesel fuel distribution infrastructure and vehicles. Can be produced from natural gas, coal, municipal solid waste, and biomass. Low sulfur content meets anticipated regulations on diesel fuels.	Fuel best suited for use in diesel engines. Conversion to gasoline increases cost, but as gasoline it is the least-expensive fuel compatible with the current fuel distribution infrastructure.

Biomass Liquids

Plants produce starches, cellulose, lignin, sugars, and oils that can be converted to fuels. Starches, cellulose, and lignin are complex solid compounds that require conversion to sugars, pyrolysis, or gasification to be converted to liquid fuels. Sugars can be processed into ethanol, but this process is not economical because the sugars are more valuable in the food market. The conversion of plant oils and animal fats into vehicular fuels was the most rapidly expanding alternative fuel source at the onset of the 21st century. However, the percentage of annual increase in production represents an insignificant market share.

Soybean oil has traditionally been the largest oil crop in the world. This is changing as palm oil is now poised to exceed soybean oil production. Animal fats and greases are less expensive than vegetable oils, but the supply is limited. Waste cooking oils and fryer grease do present disposal problems for restaurants and sewage disposal systems. These feedstocks share common chemistry for conversion to a *biodiesel* fuel, so fuels from waste fats and grease present a good disposal option.

The large fat and oil molecules (that contain 45 to 57 carbon atoms) are reacted with ethanol or methanol to form three smaller molecules containing 15 to 19 carbon atoms. Glycerin is produced as a by-product. The 15 to 19 carbon molecule liquids are less viscous than the original vegetable oils. They can be pumped and injected just like diesel fuel from petroleum. When blended with diesel, these fuels freeze at an acceptable low temperature, and they tend to form less deposits on diesel fuel injectors. Additional advantages include reductions in soot and hydrocarbon emissions.

Biodiesel can be mixed from 0% to 100% with petroleum diesel fuel without engine modification. Obviously, this makes it easy to take biodiesel from the laboratory to the commercial market. When vehicles convert to use biodiesel after extended service with petroleum diesel, the fuel filters may temporarily collect sediments that form. Aside from this, the transition to biodiesel is simple.

Biodiesel has three powerful factors working in favor of it: (1) it has political support because it provides a way to reduce excess vegetable oil supplies, (2) biodiesel and blends with petroleum diesel can be used without modifying the diesel engine, and (3) waste vegetable oil to biodiesel can eliminate a waste disposal problem and produce alternative fuels at about $1.00 per gallon. It has been shown that biodiesel enhances the lubricating factor of the fuel—a nice feature for some applications.

Vehicular Fuel Conservation and Efficiency

Defining Efficiency

Improving the efficiency of converting fuel energy to work can be as effective at reducing crude oil consumption as developing alternative fuels.It has the added advantage of reducing carbon dioxide emissions. In the 1975Energy Policy Act, the U.S. Congress set

Corporate Average Fuel Economy (CAFE) standards in the wake of the 1973 oil crisis.Increased engine and vehicle efficiencies were identified as one of the most effective ways to reduce fuel consumption.

Figure 6-2 summarizes typical losses of energy in the process of converting the chemical energy of fuel into the performance factors of changing vehicle speed, overcoming aerodynamic drag, and highway rolling losses. While the greatest losses occur in the engine as it converts chemical energy into shaft work, improvements in the other factors will reduce fuel consumption. For example, only 25 of every 100 kW of engine power may go to travel losses,

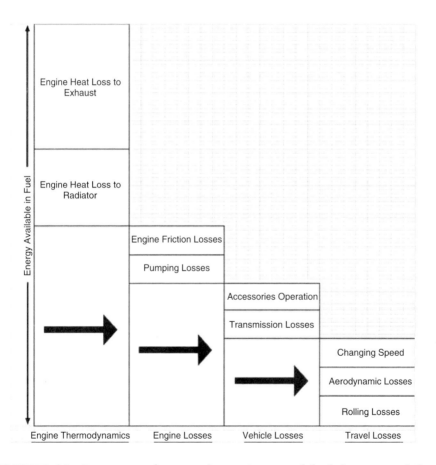

FIGURE 6-2. Summary of energy losses in use of fuel for automobile travel. Effectiveness and Impact of Corporate Average Fuel Economy (CAFÉ) Standards. National Academy Press, Washington, D.C., 2001. (http://bob.nap.edu/html/cafe/.)

but reducing the travel losses by 10 kW would indirectly lead to reductions in all the other energy expenditures except accessories operation—totaling a 25 to 40 kW reduction in the total engine power demand.

The impact of the CAFÉ standards on the average fuel consumption is shown by the increase from 13.4 to 22 miles per gallon for passenger cars between 1973 and 2000.[13] These improvements occurred even though additional weight was added for safety-related equipment required on the vehicles. The new car standards written in the legislation increased from 18 to 27.5 mpg between 1973 and 1985, with the 27.5 mpg standard continuing beyond 1985. As illustrated by Figure 6-3, the average fuel economy improved and leveled off shortly after 1985 as the older, less-fuel-efficient vehicles were replaced by vehicles meeting the 27.5 mpg fuel efficiency standard. The difference between the 22 mpg DOT report and 27.5 mpg legislated standard is partly due to the difference between public driving habits and the test drives used to certify fuel economy. Part of the difference is also due to loopholes in the law, such as getting CAFE credit for producing flexible fuel vehicles (vehicles that can run on gasoline or up to 85% ethanol in gasoline).

The fuel economy improvements of Figure 6-3 were achieved by reducing vehicle weight and improving engines. For example, a 16-valve, 4-cylinder engine wasn't even considered for small and midsized cars in 1973. In practice, going to four valves per cylinder

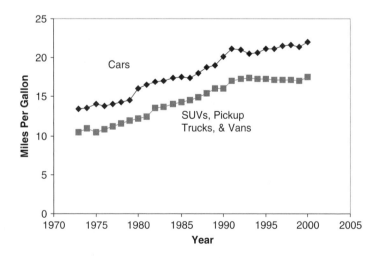

FIGURE 6-3. Average fuel economy of motor vehicles.

reduced the air-flow resistance and increased both the engine efficiency and power. A 1995, 16-valve, 4-cylinder engine could match the performance of the average 6-cylinder engine of 1973 at a fraction of the engine weight. Reducing vehicle weight with a more aerodynamic body reduces all of the vehicle travel losses (speed, aerodynamic, and road losses).

Dependence on imported crude oil in the United States certainly is greater as we enter the 21st century than it was prior to the 1973 oil crisis. Any increase in vehicle fuel economy would reduce the dependence on foreign oil and reduce the production required if we moved to an alternative fuel program. Efforts to improve vehicle fuel economy focuses on four items: (1) increased use of diesel engines or new direct-injection gasoline engines, (2) hybrid vehicles that greatly increase fuel efficiency, (3) fuel cell replacement for the internal combustion engines, and (4) move to use lower-weight vehicles.

Diesel Engines

Diesel engines are more efficient than gasoline engines for two reasons: (1) higher cylinder pressures and corresponding higher temperatures lead to improved thermal efficiency for the diesel engine and (2) the air throttling required to control the gasoline engine power output reduces engine efficiency is not used on diesel engines. Truck drivers and farmers are very aware of the reduced fuel costs associated with running a diesel engine as compared to a gasoline engine. The trucks and tractors include an extended range of gear ratios so the diesel engine operates at nearly optimum RPM as the load changes.

The thermal efficiency of an engine is computed as the ratio of work the engine delivers divided by the thermal energy available from the fuel. The Table 6-2 summary lists, by engine type, the average thermal (thermodynamic) efficiencies[14]: These data show

TABLE 6-2
Typical average thermal efficiencies for diesel and gasoline engines.

Engine Type	Thermal Efficiency
4-stroke, spark-ignition engine	30%
2-stroke, spark-ignition engine	22%
4-stroke, DI compression-ignition engine	40.3%
2-stroke, DI compression-ignition engine	43%

an improvement of over 30% in thermal efficiency for the diesel engine over the gasoline engine. In addition, a gallon of diesel fuel contains about 10% more energy than a gallon of gasoline, so the typical increase is about 40% more miles per gallon in favor of the diesel engine.

When the VW Beetle was reintroduced with the standard gasoline engine, it was rated at 24 mpg. With the new diesel engine upgrade, it is rated at 42 mpg. This represents an average fuel economy increase of about 75% based on energy available in the fuel.

There are obstacles to the adoption of the diesel engine in automobiles, including (1) poor consumer perceptions that originated with the noise and smoke of the diesel engines that were marketed in the early 1980s, (2) increased costs for diesel engines, and (3) the lack of catalytic converters for the diesel engine exhaust to reduce emissions. Catalytic converters are available for gasoline engines. The new diesel engines are quieter and do not have smoking problems when properly "tuned." It takes time for these perceptions to change. Lower diesel engine costs will occur with higher engine production volumes, and there is a sincere desire to reduce the cost of both the engine and the drive train.

Diesel engines always run with excess air in the cylinders, more air than required to burn the fuel. Excess air reduces the hydrocarbon and particulate matter in the exhaust gas, an environmental advantage.

The diesel engine exhaust contains considerable oxygen because of the excess air. This oxygen prevents traditional catalytic converters from working with diesel exhaust gas.

Diesel fuel sold at the end of the 20th century had high sulfur content that poisons catalytic converters. The new Tier 2 diesel require refiners to make low-sulfur diesel fuel available. Low-sulfur diesel and a new generation of catalytic converters will then meet the new restrictive emission standards.

The low Tier 2 emission requirements are not required in all areas of the United States, and adjustments may be possible for exempt regions. The new generation of small diesels should not cause the emissions problems in small cities and rural communities so common in metropolitan areas.

Hybrid Vehicles

Hybrid vehicles use a small internal combustion engine and a bank of batteries to supplement engine power during acceleration. The small engine runs at optimum RPM near maximum efficiency.

When the power required is more than the engine output, additional electric current is drawn from the battery pack—the wheels are either partially or totally driven by electric motors. When the vehicle is cruising, excess power that is generated charges the batteries. Hybrid vehicles can avoid idling losses, since the engine continues to operate at maximum efficiency and charges the batteries when the vehicle is at a stoplight. When the batteries are fully charged and the vehicle is stopped, the engine turns off. The "on-board" computer monitors the operator's request for acceleration, braking, stopping, and waiting, and it starts and stops the engine "on demand" for power.

The fuel that would be used during engine idling and from the engine operating at nonoptimal RPM is substantially reduced. In addition, some models have electric motors that become generators when brakes are applied to slow the vehicle, and this recovers much of the energy lost to conventional friction brakes. These features combine to boost fuel efficiency by up to 50% over conventional gasoline powered vehicles. If a diesel engine is used in a hybrid vehicle, efficiency could be higher.

In some models, like the 2005 Honda Accord hybrid, the hybrid features are used primarily to improve vehicle acceleration. In this case the hybrid option may have the same fuel economy as other vehicles. However, if a larger engine is required to deliver the same performance, the fuel economy of the hybrid would be better. The use of hybrid features allows a vehicle manufacturer to deliver improved performance without a new engine option for the vehicle.

In 1993, the U.S. Department of Energy launched the Partnership for the Next Generation Vehicle Program as described by a DOE website:

> The Hybrid Electric Vehicle (HEV) Program officially began in 1993. It was developed as a five-year cost-shared program that was a partnership between the U.S. Department of Energy (DOE) and the three largest American auto manufacturers: General Motors, Ford, and DaimlerChrysler. The "Big Three" committed to produce production-feasible HEV propulsion systems by 1998, first-generation prototypes by 2000, and market-ready HEVs by 2003.

The overall goal of the program was to develop production HEVs that achieved $2\times$ fuel economy compared to similar gasoline vehicles and had comparable performance, safety, and costs compared to similar gasoline vehicles. As the program progressed, its goals began to merge with the goals of the Partnership for a New Generation of Vehicles (PNGV). Now DOE and its partners are striving to develop vehicles that achieve at least 80 miles per gallon. PNGV is a public/private partnership between the U.S. federal government and DaimlerChrysler, Ford, and General Motors that aims to strengthen America's competitiveness by developing technologies for a new generation of vehicles.

The current PNGV goals are as follows:

1. Significantly improve national competitiveness in manufacturing.
2. Implement commercially viable innovations from ongoing research in conventional vehicles.
3. Develop vehicles that can achieve up to three times the fuel efficiency of comparable 1994 family sedans.

The PNGV program is an example of "not getting the job done." While the United States was busy talking about developing and marketing hybrid vehicles, the Japanese actually marketed these vehicles. As the 2003 deadline approached, the DOE and American auto industries essentially abandoned the program.

The current U.S. automotive industry emphasis is on fuel cell research with no advertised dates for market-ready vehicles. Commercialization of fuel cell technology is anticipated in a 15- to 20-year timeframe. Since there will be an opportunity to postpone or abandon the fuel cell program, this default U.S. DOE energy plan ensures that the United States will continue to import crude oil at increasing costs well into the future. This program does, however, create the impression that something is being done to eliminate oil imports.

If hybrid technology increases the fuel economy from 30 to 50 mpg, the fuel consumed in 100,000 miles of travel reduces from 3,333 to 2,000 gallons. The savings is about $2,300. This fuel savings is less than the additional cost of the hybrid vehicle[15] (based on net present value). However, optimistically, larger-volume production of hybrid vehicles and components will bring the annualized operating cost (including vehicle depreciation) to numbers equal or less than those of conventional vehicles. With hybrid technology there is a danger of replacing imported petroleum with imported hybrid vehicle components. An advantage of the technology is the reduced dependence on foreign petroleum.

Plug-In Hybrid

Plug-in hybrid vehicles[16] (PHEVs) are similar to hybrid vehicles with two exceptions. First, the typical 5- to 7-mile battery pack in a hybrid vehicle would be replaced with a 10- to 20-mile (and up to 60) battery pack in a hybrid vehicle. The additional batteries cost about $1,000 for every 10 miles of capacity.[17] Second, the batteries can be charged from grid electricity, allowing substantial operation without the engine running.

Whereas the HEV can reduce the vehicle's gasoline consumption up to 33% (50% increased fuel economy), the PHEV can reduce the vehicle's gasoline consumption by more than 80%. The PHEV displaces gasoline with both efficiency and grid electrical power.

PHEVs can displace the majority of the fuel they consume with electricity that is domestically produced and has stable prices. Half of the gasoline consumed in the United States is within the first 20 miles traveled by vehicles each day. For many owners, over 80% of the gasoline could be replaced with grid electricity. Here, a 30 mpg automobile would consume 3,333 gallons of gasoline for 100,000 miles of travel. A PHEV running at 80% plug-in and 20% at 50 mpg would consume 400 gallons of gasoline.

One advantage of the PHEV is a reduction in operating costs (not including depreciation). Most of the savings is in fuel costs—while $0.46/liter ($1.75 per gallon) gasoline costs about 2.9¢/km (4.6 per mile), 4 cents per kWh electricity costs about 0.9¢/km (1.42 per mile). This example is based on a 38 mpg vehicle. The savings are greater if the conventional vehicle gets 25 mpg; the PHEV saves both by increased vehicle mileage (equivalent to 38 mpg) and use of electricity rather than gasoline.

For a PHEV, displacing gasoline with grid electricity leads to greater savings per mile traveled. Chapter 9 discusses this in greater detail along with projections of lower net present costs for PHEVs compared to conventional vehicles. In addition, the money spent on electricity stays in the United States, and the hybrid vehicle component production can stay in the United States. Chapter 13 discusses in greater detail how a state can reduce the state's trade deficit by over $1 billion per year by replacing imported petroleum with indigenous electrical power.

This is where nuclear power has a major role to play in future transportation. Both HEV and PHEV options can work with gasoline or diesel engines, so these options have broad potential impact.

By plugging in at night, PHEV technology can use available off-peak electrical power (and the low 4¢/kWh cost previously cited). Increased demand for baseload power will provide the incentive to build more efficient baseload (50% efficient for new generation system power plants) rather than peak demand units (28% efficient) and can provide large amounts of transportation energy without any additional expenditure for coal and without additional pollution. Substantial reductions in greenhouse gases can be realized in this approach.

When fuel cells are used in combination with batteries on plug-in hybrid vehicles, the potential for lower-cost systems increases (lower costs are certain in a ten-year timeframe). Also, the fuel cells could eventually be refueled with hydrogen or alcohol (ethanol or methanol) fuels, which will eliminate the need for an engine on the vehicle (further reducing vehicle costs) and displacing the use of petroleum. This is likely to be the technology that closes the petroleum era. The new era would be sustainable by expanding the use of nuclear power.

References

1. http://www.sfcablecar.com/history.html, May 21, 2002.
2. Anthony Sampson. *The Seven Sisters*. New York: Bantam Books Inc., 1981, p. 184.
3. Keith Owen and Trevor Coley, *Automotive Fuel Reference Book*, 2nd ed. Warrendale, PA: Society of Automotive Engineers, 1995, p. 21.
4. Hydrogen—Fuel of the Future on Iceland, by Tom Karl Andersen, April 26, 2003. See http://www.hydro.com/en/press_room/news/varchive/no_news_view/hydrogen_iceland/hydrogen_iceland_main_en.html.
5. Owen and Coley, p. 21.
6. Ibid., p. 7.
7. Ibid., p. 269.
8. J. M. Teague, and K. K. Koyama, "Methanol Supply Issues for Alternative Fuels Demonstration Programs." SAE Paper 952771.
9. http://www.oroboros.se/syntes.htm, May 22, 2002.
10. http://www.cheresources.com/ethanolzz.shtml, May 22, 2002.
11. Owen and Coley, p. 269.
12. http://www.eia.doe.gov/pub/, May 22, 2002.
13. http://www.eia.doe.gov/emeu/mer/, May 22, 2002.
14. John B. Heywood, *Internal Combustion Engine Fundamentals*. NY: McGraw Hill, 1988, p. 887.
15. M. Duvall, "Advanced Batteries for Electric-Drive Vehicles," *Reprint Report*, Version 16, EPRI, Palo Alto, CA, March 25, 2003.
16. G. J. Suppes, S. Lopes, and C. W. Chiu, "Plug-In Fuel Cell Hybrids as Transition Technology to Hydrogen Infrastructure." *International Journal of Hydrogen Energy*, January 2004.
17. "Plug-In Hybrid Vehicles." Published by the Institute for the Analysis of Global Security. http://www.iags.org/pih.htm, 2003.

CHAPTER 7

Production of Electricity

History of Production

Discovery of Electricity and Electrical Theory

Electricity is the highest form of energy available. It can be used to produce heat and light, turn motors, or power a radio, television, or computer. It is impossible to imagine our modern society or a modern home without electricity. Essentially every U.S. business, factory, or home is connected to an electric utility just like water and sewerage. Producing all electrical power needs on-site is the last resort/default option at remote locations. While municipal water supply and sewage systems date to ancient Rome, distribution of electricity occurred in this century. Indexes that describe the standard of living in a nation show a high average use of electricity per citizen correlates with a high standard of living. Electricity use and modern lifestyle are closely coupled.

The written history of electricity probably starts with the description of an experiment recorded by the Greek philosopher Thales (640–546 B.C.). He wrote that when amber (a fossilized resin then used to make jewelry) is rubbed with fur, it can attract light objects (a feather or a thread) placed close to it. Many substances can be charged in this way. If you rub a glass rod with silk or a hard rubber rod with fur, the surface of the rods become highly charged. We now call this a *charge of electricity*, and the attraction is called an electric or *electrostatic attraction*. The Greek word for amber is *elektron* thus the source of our word *electricity* and, later, *electron*.

The curious early observers noticed a difference in the charge produced by amber and by hard rubber. They used pith balls (the dried spongy material in the stem of plants) suspended from a string. As you moved the pith balls together, they would touch and separate from each other as they were moved apart. When the pith balls were each touched with charged amber and then moved toward each other, they would push apart or repel each other. This was also true when both pith balls were touched with charged rubber. When one pith ball was touched with charged amber and the other with charged rubber, as they were brought together, they were suddenly pulled together by an attractive force.

Each of us has run a static electricity experiment. When you walk across a carpet and touch a light switch or water faucet, there is a spark that you can see if the room is dark. (This works best during the winter when the air is dry.) It works with leather or rubber-soled shoes but rarely works when you're barefoot. The explanation for these crude observations did not come until the 19th century.

The second strand of the story of electricity was the discovery of magnets. Ancient Chinese noticed that when a slender piece of a black mineral (it was later called lodestone) was suspended by a string, it would turn until one end pointed north. The same end always pointed north. When you pushed it away from north, it would swing back to point north. When a piece of lodestone was broken, the end of the small piece that was closest to the north end of the large piece again pointed north. Divide the pieces again, and each piece would point north. Even a very small piece of this material would point north. The "north-pointing" was a property of only lodestone.

When two pieces of the lodestone on strings were moved together so north-to-north-pointing ends are close together, they repelled each other. South-to-south-pointing ends also repelled each other. When a north-pointing end was brought close to a south-pointing end, they attracted each other. This is remarkably similar to the behavior of electrically charged pith balls.

We now move forward to the 18th century when we find more experiments in chemistry and physics. Benjamin Franklin did his famous (and could have been fatal) experiment flying a kite into a cloud to show that lightning was an electrical discharge. He established the convention of positive and negative to distinguish the two kinds of charge. Materials were classified by their electrical properties: Insulators were materials that retained a charge when rubbed, while conductors, mostly metals, lost their charge. It was 1767 when Joseph Priestly showed that the force between electrical charges obeyed the same mathematical form as gravity described earlier by Sir Isaac Newton.

About this time, experiments showed that it takes work to move an electric charge between two points on an electrical conductor or in an electrical circuit. This is just like moving a weight against gravity. The flow of electricity acted much like the flow of a fluid through a pipe. The difference is electricity flowed through a solid wire. Many experiments were performed to understand the flow of electricity using different metals as conductors. Metals like silver and copper were good conductors. Ceramic materials failed to conduct electricity and fell into the class called insulators. It was now clear that there must be a relation between static electric charge and the flow of electricity. It took about 20 years before experimenters were convinced that static electricity and electrical current were closely related.

Experiments using electrical current required a source of electricity, and the battery was the answer. It was about 1800 when Count Alessandro Volta invented the electric pile, or battery. The first battery was soon improved by others and became a practical source of electric current for the many experiments that followed.

There were several researchers interested in the chemistry of solutions at the end of the 18th century. Volta demonstrated that an electric current decomposed water to form hydrogen and oxygen. This was the beginning of electrochemistry. The different types of batteries were a by-product of these electrochemical experiments. As an aside, these electrochemical studies provided essential data later used to confirm the 20th-century model of the atom made up of electrons, each with a negative charge surrounding a positively charged atomic nucleus. In conductors, the electrons in the metal atoms are free to move when connected to a battery. The electrons are not free to move in an insulator.

Here is another important observation from the early experimenter's laboratories. In 1820 Hans Christian Ørstad wrapped a few turns of an insulated wire around a compass. This wire was connected to a battery, and the compass needle moved. When the wires were connected to the opposite battery terminals, the compass needle moved in the opposite direction. This demonstrated that an electric current flowing through a wire produces a magnetic field (a field is a force acting at a distance—the field is invisible and goes through objects, air, and space), which interacts with the magnetic field of the compass needle. The electric current passing through a wire produced a stronger force when the wire coil was small (closer to the compass needle) and weaker force when the wire coil is large (further from the compass needle). The needle moved faster in the strong field (small coil) and slower in the weaker field (large coil).

Michael Faraday (English) and Joseph Henry (American) independently made the next significant observation in 1831. When a wire is passed through the magnetic field between the north and south poles of a magnet, an electrical current is produced in the wire. When the wire is passed through the magnetic field in the opposite direction, the current flows in the opposite direction. When the wire stops between the poles of the magnet, the current stops. When the wire is moved directly toward the north or south magnetic pole, there is no current produced.

It was Ørstad who demonstrated that a current passing through a wire produces a magnetic field. Faraday showed that a wire passing through a magnetic field produces a current. These discoveries made possible the design of a dynamo to produce electricity from mechanical work, the electric motor to convert electricity to mechanical work, and the transformer that we use today to change the voltage on electric power transmission lines.

One of the most significant achievements of the 19th century occurred in 1864. James Clerk Maxwell published his mathematical theory that accurately represented the many experiments on electric and magnetic fields. The Maxwell theory of electromagnetism describes electrical, magnetic, and optical (this includes visible light, radio waves, x-rays, etc.) phenomena and forms the basis for many of the technological advances in the electric power and electronics industries through the 20th century. The experimental studies of electricity and magnetism available to Maxwell were the confirmation he needed for his mental picture of the interactions between electric current and magnetism and the extension to optical data.

And so, observations and understanding evolved: static electricity, batteries, the electron, conductivity in wires and lack of conductivity in insulators, magnetic fields, the relation between magnetic fields and electric current, and finally Maxwell's theory that provided a comprehensive fundamental understanding of electricity and electromagnetism. The stage was set for the commercialization of devices based on this new concept of electricity.

Commercialization of Electrical Technology

The first commercial application of electromagnets was the 1840 Samuel F. B. Morse patent on the telegraph. He used a small iron rod wrapped with a coil of insulated wire. When a current passed through the wire, the iron became a magnet. This pulled a second piece of iron to it with a "click." A spring pulled the "clicker" away from the magnet when the current was turned off. Morse developed a code of short and long time gaps between clicks that represented the letters of the alphabet. (The Morse Code distress call is... – – – ..., the letters SOS.) This Morse Code is still used today.

The U.S. Congress appropriated funds in 1844 to install wires between Washington, D.C., and Baltimore, Maryland, to demonstrate whether this invention really worked. A message was keyed in from Washington, received in Baltimore, and returned to Washington a few seconds later. With this success, telegraph wires were soon being strung, and communication between distance places was underway. The telegraph was used for communication by the railroads until after World War II.

The first electric power generators were installed in London and New York City in the same year: 1882. This is 50 years after Faraday and Henry's demonstration that passing a wire through a magnetic field produces an electric current. The generator is a machine designed to continuously pass wires through a magnetic field to produce an electric current. As illustrated by Figure 7-1, this can be achieved by rotating a wire loop between the poles of a permanent magnet (see the box "Making a Small Electric Generator"). Building upon this very simple idea, the next improvement was to have more wire loops passing through the magnetic field. Replacing the one-wire loop with a two-wire coil will double the force (the voltage) driving the electrons. The shaft work produced by steam engines, waterwheels, and the then new internal combustion engines was converted to electricity for lighting and turning electrical motors.

In practical generator designs, an iron cylinder with grooves cut into its surface to support the wire loops is mounted on a shaft. Wire is wound to form a coil using two grooves on opposite sides of the cylinder. The ends of the wire coil are attached to metal contact strips mounted on opposite sides of a smaller cylinder where a "brush" connects the coil to an external circuit. Several different loops are wound onto the large cylinder, with each contact strip separated by electrical insulation. This is the armature of the generator, which rotates between the magnet poles. This more practical configuration is illustrated by Figure 7-2. Although the magnetic field is represented as permanent magnets, in most generators they are electromagnets.

190 *Sustainable Nuclear Power*

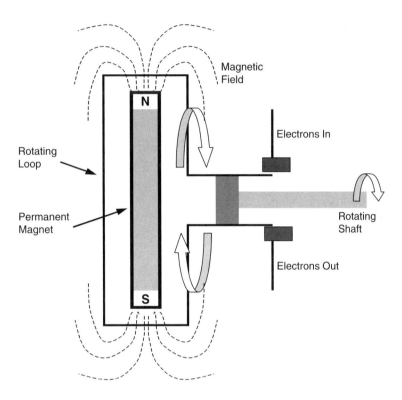

FIGURE 7-1. A simple generator.

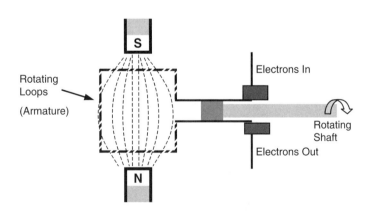

FIGURE 7-2. A generator with armature rotating inside magnetic field.

Making a Small Electric Generator

We can demonstrate how this works using a single loop of wire and the magnetic field between the north and south poles of a permanent magnet. We place a single loop of wire so the top and bottom wires are just above and below the faces of the magnet. Rotate the loop clockwise so the top half of the loop passes through the magnetic field moving down closest to the south pole of the magnet and the bottom half passes through the magnetic field moving up closest to the north pole. Since this is a wire loop, the electric current will be moving in one direction in the wire. We can use a sliding contact at each end of this loop to connect it to an external circuit.

As the wire loop continues to turn, the top wire of the loop becomes the bottom wire. The wires in this position are traveling parallel to the magnetic field, and no current is produced. As we continue to turn the loop, the direction of the current in the wire loop reverses because the two sides of the loop have exchanged positions. The contact with the external circuit at each end of the wire loop also changes, so the downward-moving wire is again connected to the same side of the external circuit. The external circuit "sees" the current flowing in one direction even though the current flow reverses in the loop. This is the way a direct current generator produces electric current.

When the wire passes through the magnetic field, the current flows through the loop and the external circuit and there will be a magnetic field around the wire. This magnetic field has the same polarity (north to north or south to south) as the magnetic field that is producing the current. This like polarity "pushes" on the wire, and a force must be applied to move the wire through the magnetic field. It is this force moving the wire through the magnetic field that is the work used to produce an electric current.

When the wire loop passes through the magnetic field quickly, more current is generated. Increasing the strength of the magnetic field also increases the current produced. Replacing the single wire loop with a coil of wire increases the number of wires passing through the magnetic field. This increases the voltage, the force that pushes the current through the loop and the external circuit.

The first commercial direct current generators combined all of these ideas. The wire loops were connected to an external circuit by two "brushes" (carbon pieces that rubbed on two or more of the contact strips attached to the wire loops) located on the centerline of the magnet poles. The wire coils, the contact strips, and the shaft are the armature of the generator. The support for the magnets, the brushes, and the bearings for the armature shaft are the essential parts of the generator. When the armature turns at constant speed, each coil connected to the external circuit is moving through the magnetic field closest to the magnet pole at maximum velocity. Each coil makes contact with the external circuit through the carbon brushes just before it gets to the centerline of the magnets and then disconnects just after passing the centerline. The next coil on the armature enters the magnetic field and is connected to the external circuit just as the previous coil disconnects. This nearly constant coil velocity through the magnetic field produces a nearly constant voltage that drives the current through the external circuit.

Modern-day generators (supplying direct current voltage), alternators (supplying alternating current voltage), and motors are a result of incremental improvements upon these generators first used at the end of the 19th century.

Thomas Edison invented the electric lightbulb in 1879 and proposed that they be used to replace the gaslights on the streets of New York. He then built the first electric power station in 1881 to supply power to the streetlights. A similar power station was built in London about this time. Steam engines powered the first electric generators.

It didn't take long for electricity to be introduced into the factories and homes. Electric lighting has better quality and is certainly more convenient than gas or oil lamps. An electric motor looks much like an electric generator. In a generator the armature's coils are "pushed" through the magnetic field by the shaft that delivers the work to produce electricity. In an electric motor, the armature's coils (also called the rotor) are "pulled" into rotary motion by the magnetic field of the stator surrounding the rotor. Electric motors can be made small or large to match the load placed on them.

In the early 1890s, belts and pulleys were used to deliver the power to a machine and to change the speed of rotation of the machine. Today, with a wide selection of electric motors and improved understanding of the technology, the motors are directly attached to the device requiring the shaft work.

The electrical power produced by a generator is consumed in many ways. A lightbulb converts the electrical energy into heat and radiation. Eventually, the radiation (light) is also absorbed by objects and is converted to heat. Electric motors convert electricity to shaft work. Eventually, this shaft work is dissipated as friction heat. There is no stopping this eventual conversion of work to heat, but the design task is to produce the maximum amount of light or useful shaft work as electricity flows through the device. Electrical distribution wires often cover tens or even hundreds of miles between the power generator and the consumer. The resistance to the flow of current in these wires represents a power loss in the electric distribution system.

Power lost to distribution lines is expensive even when using a good electrical conductor like copper. These losses increase by a factor of four when current flow through wire is doubled. Reducing the current flow is the best way to reduce power loss during distribution. Electric power is computed by multiplying the current times the voltage. To minimize distribution losses when transporting a given amount of power to a load at the end of a long transmission line, one should use a high voltage and the corresponding low current. For example, when delivering 11,000 watts of electrical power, it can be delivered as 110 volts times 100 amps or 440 volts times 25 amps. The 440-volt line has 1/16th the distribution power loss that would occur for the 110-volt line. There is a clear advantage to using high voltage to transport electric power over long distances.

Alternating current is used to transmit electrical power over national grids because it allows easy conversion to high voltage at the power plant for distribution to local transformers that reduce voltage for local use. The voltage from a direct current generator is increased by having more turns of wire in the armature coils. The insulation on the wires would also need to be improved and this high-voltage, direct-current armature soon becomes too big to be practical.

The first commercial alternating current–generating system was put in service in Germany in 1891. An alternating current generator looks much like the direct current generator. The difference is the way the current is collected from the armature coil. As the armature coil turns in the magnetic field, the current rises to a maximum and then decreases to zero and reverses direction as the coil rotates through one full turn. The current generated is connected to the load using slip rings that rub continuously on a carbon brush. The real advantage of alternating current is that one can use transformers to change the voltage for long transmission lines.

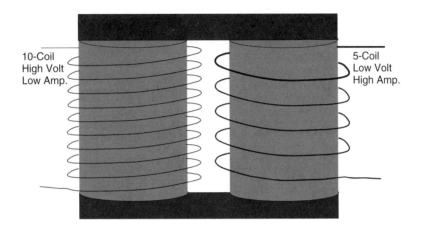

FIGURE 7-3. A transformer.

The transformer is a simple device, as shown in Figure 7-3. In this example, a ten-loop coil and a five-loop coil are wrapped around the two iron cores connected by iron bridges. As alternating current passes through the primary coil (five-loop coil), the magnetic field produced by the current in this wire builds up and magnetizes the iron core. This magnetic field is transferred to the secondary core by the iron bridges. As the magnetic field on the secondary side builds up, the secondary coil wires cut the magnetic lines as they expand outward from the magnet inducing a current flow in the secondary coil. Since there are ten turns in the secondary coil and five in the primary coil, the secondary coil voltage will be twice the primary coil voltage and the current will be halved.

When the alternating current in the primary coil reverses, the polarity of the magnetic field also reverses. Reversing the polarity of the magnetic field reverses direction of the current flow in the secondary coil. The timing of the current reversals and the voltage zeros remain the same on both sides of the transformer. Since this transfer of voltage only occurs when the voltage of the primary coil changes, this simple transformer does not work for constant-voltage direct current. The low-voltage power to the transformer is the same as the high-voltage power from the transformer. There will be a small loss due to the resistance in the wires and the little energy required as the magnet field changes direction.

We have described some key developments prior to the 20th century essential to the development of modern electrical power systems. The technology of electric power production and distribution has improved significantly during the 20th century. Special metal alloys have produced magnets with high-intensity fields that reduce the size and increase the efficiency of transformers and motors. Modern alternating current generators have the magnets located on the armature shaft, and the coils that produce the current are wound on two poles located on opposite sides of this magnetic armature. It takes a small direct current to magnetize the armature, and this spinning magnetic field produces the current in the stationary coils as the magnetic poles pass by. Spinning the armature at 3,600 revolutions per minute produces a 60-cycle-per-second current, the standard power produced in the United States. The standard in Europe is 50 cycles per second, so the armature in their turbines turns at 3,000 revolutions per minute.

Most power stations produce three-phase power using a three-phase generator. A three-phase generator has three independent single-phase coils equally spaced around the magnetized rotor of the generator. As the magnet rotates, each pair of coils produces a current independent of the other two pairs of coils. The total power of the generator is the sum of the power produced in each of the three circuits. This makes the generator smaller, and the force required to turn the armature shaft is more nearly constant. Alternating current generators vary in size. Smaller generators may provide one or two kilowatts in a portable unit powered by a small gasoline engine. Very large generators produce 1,500,000 kilowatts (1,500 megawatts) and are powered by steam turbines at large nuclear or coal-fired power plants.

Large power plants are usually located some distance from the load: cities, factories, commercial buildings, and homes. Power is now distributed over an interstate power grid with many power stations connected together. This means the electricity can move over long distances. Modern high-voltage transmission lines commonly operate between 50,000 and 150,000 volts, with some newer lines running as high as 500,000 volts. These high-voltage lines allow smaller-gauge wires to be used with lower electrical losses. Smaller wires reduce the weight of the wires and the cost of building the towers that support the distribution lines.

The voltage from the long distribution lines is reduced in steps by transformers wound to reduce the voltage. A factory or a large building may use power at 660–700 volts. Most circuits in a home operate at 110–120 volts, with the electric range and air conditioner operating at 220–240 volts. A series of transformers are used to reduce the voltage to the level required at each point of use.

The electric power industry is huge. Resources to build power-generating plants and the distribution lines are usually owned by private firms financed by issuing stocks and bonds. Large hydroelectric dams are usually owned and operated by the federal government. Large manufacturing plants often include a power plant to provide steam for processes, and electric power is usually a by-product. By-product power is the source of a substantial amount of commercial electricity.

In the year 2001, the total electrical energy used in the United States was more than 3,500 billion kilowatt-hours, about one-fourth of the total electricity used in the world. There has been about a 2% annual increase in electric power consumption since 1980, with this growth expected to continue to the year 2020. Construction of new generating plants will be necessary to replace obsolete plants and serve this increased demand. The momentum of electric power industry is immense, with the proven track record of fossil and nuclear energy providing low financial risk proven technology. Any new alternative technology that becomes available must compete with this established industry. The increased use of electricity among the emerging nations is another story. In the absence of vast national electrical power grid infrastructures, these emerging markets have been and will be the demonstration grounds for new technology without competition from the 1,000 megawatt coal and nuclear power plants in the United States.

Production of Electrical Power

As electrical power is produced, the energy in the fuel ends up as waste heat or as electrical power. When the fuel is burned to power a heat engine, there are limitations that determine the maximum amount of the fuel energy that can be converted to electrical power.

Most of the waste heat is given off at the electrical power plant by the power cycle that drives the dynamo. The thermal efficiency of the electrical power generation is defined as the electrical energy produced divided by the total energy released by the fuel consumed. The waste heat of the power cycle goes into cooling water and the exhaust gases that exit from the power plant stacks. The cooling water heat is released in cooling towers as water vapor. This is the white water vapor cloud you see coming from all electric power plants.

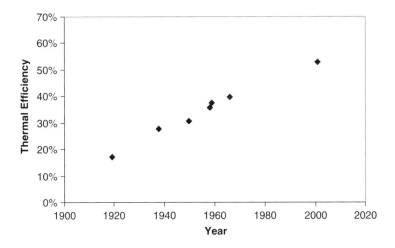

FIGURE 7-4. Increases in thermal efficiency electrical power generation during past century.

The first electrical power plants were very inefficient, with the thermal efficiency steadily increasing with improved equipment design. Data summarized by Haywood[1] and shown in Figure 7-4 illustrate this improvement in electrical power-generation efficiency over the past century. In 2002, the most efficient power plants converted 53% of the energy in natural gas to electricity. The U.S. Department of Energy was soliciting research opportunities[2] to increase the efficiency of these facilities to 60% envisioning that further increases could be achieved through incremental improvements of natural gas combined cycle technology.

Fuel cells convert the energy in chemical bonds directly to electricity. This chemical conversion to electricity can be nearly 100% efficient if the fuel cell device we build has no internal energy losses. Practical fuel cells today have maximum efficiencies over 50%. More advanced concepts envision natural gas fuel cells using high-temperature molten carbonate salts combined with a conventional steam power plant to recover heat that might achieve an overall efficiency of 70% and higher.

Figure 7-4 summarizes the "best efficiency" for any conversion cycle over the past century. Table 7-1 lists efficiencies for different conversion options leading into the 21st century. It is clear that using natural gas for peaking power is about half as efficient as the combined cycle option. These efficiencies also illustrate

TABLE 7-1
Typical thermal efficiencies for power generation options.

Type of Fossil Fuel/Atomic Energy Facility	Thermal Efficiency
Natural Gas—Peak Power Turbine	25%
Nuclear Power*,**	30%–33%
Coal-Steam Cycle	38%–45%
Nuclear—Pebble Bed Modular Reactor***	45%
Coal—Gasification Combined Cycle	48%
Natural Gas—Combined Cycle	50%–56%
Future (2010)—Natural Gas and Coal Combined Cycles	60%
Future (2015)—Fuel Cell Combined Cycles	70%

*http://www.uic.com.au/nip57.htm reports 33% thermal efficiency for nuclear power as reasonable assumption and high energy output as compared to energy input.
**http://www.nucleartourist.com/world/koeberg.htm confirms 33% efficiency after improvements.
***http://www.worldandi.com/public/2001/April/nuclear.html.

which fuels commonly used today could benefit from technology development to improve cycle efficiency.

Consider the following example. If we double the thermal efficiency of a nuclear power facility from 30% to 60%, the amount of nuclear waste generated per kilowatt-hour of electrical energy would be cut by a factor of two. Planning the next generations of nuclear facilities should be a global effort.[3]

In addition to increasing the impact of uranium reserves, increases in efficiency can also reduce costs of electrical power production. For example, if the fuel contribution to the total cost of providing nuclear power is 0.68¢/kWh at 30% efficiency, the cost is reduced to 0.34¢/kWh if the same fuel can be used to produce electricity at 60% thermal efficiency. If the nuclear reactor is improved to burn 34% of the uranium per new fuel loading rather than 3.4%, the fuel cost is reduced to 0.034¢/kWh of produced electrical power. The fuel costs for coal and natural gas are about 1.04 and 3.94¢/kWh, respectively.

As noted in Chapter 1, enough uranium has already been mined to keep present nuclear power plants in production for another 4,350 years, or all U.S. energy needs for about 400 years. These projections did not take into account increased energy consumption, nor did the increased efficiency of producing electrical power. To some extent these two tend to cancel increasing the accuracy of the 400-year projection.

Electrical Power Generation Efficiency

The evolution of electrical power generation from fuels is primarily an evolution of increasing thermal efficiency for converting heat into electrical power. These power cycles can be complex, using sophisticated methods to design and analyze the cycles.

However, evaluating the limits of electrical power efficiency is simple in terms of the temperatures used by the heat engine. This efficiency depends on three factors: (1) the average temperature at which the steam or engine's working fluid enters the engine, (2) the average temperature at which working fluid leaves the engine (going out as exhaust), and (3) the friction or bypass losses of the turbines, compressors, and pumps. The heat rejection temperature is typically near ambient temperatures and is fixed at more than 38°C (100°F). The best engine mechanical components are about 90% efficient. Improved hardware can save some of that 10% lost. Thus, the best place to improve conversion efficiency is to increase the temperature of the working fluid to the engine.

Better metals and ceramics developed during the past century have been incorporated into the design of electrical power plants and efficiencies increased. The primary reason that conventional nuclear power plants do not reach efficiencies greater than 33% is that the design has large reactor vessels with limited cooling water pressure (limiting the cooling water temperature). The water-cooled nuclear reactor temperature is limited for reasons of safety associated with the design. Nuclear power can be converted to electricity at higher efficiencies with the new pebble bed modular reactors that use helium gas as the working fluid instead of water.

Technically, energy is never lost but is converted to lower forms because the temperature decreases. Waste heat is a low form of energy and is of little use to power machines. In addition to the losses during electrical power generation, approximately 10% of the electrical power is lost in the transmission lines. To the customer, this is the same as losing 3.3% of the total energy produced by the fuel at the power plant. It is the same as waste heat and for a nuclear power plant looks like an increase in waste heat from 67% to 70.3%. Line losses become more important for a 60% efficient future natural gas combined cycle power plant. The 10% transmission line loss increases the 40%

lost during electrical power production to an effective 46% loss of the energy required to produce the electricity. All losses of electric power during distribution and inefficiency of the devices used by the customer appear as fuel lost to waste heat in the total energy budget.

Recommended Reading

1. Bailey, B. F., *The Principles of Dynamo Electric Machinery*. New York: McGraw-Hill Book Company, Inc., 1915.
2. Brown, S. L., *Electricity and Magnetism*. New York: Henry Holt and Company, 1934.
3. Franklin, W. S., and Barry MacNutt, *The Elements of Electricity and Magnetism*. London: The MacMillan Company, Ltd., 1908.
4. Hausmann, Eric, and E. P. Slack, *Physics*, 3rd ed. New York: D. Van Nostrand Company, Inc., 1948.
5. Hobart, H. M., and A. G. Ellis, *High Speed Dynamo Electric Machinery*. New York: John Wiley & Sons, 1908.
6. Patrick, D. R., and S. W. Fardo, *Electrical Distribution Systems*. Lilburn, GA: The Fairmont Press, Inc., 1999.
7. ———, *Rotating Electrical Machines and Power Systems*. Lilburn, GA: The Fairmont Press, Inc., 1997.
8. Winder, J. J., Jr., *Power Transformers, Principles and Applications*. New York: Marcel Dekker, Inc., 2002.

References

1. R. W. Haywood, *Analysis of Engineering Cycles*, 4th ed. Oxford, England: Pergamon Press, 1991.
2. "Development of Technologies and Capabilities for Developing Coal, Oil, and Gas Energy Sources." Solicitation for Financial Assistance Applications No. DE-PS26-02NT41613, U.S. Department of Energy National Energy Technology Laboratory, Pittsburgh, PA, August 21, 2002.
3. http://gen-iv.ne.doe.gov/.

CHAPTER 8

Energy in Heating, Ventilation, and Air Conditioning

The Heating, Ventilation, and Air Conditioning Industry

The energy used by American consumers for heating, ventilation, and air conditioning (HVAC) is second only to the energy used for transportation. HVAC applications consume about the same amount of energy per year as the 130 billion gallons of gasoline consumed annually in the United States (see Table 8-1). From a greenhouse gas emission perspective, about 8.8% of carbon dioxide emissions are from space heating furnaces as compared to 33.9% from electrical power production. A large fraction of the electrical power is used for HVAC.

Consumers use electricity, fuel oil, liquid petroleum gas, kerosene, and natural gas for heating. Table 8-2 shows that natural gas and fuel oil provide most of the space heating needs.

The competition created by alternative technologies and fuel choices provides relatively stable heating/cooling costs, just as with electrical power production. In the HVAC industry, energy savings result from improved building construction with better insulation and double- or triple-pane windows. The consumer can choose between investing in energy-efficient buildings with lower yearly energy costs or buying less-expensive HVAC equipment with higher annual energy expenses. Diversity in energy sources and competition among equipment manufacturers have made the

TABLE 8-1
Summary of largest applications of energy.

	Energy in Quadrillion BTU
Amount of Gasoline Consumed*	15
Electricity Produced**	12
Approximate Energy Expended on Producing Electricity***	34
Energy Consumed for HVAC (including hot water heaters)****	14

*130 billion gallons times 115,000 Btu per gallon.
**11.6 rounded up.
***Assuming average efficiency of 35%.
****Residential [5.2+.55/0.35 1.29+.0.16+0.08+0.39/0.35]+Commercial [1.7+.35/0.35+0.16/0.35 0.81/.5].

TABLE 8-2
Summary of energy sources for residential heating and air conditioning in quadrillion (1 with 15 zeros) Btus.

Residential* Air Conditioning**	0.42
Space Heating***	
Electrical	0.4
Natural Gas	3.6
Fuel Oil	0.85
Kerosene	0.06
LPG	0.26
Total	5.59

*http://www.eia.doe.gov/emeu/consumption/index.html.
**http://www.eia.doe.gov/emeu/consumptionbriefs/recs/actrends/recs_ac_trends.html#consumption.
***http://www.eia.doe.gov/fueloverview.html. See "Space Heating".

HVAC market one of the success stories in American free enterprise, but the story does not end there.

HVAC technology can provide the flexibility needed to achieve additional reductions in greenhouse gas emissions and improve the overall efficiency of electrical power generation network. The way to achieve these two objectives is best explained by examples detailed in the following sections.

Peak Load Shifting with Hot Water Heaters

In much of Europe, where electrical power costs are two to three times those in the United States, one of the most common methods to reduce electricity costs is to run the hot water heater at night. Baseload electrical power at night is produced at greater efficiency and is less expensive to the customer. The hot water heater must be large enough to supply the daily use, and some inconvenience may result in late afternoon if the system runs out of hot water. This approach benefits the customer with reduced electrical costs and reduces the greenhouse gas emissions because the electricity is produced with the most efficient power plant. Customers easily adjust to any inconvenience.

Use of Heat Pumps Instead of Fossil Fuels

The following summary compares the energy required to heat a building:

BASIS (Heat Received)	**1,000 Btu**
Fuel Burned in 80% Furnace	1,250 Btu
Fuel Burned in 90% Furnace	1,110 Btu
Electricity Consumed by a 7 HSPF Heat Pump	447 Btu
Fuel Consumed to Produce 447 Btu at 50% Thermal Efficiency Combined Cycle Plant	894 Btu
Fuel Consumed to Produce 447 Btu at 30% Thermal Efficiency Power Plant	1,490 Btu
Fuel Consumed to Produce 447 Btu at 45% New Generation Nuclear Power Plant	1,000 Btu

Of these options, use of wind energy with a 7 HSPF heat pump would consume the least fuel but would not be practical on a nationwide basis. However, the second-least energy consuming option is practical on a large scale. Using natural gas and a state-of-the-art combined cycle for electrial power production reduces the fuel consumed for heating by 20% to 30% compared to a gas furnace. The carbon dioxide emissions can be all but eliminated by using nuclear power rather than natural gas or coal. As heat pumps and power cycles improve, the quantities of fuel burned can be reduced, and this reduces the greenhouse gases released. New generations of 50% efficient coal and nuclear would provide an improved option for heating without relying on natural gas.

Use of Thermal Energy Storage for Peak Load Shifting for Heat Pumps or Air Conditioning

The peak demand for electricity occurs during daylight each day of the year when people are awake and active. The peak demand for air conditioning occurs in the afternoon of a summer day, adding to the normal power demand spike. An air conditioner is most efficient when the outside temperature is cool. It would be an advantage to run the air conditioner at night to cool an energy storage unit that would supply cool air during the day. During peak demand periods, peaking units typically generate electrical power at about 28% thermal efficiency. At night, only baseload generators would be used, such as combined cycle units operating at greater than 50% efficiency or coal-fired plants running at about 38% to 45% efficiency. Nuclear power plants operate at 30% to 33% efficiency but with zero greenhouse gas emissions. The high cost of peak demand power is due to both higher fuel and higher capital costs, since the peak-demand units are used for a small fraction of the year.

A heat pump is designed to produce warm air to heat a building during the winter heating season. The coldest part of the day is at night—the time when the difference between the inside and outside temperature is greatest. Heat pumps are most efficient when this temperature difference is small, so there is an advantage of running the heat pump during the day to heat an energy storage unit to provide heat at night. Here, the efficiency of the heat pump is at odds with the efficiency of electrical power production.

The major components of an air conditioner and a heat pump are the same. Combination units (heat pumps) are commercially available that can be used as air conditioners in summer and switched to be heat pumps in winter. These units work best in temperate climates, which is most of the United States. Development work on an efficient energy storage unit is in progress. This combination would make an energy efficient system for homes and small commercial buildings.

Potential Impact of Thermal Energy Storage for Peak Load Shifting with Heat Pumps or Air Conditioning

The peak demand of electricity relative to baseload electrical power varies during the year. Peak demand in April is typically about one and a half times the baseload. In July the peak demand can become twice the baseload. It is possible to make the demand for electricity nearly constant at a higher baseline load level by using a peak load shifting strategy. For example, a higher baseload can justify the higher cost of construction for 50% efficient combined-cycle facilities operating continuously instead of the 28% gas turbine units used to supply peak power part of the day. For every 1.0 kWh that is shifted from peak demand to baseload, there are 2.0 kWh that benefit from the improved efficiency—1.0 kWh that was shifted and 1.0 kWh during the peak demand period because the baseload has been increased to accommodate the shift load.

Before the load shift, 7.14 kWh of fuel must be consumed to provide the 2 kWh of peak power. Peak load shifting reduces this 7.14 kWh of fuel consumption to 4 kWh. This reduced cost can be shared with the customer and applied to the capital investment in the power plant. Greenhouse gas emissions are reduced by 44%.

Peak load shifting saves fuel and reduces costs for essentially every application. For those applications where the savings are passed to consumers by the local electrical providers, the consumers can realize quick paybacks for investments on energy storage devices (see the box "Example of Value of Peak Load Shifting").

Example of Value of Peak Load Shifting

Peak demand electricity is more expensive and less efficient to generate than baseload. The highest peak loads occur during the four months of high air conditioner use. The peak loads increase to a maximum at midafternoon and then decrease each day during the four months of summer. These peak-load times represent about 14% of the full year. The generators producing this peak power operate at about 10% of annual capacity. Investment costs to build these facilities are recovered by increasing the rates for peak demand electricity or by increasing the rates for all the power produced. In addition, economics dictate that capital equipment costs be minimized, and these generating units are less efficient and often use more expensive fuels.

The Northern States Power Company has programs that provide customers with incentives to reduce peak demand consumption. Their nuclear power infrastructure provides inexpensive baseload availability. Here are two of their residential rate programs:

Standard Rate Code A01:
June–September   7.35 cents/kWh
October–May   6.35 cents/kWh

This standard rate code provides easy bookkeeping, but it does not reflect the true cost and availability of electricity. Other rate codes are used to create markets for excess winter capacity and give a reduced rate of $0.0519 during the winter for electric space heating. The time of day service option (Code A04, following) reflects the difference in costs for providing peak load versus baseline load and allows the homeowner to adjust use pattern to reduce costs.

Standard Rate Code A04:
On Peak June–September   13.95 cents/kWh
On Peak October–May   11.29 cents/kWh
Off Peak   3.27 cents/kWh

When the price of electricity reflects the cost of providing peak power versus off-peak power, the peak electricity costs four times off-peak power. The definition of peak demand for this plan is 9:00 A.M. until 9:00 P.M. This plan gives a real incentive to program use of electricity to off-peak hours.

Data from http://www.xcelenergy.com/EnergyPrices/RatesTariffsMN.asp, Rate Codes A01 and A04.

EnergyGuide Labels

The U.S. government established a mandatory compliance program in the 1970s requiring that certain types of new appliances bear a label to help consumers compare the energy efficiency among similar products. In 1980, the Federal Trade Commission's Appliance Labeling Rule became effective and requires that EnergyGuide labels be placed on all new refrigerators, freezers, water heaters, dishwashers, clothes washers,

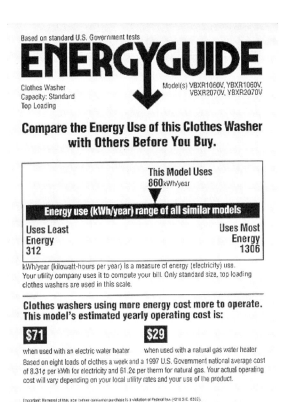

room air conditioners, heat pumps, furnaces, and boilers. These labels are bright yellow with black lettering identifying energy consumption characteristics of household appliances.

Although these labels will not tell you which appliance is the most efficient, they will tell you the annual energy consumption and operating cost for each appliance so you can compare them yourself.

EnergyGuide labels show the estimated yearly electricity consumption to operate the product along with a scale for comparison among similar products. The comparison scale shows the least and most energy used by comparable models. The labeled model is represented by an arrow pointing to its relative position on that scale. This allows consumers to compare the labeled model with other similar models. The consumption figure printed on EnergyGuide labels, in kilowatt-hours (kWh), is based on average usage assumptions, and your actual energy consumption may vary depending on the appliance usage.

> EnergyGuide labels are not required on kitchen ranges, microwave ovens, clothes dryers, on-demand water heaters, portable space heaters, and lights.
>
> From http://www.eren.doe.gov/buildings/consumer_information/energyguide.html.

Air Conditioning

Air conditioning needs are commonly met predominantly by ventilation (open windows and fans), evaporative coolers, air conditioners, and effective building design with proper landscaping. Cooling and humidity control go hand-in-hand during much of the air conditioning season. Much like the dew on the grass on a cool spring morning, the cool temperatures created by an air conditioner's evaporator cause water to condense from the air. Although modern central air systems keep the condensate out of view, it can still be seen from an automobile air conditioner as a puddle of water under a car parked after air conditioner use. The energy consumed in condensing the water from air can approach the energy consumed in reducing the air temperature.

Of all the nonarchitectural options to keep a house cool, ventilation is the least expensive. Almost as inexpensive are evaporative coolers that cool by evaporation of water to the air and are a good option in dry climates. Vapor-compression air conditioners are the most common air conditioner used by homeowners to compensate for the humid heat of summer and are responsible for most of the energy used to cool buildings. These units use the compression and condensation of refrigerants to pump heat (see the box "How Do Air Conditioners Work?").

As summarized by Table 8-2, energy consumed by air condition is about one-tenth of the energy consumed in heating. Air conditioners use less energy because they pump heat out of the house rather than converting electrical energy to "coolness." By contrast, furnaces directly convert the chemical energy in fuels to heat, and electrical resistance heaters directly convert electrical energy into heat.

The efficiency that an air conditioner pumps heat is indicated by its Seasonal Energy Efficiency Ratio, abbreviated SEER. The SEER is the Btus of cooling provided per watt-hour of electricity consumed. Federal efficiency standards require that heat pumps (air conditioners that provide both heating and cooling) have a

SEER rating of at least 10.0 with some units providing SEER values above 14.[1] Dividing the SEER rating by 3.41 puts the cooling and electrical consumption in the same units. Modern air conditioners remove three to four times more heat from the house than electrical energy consumed.

Accounting for the SEER heat pumping ability of air conditioners, the typical amount of heat added to a Midwest house during the winter is about three times as much as the heat removed in the summer. This factor of three seems about right when considering the longer heating season (six months compared to three months cooling) and the fact that outside temperatures will typically vary from about 70°F less (winter) than inside temperatures to 30°F higher (summer) than outside temperatures. The extent of coldness in the winter simply exceeds the extent of hotness in the summer.

For commercial buildings in the Midwest, about twice as much energy is consumed for heating compared to cooling. Commercial buildings are generally larger and have less outside wall per square foot of floor space. The lower outside surface areas provide better insulation. As a result, lighting, electrical office equipment, and heat given off by people will have a larger impact compared to outside weather conditions. Each of these produce heat that must be removed from the building during the summer.

How Do Air Conditioners Work?

Air conditioners use the work generated by electrical motors to pump heat out of a house. This can be expressed mathematically as follows:

3 Btu heat pumped from house + 1 Btu of electrical work

= 4 Btu heat pumped outside

This equation shows the conservation of energy in this process and gives an air conditioner a SEER rating of 3 Btu ÷ 1 Btu × 3.41 (Btu per watt-hour) = 10.23.

As shown in the figure, a vapor compression air conditioner consists of four components: evaporator, compressor, condenser, and valve. Air inside the house circulates through the evaporator. At about 40°F, the evaporator is cooler than the inside air and cools the inside air from 80°F to about 50°F. For

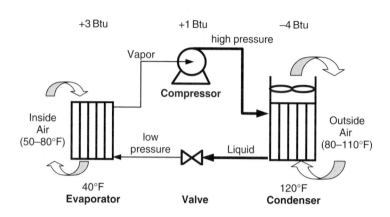

this example, 3 Btus of heat are transferred from the house air into the refrigerant circulating in the air conditioner. The heat causes the refrigerant to boil and leave the evaporator as a vapor.

With the addition of about 1 Btu of work, the vapor from the evaporator is compressed to a higher temperature and pressure. The high pressure allows warm outside air to be used to condense the refrigerant vapor (high pressures increase boiling points and favor formation of liquids) in the condenser located outside the house. The 4 Btus of heat are released from the condenser and make the warm outside air even warmer. Liquid refrigerant leaves the condenser, proceeds through a valve reducing the pressure, and causes some of the refrigerant to evaporate, which cools the liquid to 40°F in the evaporator where the cycle started.

Residential and commercial air conditioning consume 6% to 7%[i] of the electricity generated in the United States. While this number appears small, consider that this use is predominantly in the hottest one-third of the year, in the warmest and most humid part of the United States, and mostly during the daytime hours rather than at night. This 6% to 7% rapidly increases to 50%[ii] or more of the peak electrical power load during the warmest summer days.

---

[i] (0.35 + 0.42)/11.6. See table in Appendix.
[ii] 6.6% × 3 (12/4 months) × 2 (warmest half of U.S.) × 2 (warmest half of day)/0.8 (none-heating and none-cooling load).

The peak load shifting of this air conditioning load provides opportunities to justify more efficient baseload capacity and also allows for substantial reductions in greenhouse gas emissions and fossil fuel consumption.

Heating

Air conditioning arrived with the industrial revolution at the onset of the 20th century. The generation of heat is as old as civilization and often considered a trivial process. The combustion of fuels converts chemical energy into heat. The electrical resistance in wires will cause electrical energy to convert to heat.

Furnaces have historically converted wood, coal, and even cow chips (dried manure) into useful heat. The reduced soot of kerosene, fuel oil, natural gas, and liquefied petroleum gas allowed these fuels to dominate the heating fuel market into the mid-20th century. The historically low cost of natural gas combined with pipeline distribution to commercial and residential buildings make natural gas the most popular furnace fuel in the United States.

In the warmer parts of the United States, heating is often necessary in the winter, but the fuel consumption for heating does not justify the cost for natural gas pipelines. For these locations, as well as other locations, where natural gas is not available, heat pumps are now a popular alternative.

Using the same vapor-compression cycle as is used in an air conditioner, a heat pump "pumps" heat from outside air into a building. As shown in Figure 8-1, the four functional components of an air conditioner can be configured to pump heat out of a house or pump heat into a house. This can be done with the addition of two valves that reverse the flow of refrigerant (reversing the direction of the heat flow from outside to inside the house) and can be manufactured at small incremental costs above those of the air conditioner.

The heating season performance factor (HSPF) rates a heat pump performance based on the Btus of heat provided per watt-hour of electricity consumed. An HSPF rating of 6.8 or better is required by federal efficiency standards. The HSPF rating is a function of outside temperatures and decreases rapidly as outside temperatures drop below 30°F. The compressor has to work harder to generate the higher pressure difference necessary to overcome greater temperature differences between outside and inside.

212 *Sustainable Nuclear Power*

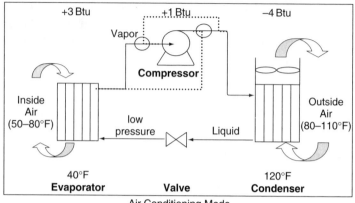
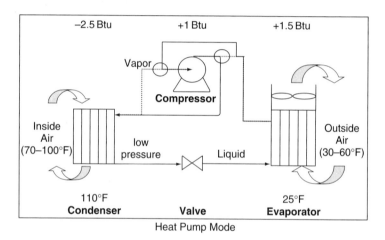

FIGURE 8-1. A heat pump operating in heating and cooling modes.

While air conditioners typically have to overcome about a 15°F temperature difference (75°F inside temperature versus a 90°F outside temperature), heat pumps will often have to overcome temperature differences in excess of 40°F (75°F inside temperature versus a 35°F outside temperature). This explains why the HSPF ratings of heat pumps are typically lower than SEER ratings of air conditioners. As outside temperatures get lower, the HSPF gets lower. At temperatures lower than the freezing point of water, the evaporator coil can freeze up like a big ice cube, and the heat flow stops.

For an incremental increase in cost above that of a conventional air conditioner (a couple of valves and minor equipment changes), the heat pump provides a significant performance advantage over electrical resistance heaters that directly convert electrical energy into heat. By tapping into the electrical power grid, the heat pump also taps into the diversity of the electrical power infrastructure, including the replacement of limited reserves of fuels using nuclear, wind, or hydroelectric energy. In addition, increased use of electrical power during nonpeak seasons (winter) and nonpeak times (night) can provide the incentive for building more efficient electrical power facilities. These improvements will become increasingly important as fossil fuel reserves are consumed and when the reduction of carbon dioxide emissions becomes a greater global priority.

At close to 10 billion gallons per year, liquid fuels used for residential space heating (heating oil, kerosene, and liquid petroleum gas) contribute about $10 billion of $200 billion in imported crude oil (and products) to the United States every year. These fuels, along with liquid fuels used for commercial heating and hot water heaters, are part of the problem and part of the solution of energy security.

The stationary nature of space heating and hot water heating applications makes electricity particularly attractive for these applications. Heat pumps used in combination with resistance heating could provide the heating demands while moving point-source emissions from inside the city to outside the city. If the electrical load created by these applications adds to the baseload during the winter, greater benefits result. Electrical costs and greenhouse gas emissions can be reduced by providing the market demand for new, efficient electrical power plants that not only provide efficient power for space heating (during the winter) but also replace less-efficient electrical power generation (peak load units operating at 28% efficiency) during the rest of the year.

For example, creating 100 additional days of baseload demand for electric heating could justify a new nuclear power plant or wind turbine farm that would provide baseload those 100 days as well as the other 265 days of the year. This is represented graphically by Figure 8-2. This would replace imported liquid heating fuels and potentially displace peaking electrical power units that consume natural gas or petroleum. These changes would generally be cost effective when all factors are considered. However, the accounting mechanisms may or may not be in place for local electrical providers to pass the savings on to the consumers who have the most control on when energy is used.

214 Sustainable Nuclear Power

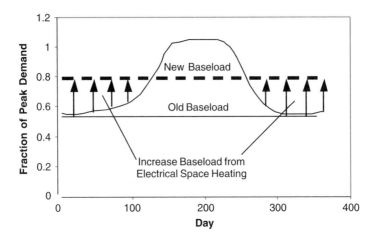

FIGURE 8-2. Impact of space heating on baseload for electrical power generation.

Peak Load Shifting and Storing Heat

Peak load shifting refers to changing the use of electricity from the middle of the day to nighttime when the electricity is in least demand. Figure 8-3 illustrates how electrical demand can vary during the 24 hours of a day. This change in demand should be of little surprise, since it is during the daytime that air conditioners run

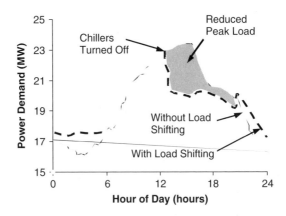

FIGURE 8-3. Example of 24-hour electricity demand during July for Fort Jackson in South Carolina. (http://www.cecer.army.mil/techreports/soh_stor/Soh_Stor-03.htm.)

when the temperature is high. It is during the day that clothes dryers run and hot water heaters recharge after the morning shower.

In the heat of a summer day, air conditioners can be responsible for over 50% of the electrical demand.[3] Peak demand power generation is both inefficient and costly. The case study at Fort Jackson illustrates the typical costs associated with peak demand electricity and illustrates how this problem can become an asset.

For large commercial or military installations, electrical power providers offer a number of different rate plans to pass on savings to these consumers as a reward for working with the electrical provider to reduce the cost for supplying electricity. These rate plans are typically based on the principal of reducing peak-demand and purchasing predictable amounts of base load electricity. Paying premium prices for all electricity purchased above the base demand is an example of such a rate plan. For the profile of Figure (8-3) at Fort Jackson, electricity consumed beyond about 19 MW is demand charge electricity for which the premium price is charged.

In 1996, the Fort Jackson Army installation paid a $5.3 million electrical bill with 51% of this ($2.7 million) as demand electricity. The nondemand portion of the electrical bill is referred to as the energy charge because it is intended to reflect the cost of all electricity consumed at baseload prices. During summer months, it was normal at Fort Jackson for 50% of the electrical bill to be demand charges. For most Army installations this demand bill exceeds one-third of the total electrical bill.

To reduce the cost of demand electricity, a chilled water storage tank was installed at Fort Jackson. During off-peak hours, air conditioners (chillers) run to chill water to about 42°F for storage. The chilled water is used to provide cooling during the day. The dashed line in Figure 8-3 illustrates how demand for electricity is typically reduced when the chillers are turned off at midday and the stored chilled water is used to provide cooling. An additional advantage of this storage system is that it allows the most efficient chiller to be used at full capacity during off-peak hours while minimizing the use of less-efficient units. Also, the chiller is operated more during the cooler hours of the day when the chiller operates more efficiently. A disadvantage of chilled water storage is that about 10% of the cooling is lost due to heat loss from the tank and mixing with the chilled water as warm water is returned to the tank.

In addition to chilled water units, ice storage is used for peak load shifting of electricity. Ice takes in considerable energy when thawing into water. Ice storage units are smaller than water storage units. Special materials, such as waxes, eutectic salts, and fat/oil derivatives have been developed to freeze at temperatures closer to room temperature, are also in use. These materials are referred to as phase change materials. Figure 8-4 shows a configuration developed in New Zealand that uses a wax phase change material.

In this configuration a phase change material (PCM) is encapsulated in a spherical nodule about the size of a golf ball. The encapsulation keeps insects and air away from the PCM and prevents the PCM from mixing with circulated air. These spherical nodules are then placed in a tank about twice the size of a building's hot water heater. Air is directed through the tank and then through the house to provide cooling. Air is circulated between the air conditioner and tank to freeze the PCM at night.

When used in Los Angeles, this peak-load shifting reduced monthly electrical bills from $19,941 to $14,943.[4] These savings were made possible because of a rate plan charging $0.11 per kWh for peak electricity and $0.061 per kWh for off-peak electricity. Through the use of phase change materials, air conditioning use went from 75% during peak demand times to 75% during off-peak times. When the tank storage is located in the building, these units can approach storage and recovery efficiencies of 100%—far better than methods for storing and recovering electrical energy.

For heating applications replacing liquid fuel with electrical power used during the winter increases demand at night. The PCM material can be used to store heat during the day to decrease peak demand at night. Even though the increase in baseload may

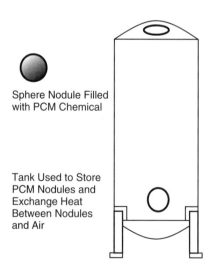

FIGURE 8-4. Example PCM device and tank for active climate control.

only be for a few months out of the year, this creates the incentive to build more efficient power generation facilities by providing a larger baseload (off-peak load) during the summer and winter.

Chilled water or ice storage systems are generally preferred for larger buildings or groups of buildings, while PCM storage tanks are preferred for small buildings. The Fort Jackson chilled water system was estimated to have a payback period of five years. A similar chilled-water storage system at the Administration Center in Sonoma County, California, cut the electrical utility bills in half and saved an estimate $8,000 per year due to reduced maintenance of the chillers that were sized at half their capacity before chilled water storage.[5] Just as peak load shifting reduces the maximum peak loads of electrical power providers, peak load shifting can also reduce the peak chiller operation demands, allowing for use of less-expensive units.

The U.S. Department of Energy reports[6] that federal government installations alone could save $50 million in electricity costs each year. Since ice or chilled water storage systems have been commercially available for over 50 years, several manufacturers and options are available.

For air conditioning, the greatest demands occur during the heat of the day when people are most active. The air conditioning inflates the daytime peaking of electrical consumption that occurs due to other activities. When heating during the winter, the coldest times are at night during off-peak hours. As a result, except for the storage of solar energy or when using ground source heat, storing heat does not have the same benefits as storing coolness.

Heat storage can eliminate energy consumption when solar heat is stored. Inevitably, solar heating systems will experience extended periods during the spring and fall when additional heating is not needed during the daytime hours, but heating may be necessary during the cooler night hours. For such systems, solar heating systems can be equipped with energy storage. Most solar heating systems offer energy storage options.

The Role of Electrical Power in HVAC to Reduce Greenhouse Gas Emissions

Energy storage and HVAC energy consumption impact strategies to provide cheap, abundant, and environmentally acceptable energy. Well-planned government programs could create the incentives necessary to build the next generation of highly efficient electrical power plants and substantially reduce the vast amounts of fossil fuels consumed for heating.

The energy consumed in the United States for HVAC applications is about the same as that burned in gasoline engines, HVAC can play a major role in reducing greenhouse gas emissions if this becomes a national priority. The 14 quadrillion Btus of energy consumed in HVAC probably underestimate the impact HVAC can have on greenhouse gas emission reduction.

The impact of near-zero greenhouse gas technologies is enhanced when electricity demand is stabilized both on the 12-month cycle through increased use of electric-based heating and on the 24-hour cycle through energy storage. Increased year-long baseload electrical demand should serve as incentive for building the most efficient combined cycle natural gas power production facilities (operating at over 50% thermal efficiency) and can lead to a new generation of more-efficient nuclear power plants.

These technologies can be used to reduce and maintain greenhouse gas emissions to 1990 levels and provide the transportation industry with time to develop new energy technologies to lower than 1990 levels. Furthermore, these technologies and conversions are cost effective when electrical providers and consumers jointly share benefits. When the electrical power providers have mechanisms to pass their peak load–reduction savings to consumers and the available technology is fully marketed, the transition can occur.

We hear messages from scientists that the dangers of global warming are real, but little has been proposed as cost-effective solutions. For electrical power generation and HVAC, the energy is abundant, the technology is available, and the knowledge and understanding to make the transition are obvious.

Example Calculations

The principle behind converting units is the mathematical axiom that any number multiplied times 1 is unchanged. Therefore, since the following is known to be true:

$$1{,}000 \text{ grams} = 1 \text{ kilogram or } 1{,}000\, g = 1\, kg$$

Then:

$$1 = \frac{1\, kg}{1{,}000\, g}$$

An example application of this conversion is as follows:

$$140\,g = 140\,g \times \frac{1\,kg}{1{,}000\,g}$$

The conversion is completed by recognizing that units cancel.

$$140\,\cancel{g} \times \frac{1\,kg}{1{,}000\,\cancel{g}} = 0.140\,kg$$

Table 8-3 summarizes commonly used conversions used in energy calculations.

TABLE 8-3
Conversion factors and abbreviations.

Conversions			Abreviations		
1 barrel	=	42 gallons	atm	atmospheres	
1 hectare	=	2.47 acres	Btu	British thermal unit	
			cal	calories	
1 kg	=	0.001 metric tons	cc	cubic centimeter	
	=	2.20462 lb$_m$	cm	centimeter	
	=	0.00110 tons	ft	feet	
			g	gram	
1 m	=	3.2808 ft	gal	gallon	
	=	39.37 in	GW	gigawatt	
			hp	horsepower	
1 m^3	=	1000 L	in	inch	
	=	35.315 ft^3	J	joule	
	=	264.17 gal	kg	kilogram	
			L	liter	
1 kJ	=	0.9486 Btu	lb$_m$	pounds mass	
	=	239.01 cal	m	meter	
			MW	megawatt	
1 W	=	1 J/s	psi	lb/in^2	
	=	0.001341 hp	s	second	
			W	watt	
			m	milli	10^{-3}
			c	centi	10^{-2}
			d	deci	10^{-1}
			k	kilo	10^{+3}
			M	mega	10^{+6}
			G	giga	10^{+9}

TABLE 8-4
Commonly used physical properties and abbreviations.

Physical Properties and Definitions			
	Biodiesel	Ethanol	Methanol
Density ($lb_{m/gal}$)	7.35	6.63	6.64
Heating Value (Btu/gal)	118,200	76,577	57,032
Gasoline (no alcohol in fuel)	115,000–119,000 Btu/gal		
Diesel	130,500–135,000 Btu/gal		
Corn	56 lb_m/bushel, 2.5 (2.6) gal ethanol/bushel		
Soybeans	18%–20% soybean oil		
SEER	Seasonal energy efficiency rating		
SEER	Btu cooling/W-hr electricity		
SEER/3.41	W-hr cooling/W-hr electricity		
HSPF	Heating season performance factor		
HSPF	Btu Heating/W-hr electricity		
Thermal Efficiency	Energy delivered/energy consumed		
Gasoline Engine (4-stroke)	30% thermal efficiency, typical		
Gasoline Engine (2-stroke)	22% thermal efficiency, typical		
Diesel Engine (4-stroke)	40% thermal efficiency, typical		

Table 8-4 summarizes commonly used physical properties and abbreviations. The physical properties of gasoline and diesel will vary based on the source of the petroleum, refining practices, and seasonal-specific formulations.

Example Calculation What is the rate of energy delivery to the driveshaft (in kW) of an automobile with a fuel economy of 30 mpg traveling 70 mph?

$$\frac{70\ miles}{3600\ s} \left| \frac{gal}{30\ miles} \right| \frac{115{,}000\ Btu}{gal} \left| \frac{kJ}{0.9486\ Btu} \right| \frac{0.3\ kJ\ shaft}{kJ\ fuel}$$

Or 23.6 kJ/s. Converting to watts, this is 23.6 kW. It would take a 23.6 kW electric motor to provide this power. Note that a typical gasoline engine efficiency of 30% was used in the calculation. This neglects frictional losses after the drive shaft.

References

1. http://www.eere.energy.gov/buildings/appliance_standards/news_detail.html/news_id=6781.
2. *Federal Register*/Vol. 69, No. 158/Tuesday, August 17, 2004/Rules and Regulations.
3. http://www.cecer.army.mil/techreports/soh_stor/Soh_Stor-03.htm. Fort Hood, Texas, example.
4. Energy-Wise News, http://www.eeca.govt.nz/, Sept. 1998.
5. http://www.eren.doe.gov/cities_counties/thermal.html.
6. http://www.pnl.gov/fta/CoolStorage/CoolStorage.htm.

CHAPTER 9

Electrical Power as Sustainable Energy

Sustainability and Electrical Power

The diversity of energy sources for producing electrical power brings with it moderate and stable prices as well as a reliable supply. In particular, nuclear, wind, and biomass sources provide sustainability and near-zero greenhouse gas emissions. It is through the electrical power grid that nuclear power can provide sustainable energy for at least the next few centuries.

Figure 9-1 summarizes how energy is used in the United States. At 38% of the energy consumed in the United States, more energy goes into electrical power production than transportation (28%), industry (22%), and buildings for commerce or residence (12%). Electricity dominates applications like lighting, air conditioning, and appliances. Over half of the electrical power is produced from coal.

The predominant fuel used for transportation in the United States is petroleum. Most of the petroleum is used in the form of gasoline. Industrial applications include use of petroleum and natural gas as chemical feedstocks and boiler fuels to provide process heat. Aside from electricity, petroleum and natural gas provide the majority of industrial energy needs with some use of coal.

In Figure 9-1, rather than including electricity in the industrial and commercial/residential energy uses, it is shown separately to demonstrate how electricity can be used to meet energy needs in the other three sectors. Most of the nonelectrical energy demands of commercial and residential buildings are for space

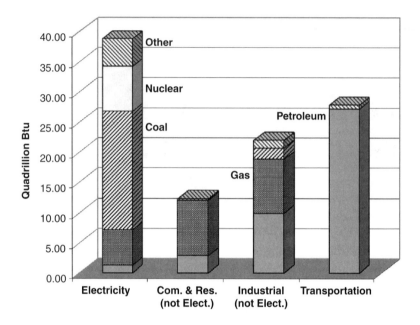

FIGURE 9-1. Energy consumption in the United States. Distribution by energy source only includes sources contributing more than 2% of the energy in each category.

heating and hot water heaters. Natural gas is the common hydrocarbon fuel delivered for these applications with some use of heating oil.

Expanded Use of Electrical Power

There are significant opportunities to expand the use of electrical power in transportation as well as in commercial and residential buildings. Fewer opportunities exist in industrial applications because natural gas and petroleum actually provide a source of carbon from which chemicals and materials are manufactured.

In industrial applications, use of biomass as feedstocks rather than petroleum and natural gas would improve sustainability. In view of this match between biomass and industrial applications, a case could be made for minimizing the use of biomass in electrical power production and space heating so it is available for industrial uses.

Electricity is more suitable for expanded use in transportation and space heating than for industrial applications. These are the two major advantages of expanded use of electrical power in transportation and commercial/residential applications:

- The distribution infrastructure is in place allowing a focus on application development (there is no question which should be first—distribution or application).
- Increased use of electricity will provide opportunities to improve our electrical power supply infrastructure. In particular, more efficient and peak load shifting approaches could further improve efficiency in current applications while expanding use to new applications.

Because of these advantages, electricity is likely to have an increasing role in providing energy demands.

A third advantage of expanded utilization of electrical power was shown by California's joint agency report entitled Reducing California's Energy Dependence.[1] The premise of this study was that petroleum sources are limited and steps should be taken to reduce the vulnerability of the economy of the state to higher prices and limited availability of petroleum. Furthermore, those alternatives to petroleum that would most benefit the local economy were identified.

Improved auto fuel economy (increasing fuel efficiency to an average of 28.5 mpg or higher) reduced total oil consumption and had the greatest benefit for the economy. The best of the improved fuel economy alternatives were those that had the lowest increase in vehicle cost per benefit of increased fuel economy. Figure 9-2 summarizes the alternative energy systems that had the greatest impact after fuel economy—options that replace petroleum with other energy sources.

The technologies summarized in Figure 9-2 did not have as great a projected dollar impact on California's economy as increasing the fuel economy of automobiles, but these technologies actually displace petroleum use and can ultimately reduce or eliminate the need to import petroleum. Efficiency increases alone will not eliminate the need to import petroleum.

The two "petroleum displacement" technologies projected to have the greatest benefit to California's economy are PHEV-20 and PHEV-60 vehicles. This is a third major advantage of increased

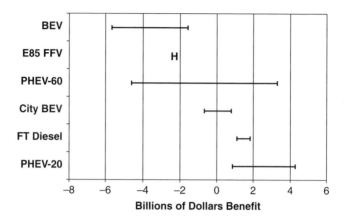

FIGURE 9-2. Top technologies for providing beneficial impact to California's economy while reducing fuel consumption. Beneficial impact is reported in projected 2001 dollars for the time period from 2002 to 2030.

use of electrical power for transportation, residences, and commercial buildings:

- Electricity can be locally produced with value added in the electrical power production process. Relative to importing natural gas, local electrical power production from low-cost fuels increases cash flows into local economies. Especially for nuclear power, where the cost of uranium is low relative to coal or natural gas, great reductions in cash flow from local economies can be realized by replacing petroleum with electrical power.

Increased Use of Electrical Power in Transportation

Battery-powered automobiles have been around for over a century, and yet little progress beyond niche markets has been made to replace petroleum. The problem is economics. Even with high-volume production of batteries, the costs are projected to be near $400 for each kWh of stored electrical power. This means it costs in excess of $25,000 to provide 200 miles of range on a compact sedan. However, new approaches in hybrid electric vehicle (HEV) technology are expected to improve the economics.

Figure 9-3 shows simplified diagrams of the series and parallel approaches to HEVs. A series design routes the engine power through the electric motor by converting the engine's mechanical

Electrical Power as Sustainable Energy 227

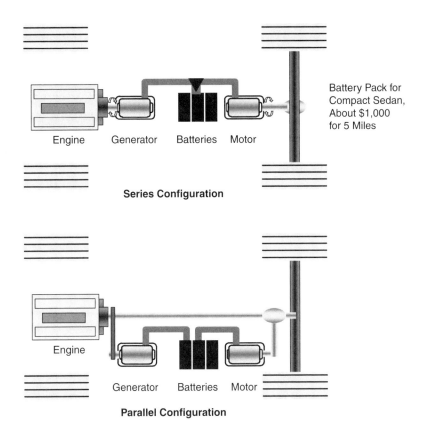

FIGURE 9-3. Simplified presentations of parallel and series HEV designs.

energy to electricity. The parallel design allows the engine or the battery pack to each power the wheels, depending on the trip demand.

The basic concept of both designs is to use a battery pack to reduce the fluctuations in power demand placed on the engine. With a level demand on engine power, a smaller engine can be used that operates at near-optimal engine speed and efficiency. When adding to the increase in engine efficiency—regenerative braking, energy storage when the engine is idling at stop lights—the overall fuel economy can be increased by about 50%. In practice, hybrid vehicle performance today ranges from delivering improved acceleration (Honda Accord) to improving fuel economy in excess of 50% (Toyota Prius).

Figure 9-4 illustrates the addition of the "plug-in" option to the HEV. The PHEV uses a larger battery pack—extending the vehicle's battery range from 3 to 5 miles to 20–40 or even 60 miles.

228 *Sustainable Nuclear Power*

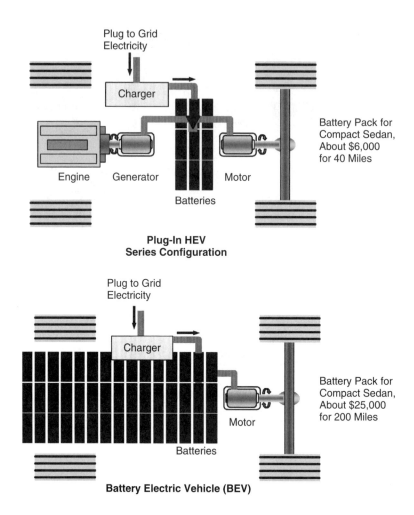

FIGURE 9-4. Comparison of PHEV and BEV designs. The PHEV has an engine and smaller battery pack. The BEV does not have a backup engine.

The addition of a battery charger to this vehicle then allows the PHEV to undertake extensive travel without engine operation. In the plug-in approach, the batteries are charged by grid electricity during the night (off-peak demand) rather than from the engine generator.

For example, a 40-mile range from the battery pack used 300 days per year provides 12,000 miles per year. The average vehicle on the highway actually travels about 20 miles on a typical day. So depending on the travel pattern, 20–40 mile PHEVs are capable of displacing about 80% of the petroleum used by the automobile.

The widespread use of PHEVs could replace the use of all imported petroleum and reduce total liquid fuel consumption to levels that could be met with ethanol and biodiesel. The PHEV can succeed where the BEV failed because the PHEV matches the size of the battery pack with the application to realize maximum use of the battery pack. In an application where the commuter needs a 40-mile battery pack, it is more cost effective to provide extended range with a backup engine (about $2,000) rather than a huge reserve of batteries (e.g., batteries for 160 mile range would cost an additional $19,000).

Figure 9-5 provides a summary of the life cycle net present cost for operating a conventional vehicle, an HEV, and a PHEV-20. A comparison of the NPV costs of the PHEV-20 to BEV illustrates that (in the right application) the PHEV can deliver the advantages of the BEV with a lower capital cost. As indicated by these cost

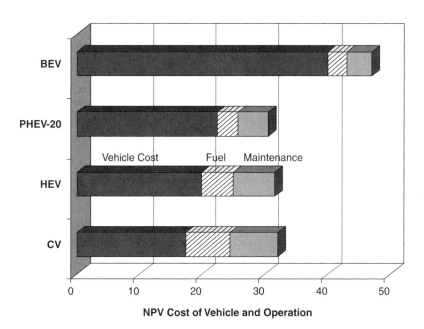

FIGURE 9-5. Comparison of net present cost for operating a conventional vehicle (CV), hybrid electric vehicle (HEV), plug-in HEV with a 20-mile range (PHEV-20), and a battery electric vehicle (BEV) with 200-mile range. Present values are based on a 7-year life cycle, $1.75 per gallon gasoline, and 6 ¢/kWh electricity. (Data on PHEV-20, HEV, and CV from A. Frank, 30 Years of HEV Research Leading to Plug-in HEVs. PHEV Workshop, 2003.)

summaries, the PHEV-20 can be more cost effective than the BEV, HEV, and conventional vehicle.

In similar analyses, the results of Frank[2] and Duvall[3] (Electrical Power Research Institute, EPRI) qualitatively agree with those of Figure 9-5: the PHEV-20 has a lower net present cost. Results are qualitatively similar for SUVs and sedans.

The primary economic advantage of the PHEV is a lower operating cost due to electricity costing less per mile than gasoline and to the lower maintenance costs. Here are the downsides of the NPV cost of the PHEV:

- The economics are based on 7 years, and many consumers have a hard time seeing past the sticker price on the vehicle.
- The capital costs are higher.
- The estimated capital costs on the PHEV are based on production of 100,000 units per year—far from the handful of prototype vehicles now being produced.

The important aspect of the PHEV is the vision that economies of scale can be attained with sustainable displacement of petroleum. It is this vision and responsibility to the future that forms the basis of sustainability.

Both the California study and the work of Professor A. Frank support the approach of using PHEVs to displace petroleum with electrical power, and they document how this can be done while saving consumers money and benefiting local economies. The city BEV (see Figure 9-6 is another vehicle that can meet these objectives. The city BEV would be a low-cost vehicle designed for niche markets such as second family cars used primarily for commuting to and from work. A range of 60 miles between charging the batteries would meet the most commuting demands.

The standing of FT diesel and E85. Fischer-Tropsch (FT) diesel fared well in the California study—probably the economic advantages of producing this fuel in California from coal or natural gas. E85 fuel (85% ethanol, 15% gasoline) did not fare well in this study. The production of the ethanol was assumed to be in the Midwest rather than California.

The PHEV is not intended to replace the conventional gasoline-powered automobile in every application. However, as depicted by Figure 9-7, most automobiles do not travel more than 30 miles in a 24-hour period. The data in Figure 9-7 imply that large sectors of the automobile market can be served with PHEV-20s (PHEV with 20 miles of range) in which the engine will be

Electrical Power as Sustainable Energy 231

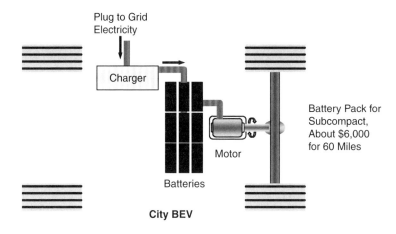

FIGURE 9-6. A city BEV. The city BEV is a compact vehicle that has a maximum range of 60 miles. It is a niche market vehicle that can meet the needs of "some" commuter needs.

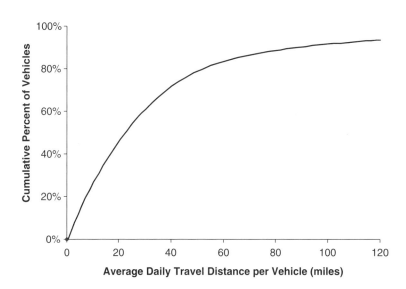

FIGURE 9-7. Miles traveled with typical automobile each day and implied ability for PHEVs to displace use of petroleum. (G. J., Suppes, S. M. Lopes, and C. W. Chiu, "Plug-In Fuel Cell Hybrids as Transition Technology to Hydrogen Infrastructure." *International Journal of Hydrogen Energy*, 29, January 2004, pp. 369–374.)

used rarely. The needs for even larger market segments would be met with PHEV-40 and PHEV-60 vehicles. Especially for two-or-more-car families, there are huge potential markets for PHEVs, and these PHEVs will substantially displace the use of petroleum in the right applications.

Additional PHEV concepts are also being studied, including the use of fuel cells on PHEVs. Suppes documented how regenerative fuel cells could be used with batteries for energy storage at a lower cost than using either batteries or fuel cells alone. An additional advantage of this approach is that the regenerative fuel cell cost curve is much lower for mass production points than the battery pack cost curve. This translates to PHEVs eventually costing less than conventional vehicles. This also points to an evolutionary path to the hydrogen economy that bypasses the need for major, risky investments into a hydrogen fuel distribution system.[4,5] Unfortunately, PHEVs using regenerative fuel cells are much further from "ready for market" than PHEVs using batteries.

PHEV technology faces many of the same commercialization barriers as all new technologies. Usually, sustainable alternatives to commercial technology are at odds with the momentum of the established industry. Any new technology faces technical and nontechnical barriers to commercialization, and these barriers are at least as great when sustainability constraints are placed on the new technology. It is rare that new sustainable alternatives present overwhelming profitability and commercialization is spontaneous.

The alternative to spontaneous commercial viabilities is a path to economic viability that includes reasonable advances in technology and overcoming reasonable risks. For the transportation sector these paths are available and that is about as good as it gets. Historically, it has taken both a path and leadership (industrial and/or political) to make new applications happen.

Battery Energy Storage

Realizing the economic benefits of PHEVs (see Figure 9-5) depends on low-cost electrical energy storage. Batteries are the preferred method of storing energy for applications ranging from watches to automobiles. However, this will not always be the case. Storing energy with hydrogen with conversion in fuel cells is already penetrating the traditional battery markets. It is all about meeting the application requirements at low cost.

Battery costs depend on the following:

- Production volume
- Type of battery (lead-acid, nickel-metal hydride, other)
- Application details like weight and recharge cycle requirements

The California Air Resources Board (ARB) has been a major driver for implementing electric vehicles. The technical assessment of batteries was considered by the Battery Technical Advisory Panel (BTAP 2000). Table 9-1 summarizes key characteristics of projected battery technology based on data from this panels' investigations that included two years of surveys and site visits with industry most active in HEV technology.[6]

Of the battery options, nickel-metal hydride batteries tend to be preferred over lead in HEV applications because of life cycle and weight advantages. Life cycle advantages are especially important if the manufacturer wishes to avoid creating the image of newly introduced HEVs as high-maintenance vehicles.

The size of battery packs varies based on vehicle size (see Table 9-2). Advocates of PHEV see the lower-cost battery pack prices of Table 9-2 as the needed advance to have PHEVs emerge with an automobile market share comparable to engine-powered automobiles. For commuting applications, the PHEVs

TABLE 9-1
Cost and capacity projections for batteries considered for HEV and PHEV applications.

Property	Lead Acid	Ni-Metal Hydride	Li Ion A	Li Ion B
Specific Energy (Wh/kg)	35	65	90	130
Operating Life (cycles)	400–1000	1,000–2,000	400–1,000	800–2,000
Cost @ 30,000–300,000 kWh Production ($/kWh)	150–200	500–840	1,000–1,350	1,000–1,700
Cost @ >300,000 kWh Production ($/kWh)	85–115	300–370	270–440	300–500
Status	mature	maturing	R&D	R&D

TABLE 9-2
Typical sizes of battery packs on EVs. Cost estimates based on entire range of Ni-metal hydride batteries in Table 9-1. Battery pack sizes are averages of values reported by Duvall.

Vehicle	Battery Size (kWh)	Battery Pack Cost $300/kWh	Battery Pack Cost $840/kWh
Mid-size PHEV-20	7.0	$2,100	$5,880
Mid-size PHEV-60	19.5	$5,850	$16,380
BEV 40 city car (micro car)	9.1	$2,730	$7,644
Mid-size BEV	27.0	$8,100	$22,680
Mid-size HEV 0	2.9	$870	$2,436

could displace 80% of the gasoline consumption with a sustainable alternative: grid electricity.

Fuel Cell Technology

As a subject of national attention, the U.S. DOE prepared a "Fuel Cell Report to Congress" in February 2003.[7] The following statement taken from this congressional report is an official rationale for fuel cell technology and targeted performance:

> Fuel cell technologies offer unique opportunities for significant reductions in both energy use and emissions for transportation and stationary power applications.
> - Efficiency improvements over conventional technologies that are inherent to fuel cells could lead to considerable energy savings and reduction in greenhouse gas emissions.
> - The use of hydrogen in fuel cells, produced from diverse, domestic resources, could result in reduced demand for foreign oil in transportation applications.
> - Widespread use of fuel cell technology could make a significant improvement in air quality in the United States. This would be a result of near-zero emission vehicles and clean power generation systems that operate on fossil fuels and zero-emission vehicles and power plants that run on hydrogen.

For the purposes of this report, the department did not attempt to quantify benefits of fuel cell commercialization and compare them to the expected public and private sector costs necessary to achieve commercialization.

Significant additional fuel cell research and development (R&D) would need to be conducted to achieve cost reductions and durability improvements for stationary and transportation applications.

Additional barriers to commercialization vary by application and fuel cell type. However, cost and durability are the major challenges facing all fuel cell technologies. (See Table 9-3.)

For fuel cell vehicles, a hydrogen fuel infrastructure and advances in hydrogen storage technology would be required to achieve the promised energy and environmental benefits. Efficient, clean, and economical processes for producing and delivering hydrogen from a variety of domestic feedstocks, including fossil, nuclear, and renewable sources, are critical to increased energy resource diversity and energy security. Five types of fuel cells dominate development efforts.

Fuel cells are expected to be suitable for a wide range of applications (see Figure 9-8). Transportation applications include vehicle propulsion and on-board auxiliary power generation. Portable applications include consumer electronics, business machinery, and recreational devices. Stationary power applications include stand-alone power plants, distributed generation, cogeneration, backup power units, and power for remote locations.

TABLE 9-3
Barriers to fuel cell commercialization.

Application	Barriers	Difficulty
Transportation	Cost	High
	Durability	High
	Fuel Infrastructure	High
	Hydrogen Storage	High
Stationary-Distributed Generation	Cost	High
	Durability	Medium
	Fuel Infrastructure	High
	Fuel Storage (Renewable Hydrogen)	Low Medium
Portable	Cost	Medium
	Durability	Medium
	System Miniaturization	High
	Fuels and Fuels Packaging	Medium

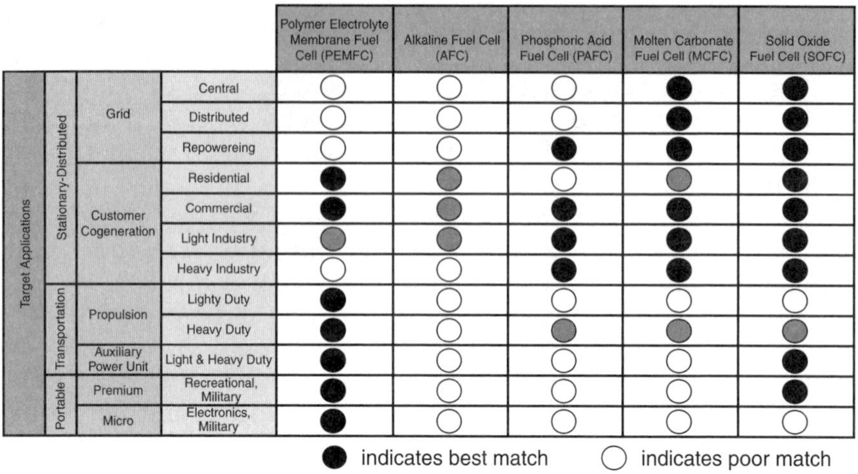

Target Applications			Polymer Electrolyte Membrane Fuel Cell (PEMFC)	Alkaline Fuel Cell (AFC)	Phosphoric Acid Fuel Cell (PAFC)	Molten Carbonate Fuel Cell (MCFC)	Solid Oxide Fuel Cell (SOFC)
Stationary-Distributed	Grid	Central	○	○	○	●	●
		Distributed	○	○	○	●	●
		Repowering	○	○	●	●	●
	Customer Cogeneration	Residential	●	●	○	●	●
		Commercial	●	●	●	●	●
		Light Industry	●	●	●	●	●
		Heavy Industry	○	○	●	●	●
Transportation	Propulsion	Lighty Duty	●	○	○	○	○
		Heavy Duty	●	○	●	●	●
	Auxiliary Power Unit	Light & Heavy Duty	●	○	○	○	●
Portable	Premium	Recreational, Military	●	○	○	○	●
	Micro	Electronics, Military	●	○	○	○	○

● indicates best match   ○ indicates poor match

FIGURE 9-8. Fuel cell technologies and their applications.

There are several different fuel cell technology paths being pursued. These divide into low-temperature and high-temperature technologies. Low-temperature technologies include phosphoric acid and polymer electrolyte membrane fuel cells (PAFCs and PEMFCs) and target transportation, portable power, and lower-capacity distributed power applications. High-temperature technologies include molten carbonate and solid oxide fuel cells (MCFCs and SOFCs) and focus on larger stationary power applications, niche stationary and distributed power, and certain mobile applications. A combination of technology developments and market forces will determine which of these technologies are successful. Currently, phosphoric acid fuel cells are the only commercially available fuel cells. More than 200 of these "first-generation" power units are now operating in stationary power applications in the United States and overseas. Most are the 200-kilowatt PC25 fuel cell manufactured by UTC Fuel Cells.

A cornerstone activity of the FE fuel cell program is the Solid State Energy Conversion Alliance (SECA), a partnership between DOE, the National Laboratories, and industry. The aim of SECA is to develop and demonstrate planar solid oxide fuel cells for distributed generation applications. Performance and cost goals for the SECA Program are shown in Table 9-4.

In addition to performance goals for the solid oxide fuel cells, the "Report to Congress" reports the Table 9-5 performance goals for PEMFCs.

TABLE 9-4
SECA performance and cost goals.

	Fuel Cell System	
Capital Cost	$400/kW	
Maintenance	3,000 hours	
Electrical Efficiency (Full Load, LHV)	Auxiliary Power Unit Stationary	50% 60%
Design Life	Auxiliary Power Unit Stationary	5,000 hours 40,000 hours
Emissions	Near Zero	

TABLE 9-5
FreedomCAR performance and cost goals (all 2010 except as noted).

	Efficiency	Power	Energy	Cost
Fuel Cell System	60% (hydrogen) 45% (w/reformer)	325 W/kg 220 W/L		$45/kW $30/kW (2015)
Hydrogen Fuel/ Storage/ Infrastructure	70% well-to-wheel		2 kW-h/kg 1.1 kW-h/L 3.0 kW-h/kg 2.7 kW-h/L	$5/kW-h $2/kW-h $1.50/gal (gas equiv.)
Electric Propulsion		>55 kW 18s 30 kW cont.		$12/kW peak
Electric Energy Storage		25 kW 18s	300 W-h	$20/kW
Engine Powertrain System	45% peak			$30/kW

Many in the scientific community consider the cost goal of $45/kW by 2010 for the PEMFC to be unrealistic. However, costs even close to $45/kW could have major implications for fuel cells, electrical power, nuclear power, and sustainability in transportation.

The most expensive component of fuel cells is the membrane electrode assemblies where reactions take place and ions are transported between the electrodes. Membrane electrode assemblies are basically sheets of the membrane material (polymeric, ceramic, or other) coated with the catalyst and designed with a microscopic system of channels to assist with flow of fuel, oxygen, and water to and from the catalyst on the membrane surface. The electric current (hence, power output) is directly proportional to the area of the membrane electrode assembly, and so, costs of a fuel cell are proportional to power output rather than total power available. On the other hand, the cost of fuel/hydrogen storage tends to be proportional to total energy stored (more fuel, large tanks, higher cost).

The fuel cell system costs of Table 9-5 are reported in kW, and the hydrogen storage is reported in kWh—consistent with fuel cell prices based on power output and fuel storage prices of the total stored energy. This means fuel cells have performance advantages over batteries when the maximum required power output is low and the required stored energy is high. For a PEMFC fuel cell stack costing $45/kW and used continuously over an 18-hour period, the power to storage ratio is 18 hours. If a 1 kW power basis is assumed, the fuel cell cost is $45 or $2.5/kWh, and the energy storage cost at $5/kWh is $90. The overall cost of stored energy for this system is $45 + $90 for a total of 18 kWh of power, or $7.5 per kWh.

When fuel cell technology matures and these costs are realized, the cost to store energy for lower-power output applications available from fuel cells could be less than for batteries. In this example—a continuous supply of power for 18 hours—the fuel cell system cost is estimated at $7.5/kWh, while the bottom of the cost curves for nickel-metal hydride and lead batteries are about $300 and $850 per kWh, respectively.

The details are important. The $45/kW PEMFC costs are projected, will require high production volumes, and may be higher for a 1 kW system compared to a 30 kW system. Both battery options provide recharging capabilities; the recharging option on the fuel cell will increase costs. Because each application brings different details to the application, there is a future for both batteries and fuel cells.

Increased Use of Electrical Power in Space Heating

Traditional Electrical Space Heating Markets

For residential and commercial buildings, electricity can be used as an alternative to natural gas, propane, and heating oil to provide space and water heating. Since none of these fossil fuels qualify as sustainable alternatives, the increased use of electricity can provide a sustainable alternative.

Historically, the use of electricity for space or water heating was an option when natural gas was not available. Since natural gas was available (before 2002) at about $3 per million Btus (or 1.02¢/kWh), with electricity prices starting at 6¢/kWh, the choice was always to use available natural gas.

For locations without natural gas, heating oil ($1.65/gallon) and propane ($1.30/gallon) have been available at about $12.6 per million Btu (4.3¢/kWh). In these locations the use of electrical resistance heating could be attractive in warm climates. The higher cost of electricity over fuel could be less than the annualized cost of installing separate propane or heating oil furnaces.

Today and in the future, these economics change due to advances in heat pumps and higher natural gas prices. In 2005 the price of natural gas exceeded $12 per MBtu (4.1¢/kWh), which translates to 4.55¢/kWh in a 90% efficient furnace. At locations where temperatures rarely go below 20°F, heat pumps can reduce the electrical resistance heating costs by 50% to 66% (COPs of 2 to 3), resulting in delivered heat at an average price of about 2.4¢/kWh. These heat pumps cost little more than central air conditioning units. Commercially available air conditioner (summer)/heat pump (winter) units are now available.

Figure 9-9 summarizes the ratios of heating costs from fuels (90% efficiency) over heating costs using a heat pump with a COP

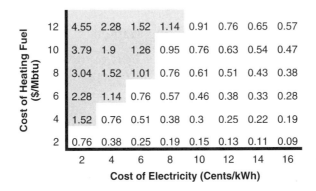

FIGURE 9-9. Ratios of costs for heating with fuel versus heating with electrical heat pump. Ratios based on COP of 2.0 and heater efficiency of 90%. Shaded regions show where heat pump is more cost effective.

of 2.0. The shaded numbers are those over 1.0 and represent combinations where a heat pump is cost effective.

In southern and southwestern states the heat pump has emerged as an economically sustainable alternative for space heating. The difficulty interpreting the data of Figure 9-9 lies in identifying the correct fuel and electricity costs. If pipeline natural gas is compared to the cost for electrical power today, $12 per MBtu natural gas and 5¢/kWh electrical production costs, the heat pump is preferred over natural gas for space heating in southern states. Provide an additional 1¢/kWh profit for electrical power distribution, it would cost about 52% more to use natural gas for space heating than it would cost for electricity.

In the southern and southwestern markets, policy alone (limiting the difference between production and delivered costs for electricity) can lock in larger and sustainable space heating markets through use of grid electricity. That policy starts with responsible decisions on fuel sources for baseload electrical power—decisions that include but are not limited to coal and nuclear power rather than natural gas to produce electrical power.

In the California electrical power market in 2005, electricity costs up to 24¢/kWh as a result of poor choices on the construction of natural gas electrical power generation. Shortages of electrical power in 2001 led to "panic" decisions to quickly install more natural gas electric power production. The natural gas-based electrical power promotes the use of natural gas that leads to dependence on natural gas with electrical power costs controlled by fluctuating gas prices.

Emerging Electrical Space Heating Markets

Paths toward expanding the use of heat pumps to cooler regions, as compared to the South, are more dependent on heat pump technology.

Figure 9-10 shows a typical performance curve for a heat pump. At temperatures above 32°F, the COP (ratio of delivered heat to consumed electricity, both in kWh) for a heat pump can be sustained above 2.0. However, the performance falls off quickly at about 32°F because ice can build up on the evaporator, which interferes with air flow through the coils and creates an increased resistance to heat flow. This happens because the evaporator coils will be about 10°F cooler than surrounding air, and these temperatures are below the dew point (or ice point) of the moist, outdoor air.

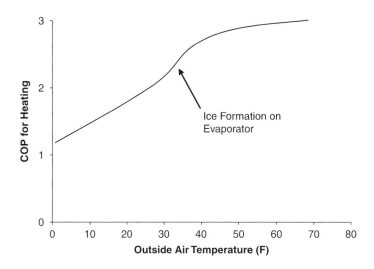

FIGURE 9-10. Dependence of a typical heat pump performance on outside temperature.

The increased thermal resistance requires even cooler evaporator temperatures to achieve the desired heat transfer and this decreases the COP. Heat must be periodically delivered to the evaporator coils to defrost them, further decreasing the effective COP.

In view of the performance curve of Figure 9-10, a good option in the Midwest, West Coast, and the southern two-thirds of the East Coast is to install both a heat pump and furnace. As temperatures go below about 20°F, programmable thermostats are available that will automatically switch from the heat pump to the furnace. The overall impact of this approach is an incremental substitution of fossil fuels for electrical power. At locations where air conditioners are common, the incremental cost for upgrading a central air conditioning unit to a heat pump can be small. In fact, the combination of a heat pump and an 80% efficient furnace can be less costly and more sustainable than an air conditioner and a 90% efficient furnace.

Hinrichs and Kleinbach[8] report that reducing the thermostat setting from a constant 72°F to 68°F during the day and 55°F at night can reduce heating fuel consumption (costs) by 25% to 50% (Dallas versus Minneapolis). An extension of this approach to systems that use both heat pumps and furnaces can provide further reductions in natural gas use.

Temperatures often fluctuate by 20°F or more between day and night, and this can translate to the COP increasing from 1.8 (20°F) to 2.7 (40°F). The combination of the thermal mass of the house (i.e., 13°F decrease in temperature during the night) and effective insulation can reduce periods of inefficient heat pump operation.

Programming of thermostats to take advantage of fluctuations in temperatures between day and night can also extend the useful range of heat pumps to more northerly regions. By heating the house during the day (when the outside temperature is 20°F) rather than at night (when the outside temperature may be 0°F) the heat pump goes from an unacceptable COP of less than 1.5 to a COP of 1.7 or higher.

The use of lower nighttime thermostat settings with heat pumps is a cost-effective approach to reduce heating costs while converting to more sustainable energy options. For new construction or when replacing an air conditioning unit, the savings start at once and extend into the future.

The downside of this approach is that heat must be built up in the thermal mass of the house during the warmest part of the day (after about 10:00 A.M.), which implies that the house will be cool during the morning hours. To avoid taking a shower in a 55°F bathroom, resistance heaters can be used to heat the bathrooms without significant compromise of the savings. Alternatively, phase change materials in the bathrooms could substantially reduce temperature fluctuations in these rooms—a phase change material approach that is not cost effective for the entire house may be cost effective on a room-by-room basis.

Ground Source Heat Pumps

As indicated by the performance curve in Figure 9-10, heat pumps are particularly advantageous at warmer evaporator temperatures. Warmer evaporator temperatures can be accomplished even in northern states using ground source heat pumps.

Ground source heat pumps employ ground, groundwater, or surface water as the heat sink rather than outside air. Water circulation from the ground source over the evaporator coils can be from open or closed water cycles. The use of a closed cycle ground system is more widely used and debatably has the least impact on the environment. Only the use of the ground system will be discussed.

Figure 9-11 compares a closed ground source heat pump to the traditional air heat pump. By placing the network of piping below the frost line of the ground, a water supply at 32°F or higher

Electrical Power as Sustainable Energy 243

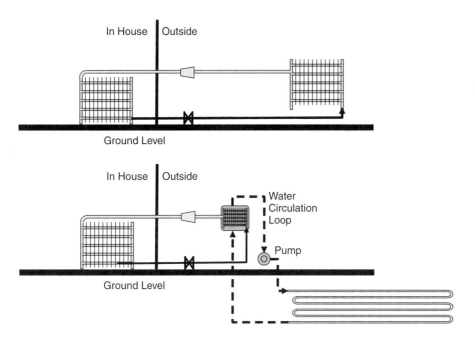

FIGURE 9-11. Comparison of heat pump using air heat sink to ground source unit.

can typically be supplied to the evaporator throughout the winter, even with sustained outside air temperatures below 0°F.

Properly designed ground source heat pumps eliminate the need for backup furnaces and increase the efficiency of both heating and air conditioning. The major drawback of ground source heat pumps is the cost. A typical cost for a conventional heat pump is about $1,000 per ton of air conditioning capacity ($2,000 for a 2,000-square-foot home—will vary by contractor and location), the cost for a comparable ground source system is about $3,000 per ton of capacity.[i] Polyethylene U-tube pipes are often used for heat exchange with the ground. Horizontal systems (as illustrated in Figure 9-11) are typically less costly than vertical systems (typically 150–250 feet deep) but require larger land area for the layout.

The payback period for most ground source systems is around 15 years or longer. The high costs of ground source heat units tend

---

[i] This is based on DOE EERE example for installation at Fort Polk, Louisiana, where 6,600 tons of cooling capacity for 4,000 homes were supplied at a cost of $19 million.

to limit their use in the Midwest to federal buildings and schools large enough to negotiate reduced electrical costs during the heating season (e.g., programs that avoid use of peaking electrical power in the summer) and qualify for federal programs that cover much of the up-front costs.

In the Great Lakes region, open cycle lake water (or groundwater effectively connected to the lakes) is available and can reduce the costs of ground source units. The combination of lower costs (lake water) and high heating requirements (northern states) makes a ground source more common in the Great Lakes region.

Hybrid Heat Pump Systems

The commercialization of HEV vehicles is successful because the use of engines and batteries to propel vehicles has advantages beyond the use of either engines or batteries alone. The combined systems tend to have smaller engines and battery packs than the conventional vehicles or BEVs. Similar opportunities exist in the heat pump market.

Could a $1,500-per-ton heat pump that uses both above- and below-ground heat sources be more cost effective than the nonhybrid unit? In the wide range of climates, the answer depends on location.

In summary:

- Heat pumps are in use today with applications ranging from moderate heating demands in southern states to intensive ground source units in the North.
- With increasing natural gas and heating oil prices, larger regions of the United States will benefit from the lower relative annualized costs using heat pumps.
- There are regions where a heat pump can be used in combination with furnaces to provide space heating at a lower cost than furnaces alone—especially where air conditioners are already used (the heat pump option on the air conditioner system costs little more than the air conditioner system alone).
- Programming thermostats to operate during the warmer daytime hours can extend the use of heat pumps for consumers willing to put up with cooler houses during the morning hours. This works best if there is a demand for electric power at night (e.g., charging of PHEVs) without producing new peak demands for electricity.

- Hybrid heat pump systems have the potential to improve the heat pump economics.

Switching from fossil fuels to electrical power can bring sustainability and increased cash flows in local economies. The PHEV and heat pump technologies are available and can give consumers savings.

Increased Use of Electrical Power for Hot Water Heating

Energy used to heat water can be a larger fraction of energy demand in southern states and a small fraction of space heating energy in northern states. Hot water heating can often be the highest or second-highest nonelectrical energy cost in residences and commercial buildings.

Heat pumps are capable of efficiently pumping heat from ambient (80°F) temperatures to 160°–180°F hot water, but such units are not available. Water often enters a house at temperatures lower than 70°F. Even heating to an intermediate temperature of 120°F with a heat pump might reduce fossil fuel heating of water by half.

Heating water with heat from the condensing coil of an air conditioner will require a second heat exchanger. When the water heater temperature reaches 120°–130°F, the air conditioner efficiency drops. The conventional outside air heat exchanger must also be in place to help the air conditioner run when the water is hot. The cost of buying and maintaining two heat exchangers will not be justified in northern state climates where there are 30 days or less when air conditioners operate.

Topics of National Attention

The emphasis of this chapter has been on replacing fossil fuels with electricity for transportation, space heating, and hot water heaters. These transformations would allow nuclear power to provide a sustainable energy future for the United States as well as the rest of the world.

A lack of emphasis on other topics does not imply lack of merit for other approaches. Insulation, good construction practices, effective use of vegetation/trees, and passive solar heating should all be used to promote sustainable energy practices and all will be justified on the basis of cost effectiveness.

Example Calculations

Automobile Cruising kWh Calculation

Estimate the maximum sustained energy requirement of an automobile cruising at 80 mpg with a fuel economy of 38 mpg. Assume 30% of the energy in the fuel is delivered to the wheels and the gasoline has an energy density of 115,000 Btu/gallon. This 30% efficiency is referred to as the powertrain efficiency (also known as drivetrain efficiency); it includes the engine, transmission, driveshafts, differentials, and turning the wheels.

Solution

Part 1—Calculate the rate of fuel consumption in kW.

Fuel Consumption = 80 mph ÷ 38 mpg × 115,000 Btu/gallon

= 242,105 Btu/h

Conversions:

1 h/3,600 sec

1 kJ/0.9486 Btu

1 kW = 1 kJ/sec

Therefore:

Fuel Consumption = 242,105 Btu/h ÷ 3,600 sec/h

× 1 kJ/0.9486 Btu × 1 kW/(kJ/sec)

= 70.9 kW

Part 2—Assume that 30% of the fuel's energy makes it to the wheels.

Power to Wheels = 70.9 kW × 0.30 Wheel:Fuel Power

= 21.3 kW

Alternative Automobile Cruising kWh Calculation

For the previous example, repeat the estimate for 30 mpg, and at 30 mpg, estimate the energy economy in kWh wheel energy requirement per mile. Assume 30% of the fuel's energy is delivered to the wheels.

Solution

Power to Wheels = 21.3 kW × 38/30 (mpg/mpg)

= 27 kW

Conversions:

0.002778 kWh/9.486 Btu

Energy Economy = 115,000 Btu/gallon ÷ 30 mpg

× 0.002778 kWh/9.486 Btu

× 0.30 kWh to wheel/kWh in fuel

= 0.337 kWh/mile

Battery Pack Sizing Calculation

For the previous example, estimate the size of a battery pack for a 60-mile range and compare this to the numbers in this chapter.

Solution

Assume the battery is specified in delivered power and the electric motor is 90% efficient:

Ideal Motor Battery Pack = 0.337 kWh/mile × 60 miles

= 20.2 kWh

Batter Pack (90% motor) = 20.2 kWh ÷ 0.9 Wheel:Battery Power

= 22.45 kWh

Comparison: Table 9-2 estimates 19.2 kWh for a mid-sized PHEV-60. Since a PHEV is both an HEV and a PHEV, it is reasonable that 30 mpg is consistent with a large sedan rather than a mid-sized sedan. Thirty mpg is low for a mid-sized HEV sedan. In view of this, the calculation agrees with the sizing reported in Table 9-2.

Recommended Reading

1. "Fuel Cell Report to Congress." Report ESECS EE-1973, U.S. Department of Energy, February 2003. Available at http://www.eere.energy.gov/hydrogenandfuelcells/pdfs/fc_report_congress_feb2003.pdf.
2. Suppes, G. J., "Plug-In HEV Roadmap to Hydrogen Economy." SAE Paper Number 2005-01-3830, 2005.

References

1. "Reducing California's Energy Dependence." Joint Agency Report of the California Energy Commission and California Resources Board, Report P600-03-005F, August 2003.
2. A. Frank, 30 Years of HEV Research Leading to Plug-In HEVs. PHEV Workshop, 2003.
3. M. Duvall, "Advanced Batteries for Electric Drive Vehicles." EPRI Report 1009299, May 2004.
4. G. J. Suppes, "Plug-In Hybrid Electric Vehicle Roadmap to Hydrogen Economy." SAE Paper 2005-01-3830.
5. G. J. Suppes, "Plug-In Hybrid with Fuel Cell Battery Charger." *Journal of Hydrogen Energy*, 30, 2005, pp. 113–121.
6. M. Anderman, "Brief Assessment of Progress in EV Battery Technology Since the BTAP." June 2000 Report. Presentation available at www.arb.ca.gov/msprog/zevprog/2003rule/03board/anderman.pdf.
7. "Fuel Cell Report to Congress." Report ESECS EE-1973, U.S. Department of Energy, February 2003.
8. R. A. Hinrichs, and M. Kleinbach, *Energy—Its Use and the Environment*, 3rd ed. New York: Brooks/Cole, 2002.

CHAPTER 10

Atomic Processes

Energies of Nuclear Processes

The huge quantities of energy liberated in nuclear power plants originate from the nuclei of atoms. In the fission process, relatively stable nuclei are induced into excited states that release energy as they form new stable nuclei. The heat produced in the nuclear reactor is converted to work through a heat engine power cycle. (See box for summary of terms used in this chapter.)

Atomic Nature of Matter—Terms

Here is a short summary of the terms used in the physical theory of matter to describe the nature of chemical compounds and the processes involving the nucleus of atoms that occur in a nuclear reactor.

Atoms consist of three basic subatomic particles. These particles are the proton, the neutron, and the electron.

Protons are particles that have a positive charge and a mass about the same as the mass of a hydrogen atom. Protons exist in the nucleus of an atom. The nucleus of the hydrogen atoms is one proton, and the mass of the hydrogen atom defines the atomic mass unit (amu) used in nuclear calculations.

Neutrons are particles that have no electrical charge and have a mass about the same as a hydrogen atom (approximately 1 amu). Neutrons exist in the nucleus of an atom.

Electrons are particles with a negative charge and have a mass about 1/1,837 of the mass of a hydrogen atom. Each electron exists in a well-defined, unique orbital shell around the nucleus of an atom.

The **atomic number** of an atom is the number of protons in the nucleus.

Nuclides are atoms that contain a particular number of protons and neutrons.

Isotopes are nuclides that have the same atomic number of protons (and electrons, therefore the same chemical properties) but differ in the number of neutrons.

The **mass number** of an atom is the total mass number of nucleons (protons and neutrons) in the nucleus.

The **stability** of a nucleus is determined by the different forces interacting within it. There is the long-range repulsive electrostatic force that acts between the protons and is very strong at the close distances in the nucleus. The gravitational force between the nucleons in the nucleus is negligible. The nuclear force is a short-range, strong attractive force, independent of charge, that acts between all of the nucleons holding the nucleus together.

The **radius** of a nucleus ranges from 1.25 to 7.74×10^{-13} cm (for hydrogen and uranium-238). The average diameter of an atom, except for a few very light atoms, is about 2×10^{-8} cm, making the atom more than 25,000 times as large as the nucleus. The nucleus is very small and very dense and contains nearly all of the mass of the atom.

Radioactive Nuclides. Atoms that disintegrate by the emission of a particle of electromagnetic radiation, most commonly an alpha or beta particle, or gamma radiation. There are three classes of radio nuclides:

1. Primary—with half-lives greater than 10^8 years. These may be alpha or beta emitters.
2. Secondary—formed by the radioactive transformation of uranium-235, uranium-238, or thorium-232.
3. Induced—have geologically short half-lifetimes and are formed by induced nuclear reactions. All of these reactions result in transmutation with a new (radioactive or non-radioactive) nuclide formed.

Nuclear Fission is a nuclear reaction that splits the atom nucleus, forming two new atoms each with about half of the original mass. There is the emission of a great quantity of energy, since the mass of the new atoms is slightly less than the parent atom, the mass loss is converted to energy by the Einstein equation.

The natural stability of atoms is characterized by their half-lives. The half-life of U-238 is 4.5×10^9 years. If a 1-pound meteor of pure U-238 were flying through outer space today, in 4.5 billion years that meteor would have a total mass slightly less than 1 pound—one-half would be U-238, and the other half would be fission products, mostly lead. Since Earth is about 5 billion years old, the U-238 present today is about half of that present when the Earth was formed.

While the atomic stability is typically discussed in relative rather than absolute terms, atoms with half-lives greater than 4.5 billion years are generally recognized as stable. Lead (Pb-206) is stable; the change in concentration of a 1-pound lead meteor would be negligible over a 4.5-billion-year period.

The decay of U-238 is a natural process (specifically, an α-decay process).[i] An unnatural decay of a nucleus by nuclear fission can be induced by collision with a neutron. A nuclear reactor environment is designed to sustain a critical concentration of free neutrons that gives a constant and controlled source of heat from induced fission. U-235 is the isotope that provides most of the reactor energy. Figure 10-1 illustrates the overall process by which U-235 releases heat through neutron-induced fission.

The energies released from the excited nuclei are not commonly observed in nature. However, analogous electron processes are often observed. For example, an incandescent lightbulb operates on the principle of using electrical power to increase the energy of the metal filament (high temperature). Some of this energy

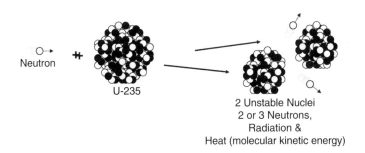

FIGURE 10-1. A neutron-induced fission of U-235.

---

[i] The α-decay half-life for U-238 is 4.5×10^9 years. The fission decay half-life is 8.0×10^{15} years.

TABLE 10-1
Examples of different emissions from nuclei and electrons.

Type of Emission	Source	Energy Level
Nuclear		
Beta	Atomic decay in nuclear reactor	Disintegration energy of S-38: 2.94 MeV
Alpha	Atomic decay in nuclear reactor	—
Neutron	Atomic decay in nuclear reactor	Fission release: ~2 MeV Fast neutron: >1 MeV Thermal neutron: 0.025 eV
γ-Ray	Nuclear transition from excited state to lower-energy state	Relaxing of excited states of Ni-60: 1.174, 2.158, and 1.332 MeV U-234 decay to Th-230: 0.068 MeV
Electron		
X-ray		Typically from 5–100,000 eV

produces excited states of the electrons that surround the metal nuclei. These *excited* electron states emit visible radiation (light) as they pass to lower energy, more stable states, or *ground* states.

Table 10-1 provides several example emissions that occur when electrons and nuclei go from *excited* states to *ground* states (referred to as stable states for nuclei). Electrons and nuclei have multiple excited states and one or two stable/ground states. The energies are reported in electron volts; one electron volt is equivalent to 1.602×10^{-19} joules or 1.18×10^{-19} foot-pounds.

Among the lowest energy emissions from electrons is visible light resulting from electricity flowing through an incandescent lightbulb. Among the highest energy emissions are neutrons emitted as part of atomic decay (2,000,000 eV). As illustrated by the comparison of Table 10-2, the energies associated with atomic processes are much larger than with electron processes. Atomic processes tend to be useful in large power plants, while electronic processes tend to have applications in homes and offices.

TABLE 10-2
Examples of energy levels in electron volts for different processes.

Other Processes	
U-235 fission to Rb-93 + Cs-140	200 MeV
Ionization	Remove outer electron from lead: 7.38 eV
	Remove inner electron from lead: 88,000 eV
Mass defect	Mass defect of Li-7: 931.5 MeV
Binding energy	Binding energy of Li-7: 1784 MeV

Electron emissions tend to be photonic (light, energetic x-rays); nuclear emissions may be photonic (γ-rays) or have white particle mass (α, β, and neutron) as energy components.

Of the emissions in Table 10-1, only the neutrons can collide and combine with a nucleus—often leading to an unstable state of the nucleus. In some physics laboratories atomic accelerators are able to increase the energy of particles that collide to produce excited nuclei or new elements.

If a neutron loses enough energy through collisions, it will at sufficiently low energy become atomic hydrogen (one proton and one electron). Beta (β) particles become electrons, and alpha (α) particles become helium (He-4). These transitions are summarized in Table 10-3.

The U.S. Department of Energy publication, entitled *DOE Fundamentals Handbook, Nuclear Physics and Reactor Theory*,[1] provides a detailed summary of key aspects of nuclear physics

TABLE 10-3
While electronic emissions dissipate, nuclear emissions do not dissipate.

Type of Emission	Product of Emission
Nuclear	
Beta	Is a newly produced, excited electron
Alpha	Is a newly produced, excited helium molecule
Neutron	Is a newly produced, excited atomic hydrogen (but remains a neutron if incorporated into a nucleus)

including the following excerpts that describe nuclides, nuclear stability, and conventions for reporting the atomic information.

Chart of the Nuclides

A tabulated chart called the Chart of the Nuclides lists the stable and unstable nuclides in addition to pertinent information about each one. Figure 10-2 shows a small portion of a typical chart. This chart plots a box for each individual nuclide, with the number of protons (Z) on the vertical axis and the number of neutrons (N = A − Z) on the horizontal axis.

The completely gray squares indicate stable isotopes. Those in white squares are artificially radioactive, meaning that they are produced by artificial techniques and do not occur naturally. By consulting a complete chart, other types of isotopes can be found, such as naturally occurring radioactive types (but none are found in the region of the chart that is illustrated in Figure 10-2).

Located in the box on the far left of each horizontal row is general information about the element. The box contains the chemical symbol of the element in addition to the average atomic weight of

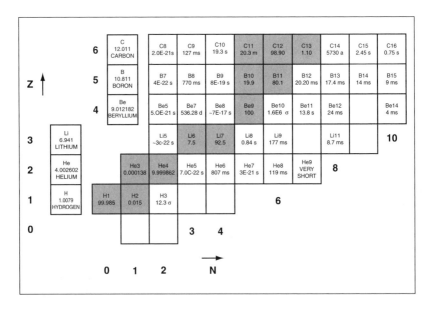

FIGURE 10-2. Excerpt from the Chart of the Nuclides.

Atomic Processes 255

the naturally occurring substance and the average thermal neutron absorption cross section, which will be discussed in a later module. The known isotopes (elements with the same atomic number Z but different mass number A) of each element are listed to the right.

Information for Stable Nuclides For the stable isotopes, in addition to the symbol and the atomic mass number, the percentage of each isotope in the naturally occurring element is listed, as well as the thermal neutron activation cross section and the mass in atomic mass units (amu). A typical block for a stable nuclide from the chart of the nuclides is shown in Figure 10-3.

Information for Unstable Nuclides For unstable isotopes the additional information includes the half-life, the mode of decay (for example, β^-, α), the total disintegration energy in MeV (million electron volts), and the mass in amu when available. A typical block for an unstable nuclide from the chart of the nuclides is shown in Figure 10-4.

Neutron-to-Proton Ratios Figure 10-5 shows the distribution of the stable nuclides plotted on the same axes as the Chart of the Nuclides. It provides the skeleton of the complete chart. As the mass numbers become higher, the ratio of neutrons to protons in the nucleus becomes larger. For helium-4 (2 protons and 2 neutrons) and oxygen-16 (8 protons and 8 neutrons) this ratio is unity. For indium-115 (49 protons and 66 neutrons) the ratio of neutrons to protons has increased to 1.35, and for uranium-238 (92 protons and 146 neutrons) the neutron-to-proton ratio is 1.59.

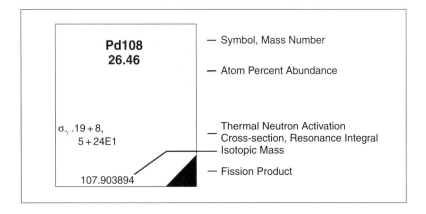

FIGURE 10-3. Presentation format for stable isotopes in chart of nuclides.

256 *Sustainable Nuclear Power*

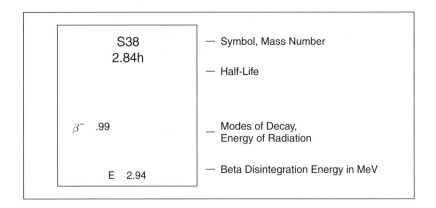

FIGURE 10-4. Presentation format for unstable isotopes in chart of nuclides.

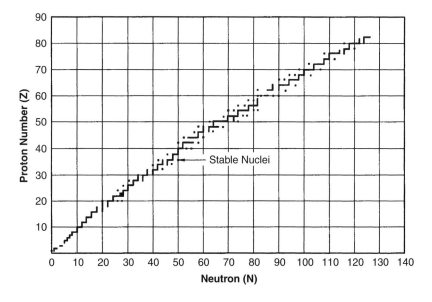

FIGURE 10-5. Skeleton of complete chart of nuclides illustrating stable nuclei.

If a heavy nucleus were to split into two fragments, each fragment would form a nucleus that would have approximately the same neutron-to-proton ratio as the heavy nucleus. This high neutron-to-proton ratio places the fragments below and to the right of the stability curve displayed by Figure 10-5. The instability caused by this excess of neutrons is generally rectified by successive **beta**

emissions, each of which converts a neutron to a proton and moves the nucleus toward a more stable neutron-to-proton ratio.

Careful measurements have shown that the mass of a particular atom is always slightly less than the sum of the masses of the individual neutrons, protons, and electrons of which the atom consists. The difference between the mass of the atom and the sum of the masses of its parts is called the **mass defect** (Δm).

The loss in mass, or mass defect, is due to the conversion of mass to binding energy when the nucleus is formed. Binding energy is defined as the amount of energy that must be supplied to a nucleus to completely separate its nuclear particles (nucleons). It can also be understood as the amount of energy that would be released if the nucleus was formed from the separate particles. **Binding energy** is the energy equivalent of the mass defect. Since the mass defect was converted to binding energy (BE) when the nucleus was formed, it is possible to calculate the binding energy using a conversion factor derived by the mass-energy relationship from Einstein's Theory of Relativity.

Energy Levels of Atoms   The electrons that circle the nucleus move in well-defined orbits. Some of these electrons are more tightly bound in the atom than others. For example, only 7.38 eV is required to remove the outermost electron from a lead atom, while 88,000 eV is required to remove the innermost electron. The process of removing an electron from an atom is called ionization, and the energy required to remove the electron is called the ionization energy.

In a neutral atom (number of electrons = Z), it is possible for the electrons to be in a variety of different orbits, each with a different energy level. The state of lowest energy is the one in which the atom is normally found and is called the ground state. When the atom possesses more energy than its ground state energy, it is said to be in an excited state.

An atom cannot stay in the excited state for an indefinite period of time. An excited atom will eventually transition to either a lower-energy excited state, or directly to its ground state, by emitting a discrete bundle of electromagnetic energy called an **x-ray**. The energy of the x-ray will be equal to the difference between the energy levels of the atom and will typically range from several eV to 100,000 eV in magnitude.

Energy Levels of the Nucleus   The nucleons in the nucleus of an atom, like the electrons that circle the nucleus, exist in shells that correspond to energy states. The energy shells of the nucleus are less defined and less understood than those of the electrons. There is a state of lowest energy (the ground state) and discrete possible excited states for a nucleus. Where the discrete energy states for the electrons of an atom are measured in eV or keV, the (k = 1,000) energy levels of the nucleus are considerably greater and typically measured in MeV (M = 1,000,000).

A nucleus that is in the excited state will not remain at that energy level for an indefinite period. Like the electrons in an excited atom, the nucleons in an excited nucleus will transition toward their lowest energy configuration and in doing so emit a discrete bundle of electromagnetic radiation called a gamma ray (γ-ray). The only differences between x-rays and γ-rays are their energy levels and whether they are emitted from the electron shell or from the nucleus. The ground state and the excited states of a nucleus can be depicted in a nuclear energy-level diagram. The nuclear energy-level diagram consists of a stack of horizontal bars, one bar for each of the excited states of the nucleus. The vertical distance between the bar representing an excited state and the bar representing the ground state is proportional to the energy level of the excited state with respect to the ground state. This difference in energy between the ground state and the excited state is called the excitation energy of the excited state. The ground state of a nuclide has zero excitation energy. The bars for the excited states are labeled with their respective energy levels. Figure 10-6 is the energy level diagram for nickel-60.

Stability of Nuclei As mass numbers become larger, the ratio of neutrons to protons in the nucleus becomes larger for the stable

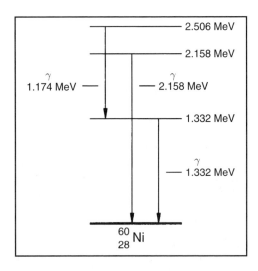

FIGURE 10-6. Energy level diagram for nickel-60.

nuclei. Nonstable nuclei may have an excess or deficiency of neutrons and undergo a transformation process known as beta (β) decay. Nonstable nuclei can also undergo a variety of other processes such as alpha (α) or neutron (n) decay. As a result of these decay processes, the final nucleus is in a more stable or more tightly bound configuration.

Natural Radioactivity In 1896, the French physicist Becquerel discovered that crystals of a uranium salt emitted rays that were similar to x-rays in that they were highly penetrating, could affect a photographic plate, and induced electrical conductivity in gases. Becquerel's discovery was followed in 1898 by the identification of two other radioactive elements, polonium and radium, by Pierre and Marie Curie.

Heavy elements, such as uranium or thorium, and their unstable decay chain elements, emit radiation in their naturally occurring state. Uranium and thorium, present since their creation at the beginning of geological time, have an extremely slow rate of decay. All naturally occurring nuclides with atomic numbers greater than 82 are radioactive.

Nuclear Decay

Table 10-4 provides examples of different types of nuclear transitions. These transitions can be from a highly unstable nucleus or from a relatively stable nucleus. A highly unstable nucleus has a short half-life (time in which the concentration of that isotope is reduced by 50% due to atomic transition), while stable molecules have long half-lives.

During and immediately after a nuclear reactor is shut down, radiation levels are very high due to the large number of radio nuclides with short half-lives. These short half-life nuclei rapidly undergo nuclear transitions. For a given nuclei, this process continues in a **decay chain** until a molecule with a stable or long-half-life nucleus is formed. The following decay of rubidium-91 to zirconium-91 illustrates a decay chain. The numbers under the arrows indicate half-lives in seconds, hours, days, and years.

$$^{91}_{37}\text{Rb} \xrightarrow[58.0 \text{ s}]{\beta^-} {}^{91}_{38}\text{Sr} \xrightarrow[9.5 \text{ hrs}]{\beta^-} {}^{91}_{39}\text{Y} \xrightarrow[58.5 \text{ d}]{\beta^-} {}^{91}_{40}\text{Zr}$$

These decay chains are important when treating fission products. The short-lived products will rapidly decay. If spent fuel is

TABLE 10-4
Example notations of nuclear processes.

Process	Formula	Description
Alpha Decay	$^{234}_{92}U \to {}^{230}_{90}Th + {}^{4}_{2}\alpha + \gamma + KE$	KE is kinetic energy of the α-particle is helium nucleus.
Beta Decay	$^{239}_{93}Np \to {}^{239}_{94}Pu + {}^{0}_{-1}\beta + {}^{0}_{0}\nu$	Neutron converted to proton. ν stands for neutrino—interacts little with atoms and escapes at speed of light.
Beta Decay	$^{13}_{7}N \to {}^{13}_{6}C + {}^{0}_{+1}\beta + {}^{0}_{0}\nu$	Proton converted to neutron through positron formation.
Electron Capture	$^{7}_{4}Be + {}^{0}_{-1}e \to {}^{7}_{3}Li + {}^{0}_{0}\nu$	Proton converted to neutron through electron capture.

DOE Fundamentals Handbook, Nuclear Physics and Reactor Theory, Vol. 1 of 2, U.S. Department of Energy, Washington, D.C., DOE-HDBK-1019/1-93, 1993.

stored for 30 years at the nuclear power plant, this will reduce the concentration of all nuclides with half-lives less than three years to less than 0.1% of the initial concentration.

Fortunately, the majority of the short-lived isotopes decay to stable nuclides. About 10% of the fission products remain as high-level waste after 30 years of storage; the remainder have decayed to stable nuclides.

Conditions for Successful Nuclear Fission

For nuclides to successfully undergo neutron-induced fission, a number of conditions must be met that are analogous to a chemical reaction. Table 10-4 summarizes and compares the factors that lead to fission with those conditions that promote chemical reactions.

Uranium and Other Fertile Materials

A nuclear reactor is designed to provide a flux of neutrons with the right energy to provide a constant and steady rate of nuclear fission. Each U-235 yields about 200 MeV per atom of uranium that fissions.

U-235 is referred to as a **fissile** material because U-235 will absorb a neutron with down to zero kinetic energy (referred to as

Atomic Processes 261

TABLE 10-5
Factors impacting the rate of nuclear fission versus analogous factors for chemical reaction.

Factor	Nuclide	Chemical Reagent
Materials must have a propensity to react	A low critical energy that corresponds to classifications as fissile or fissionable	A low activation energy
Materials must have ability to go to lower energy state	Products must have a higher binding energy	Products must have a lower Gibbs-free energy
Degree of molecular excitement should be optimal	The energy of the neutron must be correct—high (fast) or low (thermal) energy level may be optimal	Temperature must be high enough to react but low enough to stabilize the products
Events must be concentrated rather than disperse	Concentrations of reacting materials (e.g., U-235) must be high enough to sustain reaction but not so high to run away (explode)	High concentrations are needed for reasonable reactor size, or a solvent must be used to avoid runaway

thermal neutrons), and this results in fission. Table 10-5 summarizes the three types of materials that are of interest in nuclear fission fuels. Fissile atoms undergo fission because a neutron of zero kinetic energy can induce fission. Nuclides are **fissionable** if neutrons of zero or higher kinetic energies are able to induce fission.

When a neutron combines with a stable nucleus, a binding energy corresponding to that neutron addition is released. When that binding energy is greater than a **critical energy** (specific to the nuclide before addition of the neutron), the nuclide can undergo fission. Table 10-7 provides the binding energies (MeV/nucleon) and critical energies of the five **fissile** and **fissionable** materials. Th-232 and U-238 are fissionable but not fissile because higher-energy neutrons must bring sufficient kinetic energy so that the sum of kinetic and binding energies exceeds the critical energy.

When U-238 or Th-232 absorb neutrons and fission does **not** occur, they can undergo the decay chain summarized by

TABLE 10-6
Definitions and examples of nuclear fission fuels.

Material Category	Definition	Examples
Fissile	Nuclides for which fission is possible with neutrons of any energy level	U-235, U-233, and Pu-239
Fissionable	Nuclides for which fission is possible with neutron collision	U-235, U-233, Pu-239, U-238, and Th-232
Fertile	Materials that can absorb a neutron and become fissile materials	U-238 and Th-232.

TABLE 10-7
Critical energy versus energy released with absorption of additional neutron.

Target Nucleus	Critical Energy E_{crit}	Binding Energy of Last Neutron BE_n	$BE_n - E_{crit}$
Th-232	7.5 MeV	5.4 MeV	−2.1 MeV
U-238	7.0 MeV	5.5 MeV	−1.5 MeV
U-235	6.5 MeV	6.8 MeV	+0.3 MeV
U-233	6.0 MeV	7.0 MeV	+1.0 MeV
Pu-239	5.0 MeV	6.6 MeV	+1.6 MeV

DOE Fundamentals Handbook, Nuclear Physics and Reactor Theory, Vol. 1 of 2, U.S. Department of Energy, Washington, D.C., DOE-HDBK-1019/1-93, 1993.

Figure 10-7, resulting in formation of U-233 and Pu-239. U-238 and Th-232 are referred to as **fertile materials** because absorption of a neutron can produce a fissile material. Because of these nuclear processes, it is possible for a nuclear reactor to produce more fuel than is consumed. Reactors designed to do this are called **breeder reactors**. In light water **converter reactors** (also referred to as burner reactors), which consume more fuel than produced, about one-third of the energy produced is a result of Pu-239 production with subsequent fission. At the end of the nuclear fuel burn in a light-water reactor, about 0.9% Pu-239 is in the fuel. New fuel contains 3.4% U-235 (0% Pu-239).

Atomic Processes 263

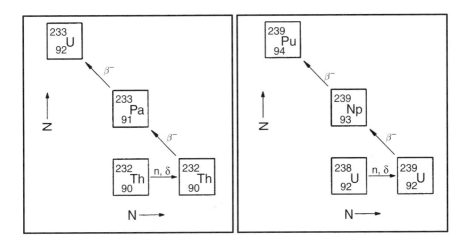

FIGURE 10-7. Decay tracks for fertile collisions with Th-232 and U-238.

The decay tracks of Figure 10-7 (absorption without fission) are broadly referred to as transmutation processes. Transmutation is important for converting fertile fuel to fissile fuel. The susceptibility of materials to transmutation is covered in the next section under the topic of absorption cross section.

Transmutation is important for creating fissile materials and for converting problem radioactive wastes into more benign materials. Not all nuclides in nuclear waste present the same degree of waste-handling problems. For example, nuclides with short half-lives (less than five years) can be stored until the radioactivity decays to benign levels. Wastes with very long half-lives tend to be less hazardous than the uranium mined to make the nuclear fuel. However, wastes with intermediate half-lives are more hazardous than natural ores and take too long to decay in 30–60 years of temporary storage. Transmutation can be used to transform some of these waste materials into new nuclides that can decay quickly or are stable. This topic is discussed in Chapter 13 from the perspective of sustainable management of spent fuel inventories.

Binding Energy Constraints

Available technology limits sustainable fission power to fissionable materials originating from natural uranium and thorium. For fission to occur, the nuclei produced from the nuclear transformation must have a higher binding energy than the nuclei undergoing fission. The higher binding energies represent more permanent nuclei. The most stable nuclei are more like iron which has the highest binding energies.

The binding energy trends in Figure 10-8 illustrate that nuclei with atomic weights greater than about 60 can undergo fission to produce nuclei more tightly bound. Nuclei with atomic weights less than about 60 can undergo fusion to produce more tightly bound nuclei.

The total energy release from the fission of U-235 is about 200 MeV. About 187 MeV of the energy is immediately released in the form of kinetic energy of the fission fragments, the fission neutrons, and γ-rays. The excited product nuclei will release the remaining 13 MeV in the form of kinetic energy of delayed beta particles and decay γ-rays. Tables 10-8 and 10-9 report average quantities of instantaneous and delayed energy release from U-235

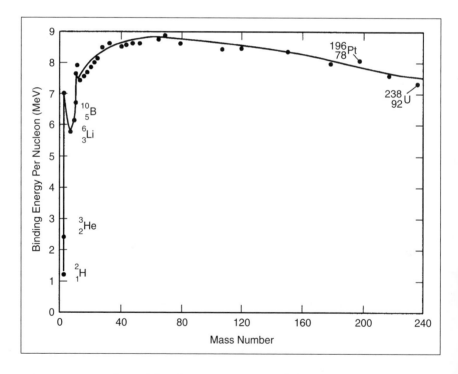

FIGURE 10-8. Plot of binding energies as function of mass number. Higher values reflect more stable compounds. The values are the binding energy per nuclei release of energy if free protons, neutrons, and electrons combine to form the most stable nuclei for that atomic number.

TABLE 10-8
Instantaneous energy from fission.

Kinetic Energy of Fission Products	167 MeV
Energy of Fission Neutrons	5 MeV
Instantaneous γ-Ray Energy	5 MeV
Capture γ-Ray Energy	10 MeV
Total	187 MeV

DOE Fundamentals Handbook, Nuclear Physics and Reactor Theory, Vol. 1 of 2, U.S. Department of Energy, Washington, D.C., DOE-HDBK-1019/1-93, 1993.

TABLE 10-9
Delayed energy from fission.

β-Particles from Fission Products	7 MeV
γ-Rays from Fission Products	6 MeV
Neutrinos	10 MeV
Total	23 MeV

DOE Fundamentals Handbook, Nuclear Physics and Reactor Theory, Vol. 1 of 2, U.S. Department of Energy, Washington, D.C., DOE-HDBK-1019/1-93, 1993.

fission by a thermal neutron. Ten MeV of energy from the neutrinos escape the reactor system.

Nuclear Cross Sections

Nuclear cross sections are tabulated for atoms and characterize the probability the nuclide will interact with a neutron. Different representative cross sections are reported for different types of interaction. While fissile, fissionable, and fertile classifications indicate what can happen if a neutron is absorbed by a nuclide, the cross sections represent the size of the target for neutron capture.

The cross sections are dependent on the energy of the neutron and the properties of the nuclide. These microscopic cross sections may be viewed as the area available for a neutron to hit the nucleus to induce reaction. A larger cross section provides an increased probability for reaction.

Table 10-10 provides example nuclear cross sections for U-235 and U-238. Cross sections (reported in barns, 1 barn = 10^{-24} cm^2)

TABLE 10-10
Example cross section areas.

Nuclide	Kinetic Energy of Neutron (eV)	Fission Cross Section (barns)	Capture Cross Section (barns)
U-235	0.5	50	7
U-235	1,000,000	2	0.15
U-238	0.5	0.6	N/A
U-238	1,000,000	0.1	0.02

Source: Y. Kadi, CERN. From R. L. Garwin and G. Charpak, *Megawatts and Megatons*. New York: Alfred A. Knopf, 2001.

for both fission and capture are provided. The thermal neutrons (<1eV) typically have cross sections 20 to 30 times larger than fast neutrons (1–2 MeV). It is the large cross section for U-235 and thermal neutrons that made it the fuel of choice for commercial nuclear reactors.

Fission and capture cross sections are two of the four cross sections that dominate nuclear reactor behavior. Table 10-11 illustrates these and includes elastic and inelastic cross sections.

The fission cross section for U-235 is about 50 barns for the thermal neutron versus 2 barns for the fast neutron. For U-235 (fissile), the fission cross section is greater than the capture cross section, fission will occur more often than capture.

Fertile nuclides like U-238 have a small fission cross section for low neutron energy levels. The fission cross section for U-238 is equal to the capture cross section at 1.3 MeV. Higher energy neutrons will tend to cause fission, while lower-energy neutrons will tend to cause transmutation, the pathway to forming plutonium.

Fission, capture, scatter, and total cross sections are a few of the different types of cross sections that are characterized. Figure 10-9 shows a typical plot of total nuclear cross section versus the energy level of the neutron. The complex nature of the free neutron interaction with nuclei goes beyond the scope of this text with much yet to be learned. Key points have been covered. Especially important is the distinction between thermal neutrons (<1 eV) and fast neutrons (typically >1 MeV). The thermal neutron is key in propagating reactions in the Generation II nuclear reactors including current commercial light water reactors. For fast-spectrum reactors (the Generation IV designs) fast neutrons are

Atomic Processes 267

TABLE 10-11
Prominent cross sections in nuclear reactors.

Process	Description
Fission	Neutron + Target Nucleus → 2 Unstable Nuclei
Transmutation	Neutron + Target Nucleus → Nucleus with Additional Neutron
Scattering (Elastic)	Neutron + Target Nucleus → Neutron Same Energy + Nucleus Same Energy
Scattering (Inelastic)	Neutron + Target Nucleus → Neutron Lower Energy + Excited Nucleus

key to the performance. Fast neutrons can directly induce fission in U-238 and fissile actinides.

Actinides are nuclides with atomic numbers between 89 and 104 (with an atomic number of 92, uranium is an actinide). Actinides such as plutonium (Pu), neptunium (Np), americium (Am), and curium (Cm) are formed in nuclear reactors (see Table 10-12). Once formed, they can keep absorbing thermal neutrons, slowly reducing the number of neutrons available to promote fission. Fast neutrons tend to produce fission. So, fast neutrons tend to cause actinides to fission releasing energy rather than inhibit fission by absorbing neutrons.

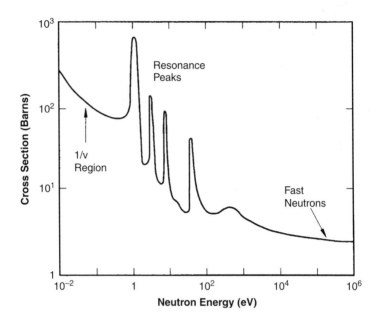

FIGURE 10-9. Typical neutron absorption cross section vs. neutron energy.

TABLE 10-12
Transuranics of primary interest to AFCI program including uranium as reference.

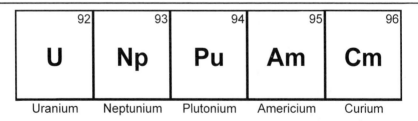

Why do transuranics matter?
- Transuranics affect repository performance by dominating long-term heat load and long-term radiotoxicity.
- Transuranics and enriched uranium are the only materials of concern for weapons proliferation.
- Transuranics can be destroyed while producing extra energy if recycled in (fast-spectrum) nuclear reactors.

Report to Congress Advanced Fuel Cycle Initiative. "Objectives, Approaches, and Technology Summary." Prepared by U.S. DOE, Office of Nuclear Energy, Science, and Technology, May 2005.

Fast-spectrum reactors are important for sustainable nuclear power. Fast-spectrum reactors eliminate the need to separate the actinides when reprocessing nuclear fuel. This should reduce the cost and promote sustainable economics. Using all the actinides as fuel removes them from the waste stream.

Concentrated Events

Fissile materials U-235 and Pu-239 meet the constraints of fission and release energy as they transform to smaller, more stable nuclei. The chain reaction is maintained by the neutron flux. The final components of the controlled release of the nuclear energy are the initiation of the neutron flux and maintaining of the neutron flux. The neutron flux is the number of neutrons passing through an area of one cm^2 per second. Since the neutrons tend to be moving through solids and liquids (being stopped or scattered only by the dense nuclei), the energy of the neutrons decreases (they slow down) as they travel through the reactor core. A discussion of materials for initiating the neutron flux is beyond the scope of this text. There are such materials that are used to start the reactor by initiating fission.

Sustaining the neutron flux is one of the most important criteria in nuclear reactor design. The neutron flux is depleted by neutron capture and by scattering out of the reactor core volume. The neutron poisons (e.g., boron) are used to maintain the neutron flux for constant energy production.

In a controlled reactor environment, the neutron flux achieves a steady-state consistent with the desired heat release. This is achieved by having the right concentration of U-235 or Pu-239 present, which is done by concentrating these in the fuel rods and properly spacing the fuel rods. The right fuel rod concentration is typically between 2.6% and 4.0% U-235 in a light-water reactor. Some of the proposed Generation IV designs may use concentrations up to 20%. The spacing of the fuel rods in the reactor and the fissile isotope concentration in the fuel provide controlled release of energy. Since the medium between the fuel rods changes the kinetic energy of the neutrons, it is necessary to match the medium with fuel composition and spacing.

Light-water reactors are designed for controlled delivery of thermal neutrons (<1 eV) to the fuel rods. Liquid water (as opposed to water vapor) between the fuel rods provides an average of 12 scattering collisions with water to produce the thermal neutrons that will successfully fission another U-235 nucleus.

If water is absent, the energy level of the neutrons is too high, and the lower nuclear fission cross section leads to fewer successful fissions and the neutrons escape from the reactor core. In light-water reactors this happens if water vapor is present between the fuel rods, and this will lead to a "passive" shutdown of the reactor. A flow of cooling water must be maintained to remove the decay heat from the fission products in the fuel.

In fast flux Generation IV reactors, the reactor fuel configuration and isotope concentrations are such that collisions of fast neutrons maintain the nuclear fission process. The presence of higher-energy neutrons allows use of most fissionable materials to propagate the nuclear fission process.

Fast neutron reactors bombard the actinides with fast neutrons and they undergo fission. In thermal flux reactors, the higher actinides accumulate and present a radioactive "waste" problem. The excitation and fission (energy release) of all actinides in Generation IV reactors represent an important step toward sustainable nuclear energy. This process reduces waste, makes chemical separations for fuel recycling easier, and allows total use of uranium fuel. This includes the stockpiles of depleted uranium left from producing military, highly enriched U-235 and from production of domestic nuclear fuel.

Transmutation

For every 100 kg of fuel introduced into a light-water reactor, about 3.4 kg of fission products are formed. Of these fission products, about 0.4 kg is high-level radioactive waste after about 30 years of storage at the nuclear power plant. This 0.4 kg can be placed in a repository as high-level waste for 1,000 years to become stable, or some of it could be transmuted. The transmutation of I-125 (see the box "What Is Transmutation?") is considered viable with existing methods. Iodine is about 0.1 kg of the high-level waste. Other techniques could be developed for the other 0.3 kg.

What Is Transmutation?

Transmutation transforms one atom into another by changing its nuclear structure. This is accomplished by bombarding the atoms of interest with neutrons either in an accelerator or a nuclear reactor. The nucleus of the radioactive atom absorbs a neutron, this new nucleus is stabilized by emitting energy (a beta particle) leaving a nucleus of a nonradioactive atom.

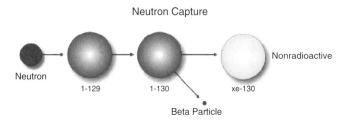

Source: Report to Congress Advanced Fuel Cycle Initiative: "The Future Path for Advanced Spent Fuel Treatment and Transmutation Research." Prepared by the U.S. DOE, January 2003.

While plutonium-based fuels have been manufactured on a commercial basis, almost no work has been done to create or irradiate fuels that contain neptunium, americium, or curium.

Transmutation fuels that can significantly destroy the higher actinides should be capable of very high burnups. This will minimize the number of reprocessing cycles required, and reduce material losses during the separation and refabrication processes. Recycled fuel would be fabricated in hot cells or other remote environment due to the gamma radiation from the fission products that remain in this fuel. If these advanced fuels are to be candidates for deployment with Generation IV–reactor systems, research, development, and testing would be needed beyond Phase II. AFCI (Advanced Fuel Cycle Initiative) Series Two would include effort to evaluate the various fuel types that might serve as fuel for fast-spectrum reactors or accelerator-driven transmutation systems.

Determination of the optimum fuel form for transmutation will be fabricated using remote handling technologies. Transmutation reactors are subcritical ensuring safe operation and leaves a final waste form acceptable for Yucca Mountain — a major research objective of the program.

Oxide, nitride, metallic, dispersion, ceramic, and coated particle fuel forms are currently under investigation. Fabrication of several test fuel specimens of these fuel forms containing plutonium mixed with minor actinides is underway. The department plans to irradiate these fuels in the ATR in Idaho in fiscal year 2003, with a more ambitious follow-on irradiation program to be carried out in France by European partners. A consortium of institutions is planning the construction of an experimental assembly containing minor actinide fuels that would come from several countries; this assembly would be irradiated in a French fast-spectrum reactor (PHENIX).

Successful testing in the ATR and initiation of the French PHENIX reactor tests during Phase II would permit DOE to select the most promising path forward for AFCI Series Two transmutation fuels. Planning for potential Phase III scaled-up fast-spectrum irradiations in foreign facilities will follow.

Fast-spectrum systems can be either fast reactors (which employ critical reactor cores that operate 12 to 18 months between refueling cycles) or accelerator-driven systems that employ reactor cores that are subcritical (i.e., they need a constant source of neutrons to maintain fuel fission). The external source of neutrons is produced by a particle accelerator and a target system. Both designs employ fast neutrons. The accelerator system has the advantage that it can transmute all radioactive elements without producing plutonium.

Accelerator systems are more expensive than fast reactors and require significantly more research and development. The fuel technology is basically the same.

Based on the systems analysis carried out in Phase I of this research, it does not appear accelerator transmutation systems will be used as the primary transmuters of the long-lived toxic materials in spent fuel. The important role is assuring the very low levels of radio toxicity of the fission product waste, the goal of this activity. The high construction and operating costs of accelerator-based systems make them unsatisfactory for wide application as commercial-scale transmuters. Fast reactor systems should prove sufficiently economic to justify their deployment. This is a key element of evaluation in the multinational Generation IV Nuclear Energy Systems Initiative.

Accelerator-Driven Systems Physics and Materials Research and Development

Many countries are considering ADSs (Accelerator Driven Systems) as a viable approach to transmutation because these systems may be capable of destroying long-lived radioactive isotopes of all types without making plutonium. An accelerator-driven system consists of an accelerator that produces high-energy protons that strike a heavy metal target to produce high-energy (fast) neutrons through a spallation process to drive a subcritical reactor assembly.

Accelerator-driven transmutation (see the box "Making New Molecules in the Lab") has been an important part of nuclear physics research for decades. This is the method used to produce the high atomic number elements that do not occur in nature.

Making New Molecules in the Lab

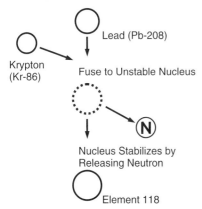

On small scales, scientists are able to make new molecules similar to the manner in which supernovas combine two smaller molecules to form a larger molecule. As shown in the figure, krypton and lead combine to form a compound nucleus. This nucleus is very unstable and rapidly degrades to a more stable atom indicated as Element 118.[2]

This work was performed at the Lawrence Berkeley National Lab. During this laboratory synthesis, two atoms are joined to actually form a less stable molecule. This is possible in a particle accelerator that puts kinetic energy (high speed) into the krypton. This high speed provides the extra energy needed to fuse the nuclei. The atomic rearrangement resulting in the release of a neutron helps to lock in a final element that is stable.

Nuclear Fusion

It is difficult to predict what energy options will be available beyond 30 years. Nuclear fusion, which is the primary source of energy in our corner of the universe (see the box "From Hydrogen to Helium"), may be an option.

The potential benefits of controlled fusion are great. Successful development of a fusion power plant is proving to be a most difficult scientific and technological challenge. Although progress has been steady, it may be 50 years before a demonstration fusion reactor that produces energy is built.

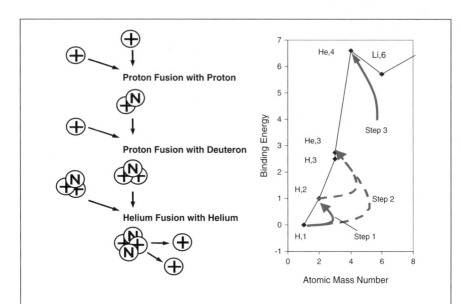

From Hydrogen to Helium

The most abundant atom in the universe is hydrogen. Hydrogen is the fuel of the stars. The diagram[3] illustrates how four protons interact to produce a new stable molecule: helium. Considerable energy is released in the process—about 25 MeV of energy for every helium atom formed (570 million Btus per gram helium formed).

Source: D. Morgan, *Magnetic Fusion: The DOE Fusion Energy Sciences Program*. CRS Issue Brief for Congress, Order Code IB91029, January 15, 2002.

Fusion occurs when the nuclei of light atoms, such as isotopes of the element hydrogen (deuterium and tritium), collide with sufficient energy to overcome the natural repulsive forces that exist between such nuclei (see Figure 10-10). When this collision takes place, a D-T reaction is said to have occurred. If the two nuclei fuse, a heavier element, a form of helium (an alpha particle), is created, along with release of a large quantity of energy. For the fusion reaction to take place, the nuclei must be heated to a very high temperature. In a hydrogen bomb, this is done by exploding a fission bomb, uranium or plutonium, forcing the deuterium and

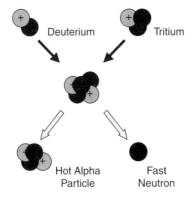

FIGURE 10-10. Laboratory fusion.

tritium together at very high temperature and pressure. The fusion reactions release the energy producing the hydrogen bomb explosion. A fusion power reactor must "tame" this energy source so the energy can be collected.

Fusion reactions are possible between a number of light atoms, including deuterium alone (a D-D reaction); deuterium and helium-3, an isotope of the element helium (a D-3He reaction); and hydrogen and the element lithium, a light metal. All of these reactions occur much less frequently at a given temperature than the D-T reaction. For instance, the fusion energy produced from D-T reactions in a mixture of deuterium and tritium will be about 300 times greater than that from D-D reactions. For this reason, research into controlled fusion has concentrated on developing deuterium-tritium-fueled reactors.

The potential benefits of fusion energy are great and international financial support for fusion energy research should continue. The sun is the nearest operating fusion reactor. It is impossible to duplicate the conditions that exist there in an earthbound fusion reactor that allows the collection of the energy. It will take an apparatus that can bring the hydrogen isotopes together, allow fusion to occur, and a way to harvest the energy released by the reaction. Since experts say it may take 50 years to reach this goal, it is beyond the scope of this book.

Radiological Toxicology

The radioactivity of uranium ore is often considered a threshold level of acceptable radiation. In practice, a concentrated uranium ingot can be handled with little concern of radioactive toxicology. Handling fuel pins need not be performed remotely when preparing fuel for nuclear reactors (one of the downsides of advanced reprocessing is the fuel will be radioactive and require remote handling).

A brief introduction to radiation poisoning is needed to understand the risks of radiation and methods for reducing risks. Both the U.S. Environmental Protection Agency and U.S. Nuclear Regulatory Commission (NRC) have websites that detail how one is exposed to radiation poisoning and the impact of that exposure.

The following is an EPA summary on sources of radiation and radiation poisoning.[4]

What Is Radiation? Radiation is energy that travels in the form of waves or high speed particles. When we hear the word *radiation*, we generally think of nuclear power plants, nuclear weapons, or radiation treatments for cancer. We would also be correct to add microwaves, radar, electrical power lines, cell phones, and sunshine to the list. There are many different types of radiation that have a range of energy forming an electromagnetic spectrum. However, when you see the word *radiation* on this website, we are referring to the types of radiation used in nuclear power, nuclear weapons, and medicine. These types of radiation have enough energy to break chemical bonds in molecules or remove tightly bound electrons from atoms, thus creating charged molecules or atoms (ions). These types of radiation are referred to as ionizing radiation.

What's the Difference Between Radiation and Radioactivity? Radiation is the energy that is released as particles or rays during radioactive decay. Radioactivity is the property of an atom that describes spontaneous changes in its nucleus that create a different element. These changes usually happen as emissions of alpha or beta particles and often gamma rays. The rate of emission is referred to as a material's "activity."

Each occurrence of a nucleus throwing off particles or energy is referred to as a "disintegration." The number of disintegrations per unit time (minutes, seconds, or hours) is called the activity of a sample. Activity is expressed in curies. One curie equals 37 billion disintegrations per second. (Since each disintegration transforms the atom to a new nuclide, transformation is often substituted for disintegration in talking about radioactive decay and activity.)

Exposure from radiation can occur by direct exposure, inhalation, and ingestion.

Direct (External) Exposure The concern about exposure to different kinds of radiation varies:

- Limited concern about alpha particles. They cannot penetrate the outer layer of skin, but if you have any open wounds, you may be at risk.
- Greater concern about beta particles. They can burn the skin in some cases or damage eyes.
- Greatest concern is about gamma radiation. Different radionuclides emit gamma rays of different strength, but gamma rays can travel long distances and penetrate entirely through the body.

Gamma rays can be slowed by dense material (shielding), such as lead, and can be stopped if the material is thick enough. Examples of shielding are containers; protective clothing, such as a lead apron; and soil covering buried radioactive materials.

Inhalation Exposure by the inhalation pathway occurs when people breathe radioactive materials into the lungs. The chief concerns are radioactively contaminated dust, smoke, or gaseous radionuclides such as radon.

Radioactive particles can lodge in the lungs and remain for a long time. As long as it remains and continues to decay, the exposure continues. For radionuclides that decay slowly, the exposure continues over a very long time. Inhalation is of most concern for radionuclides that are alpha or beta particle emitters. Alpha and beta particles can transfer large amounts of energy to surrounding tissue, damaging DNA or other cellular material. This damage can eventually lead to cancer or other diseases and mutations.

Ingestion Exposure by the ingestion pathway occurs when someone swallows radioactive materials. Alpha- and beta-emitting radionuclides are of most concern for ingested radioactive materials. They release large amounts of energy directly to tissue, causing DNA and other cell damage.

Ingested radionuclides can expose the entire digestive system. Some radionuclides can also be absorbed and expose the kidneys and other organs, as well as the bones. Radionuclides that are eliminated by the body fairly quickly are of limited concern. These radionuclides have a short biological half-life.

Minimizing direct exposure to radiation is achieved by shielding and distance between the radiation-emitting object and the person and by reducing the time in the presence of the radiation-emitting object. Minimizing of inhalation and indigestion is achieved by keeping radioactive isotopes out of the environment. Once radiation is in the environment, the materials can be removed or isolated so the isotopes do not get into water, air, or vegetation.

The following NRC summary describes health effects on radiation exposure.[5]

Biological Effects of Radiation We tend to think of biological effects of radiation in terms of their effect on living cells. For low levels of radiation exposure, the biological effects are so small that they may not be detected. The body has repair mechanisms against damage induced by radiation as well as by chemical carcinogens. Consequently, biological effects of radiation on living cells may result in three outcomes: (1) injured or damaged cells repair themselves, resulting in no residual damage; (2) cells die, much like millions of body cells do every day, being replaced through normal biological processes; or (3) cells incorrectly repair themselves, resulting in a biophysical change.

The associations between radiation exposure and the development of cancer are mostly based on populations exposed to relatively high levels of ionizing radiation (e.g., Japanese atomic bomb survivors and recipients of selected diagnostic or therapeutic medical procedures). Cancers associated with high-dose exposure (greater than 50,000 mrem) include leukemia, breast, bladder, colon, liver, lung, esophagus, ovarian, multiple myeloma, and stomach cancers. Department of Health and Human Services literature also suggests a possible association between ionizing radiation exposure and prostate, nasal cavity/sinuses, pharyngeal and laryngeal, and pancreatic cancer.

The period of time between radiation exposure and the detection of cancer is known as the latent period and can be many years. Those cancers that may develop as a result of radiation exposure are indistinguishable from those that occur naturally or as a result of exposure to other chemical carcinogens. Furthermore, National Cancer Institute literature indicates that other chemical and physical hazards and lifestyle factors (e.g., smoking, alcohol consumption, and diet) significantly contribute to many of these same diseases.

Although radiation may cause cancers at high doses and high dose rates, currently there are no data to unequivocally establish the occurrence of cancer following exposure to low doses and dose rates—below about 10,000 mrem (100 mSv). Those people living in areas having high levels of background radiation—above

1,000 mrem (10 mSv) per year—such as Denver, Colorado, have shown no adverse biological effects.

Even so, the radiation protection community conservatively assumes that any amount of radiation may pose some risk for causing cancer and hereditary effect and that the risk is higher for higher radiation exposures. A linear, no-threshold (LNT) dose response relationship is used to describe the relationship between radiation dose and the occurrence of cancer. This dose-response model suggests that any increase in dose, no matter how small, results in an incremental increase in risk. The LNT hypothesis is accepted by the NRC as a conservative model for determining radiation dose standards, recognizing that the model may overestimate radiation risk.

High radiation doses tend to kill cells, while low doses tend to damage or alter the genetic code (DNA) of irradiated cells. High doses can kill so many cells that tissues and organs are damaged immediately. This in turn may cause a rapid body response often called Acute Radiation Syndrome. The higher the radiation dose, the sooner the effects of radiation will appear and the higher the probability of death. This syndrome was observed in many atomic bomb survivors in 1945 and emergency workers responding to the 1986 Chernobyl nuclear power plant accident. Approximately 134 plant workers and firefighters battling the fire at the Chernobyl power plant received high radiation doses—80,000 to 1,600,000 mrem (800 to 16,000 mSv)—and suffered from acute radiation sickness. Of these, 28 died within the first three months from their radiation injuries. Two more patients died during the first days as a result of combined injuries from the fire and radiation.

Because radiation affects different people in different ways, it is not possible to indicate what dose is needed to be fatal. However, it is believed that 50% of a population would die within 30 days after receiving a dose to the whole body, over a period ranging from a few minutes to a few hours, between 350,000 to 500,000 mrem (3,500 to 5,000 mSv). This would vary depending on the health of the individuals before the exposure and the medical care received after the exposure. These doses expose the whole body to radiation in a very short period of time (minutes to hours). Similar exposure of only parts of the body will likely lead to more localized effects, such as skin burns.

Conversely, low doses—less than 10,000 mrem (100 mSv)— spread out over long periods of time (years to decades) don't cause an immediate problem to any body organ. The effects of low doses of radiation, if any, would occur at the level of the cell, and thus changes may not be observed for many years (usually 5–20 years) after exposure.

Genetic effects and the development of cancer are the primary health concerns attributed to radiation exposure. The likelihood of cancer occurring after radiation exposure is about five times greater than a genetic effect (e.g., increased stillbirths, congenital abnormalities, infant mortality, childhood mortality, and decreased birth weight). Genetic effects are the result of a mutation produced in the reproductive cells of an exposed individual that are passed on to their offspring. These effects may appear in the exposed person's direct offspring or may appear several generations later, depending on whether the altered genes are dominant or recessive.

Although radiation-induced genetic effects have been observed in laboratory animals (given very high doses of radiation), no evidence of genetic effects has been observed among the children born to atomic bomb survivors from Hiroshima and Nagasaki.

Example Calculations

Example calculations are from *DOE Fundamentals Handbook: Nuclear Physics and Reactor Theory*, Volume 1 of 2.

Calculation of Mass Defect The mass defect can be calculated using Equation 10-1. In calculating the mass defect, it is important to use the full accuracy of mass measurements because the difference in mass is small compared to the mass of the atom. Rounding off the masses of atoms and particles to three or four significant digits prior to the calculation will result in a calculated mass defect of zero.

$$\Delta m = [Z(m_p + m_e) + (A - Z)m_n] - m_{atom} \qquad (10\text{-}1)$$

where:

Δm = mass defect (amu)
m_p = mass of a proton (1.007277 amu)
m_n = mass of a neutron (1.008665 amu)
m_e = mass of an electron (0.000548597 amu)
m_{atom} = mass of nuclide $^A_Z X$ (amu)
Z = atomic number (number of protons)
A = mass number (number of nucleons)

Example: Calculate the mass defect for lithium-7. The mass of lithium-7 is 7.016003 amu.
Solution:

$$\Delta m = [Z(m_p + m_e) + (A - Z)m_n] - m_{atom}$$
$$\Delta m = [3(1.007826\,\text{amu}) + (7 - 3)1.008665\,\text{amu}] - 7.016003\,\text{amu}$$
$$\Delta m = 0.0421335\,\text{amu}$$

Calculation of Binding Energy Since the mass defect was converted to binding energy (BE) when the nucleus was formed, it is possible to calculate the binding energy using a conversion factor derived by the mass-energy relationship from Einstein's Theory of Relativity. Einstein's famous equation relating mass and energy is $E = mc^2$, where c is the velocity of light ($c = 2.998 \times 10^8$ m/sec). The energy equivalent of 1 amu can be determined by inserting this quantity of mass into Einstein's equation and applying conversion factors.

$$\begin{aligned}
E &= mc^2 \\
&= 1\,amu \\
&= 1\,amu \left(\frac{1.6606 \times 10^{-27}\,kg}{1\,amu}\right) \left(2.998 \times 10^8 \tfrac{m}{\sec}\right)^2 \times \left(\frac{1\,N}{1\,\frac{kg \bullet m}{\sec^2}}\right) \left(\frac{1\,J}{1\,N \bullet m}\right) \\
&= 1.4924 \times 10^{-10}\,J \left(\frac{1\,MeV}{1.6022 \times 10^{-13}\,J}\right) \left(2.998 \times 10^8 \tfrac{m}{\sec}\right)^2 \\
&= 931.5\,MeV
\end{aligned}$$

Conversion Factors

$$1\,\text{amu} = 1.6606 \times 10^{-27}\,\text{kg}$$
$$1\,\text{Newton} = 1\,\text{kg-m/sec}^2$$
$$1\,\text{Joule} = 1\,\text{Newton-meter}$$
$$1\,\text{MeV} = 1.6022 \times 10^{-13}\,\text{Joules}$$

Since 1 amu is equivalent to 931.5 MeV of energy, the binding energy can be calculated using Equation 10-2.

$$\text{B.E.} = \Delta m \left(\frac{931.5\,MeV}{1\,amu}\right) \tag{10-2}$$

Example Calculate the mass defect and binding energy for uranium-235. One uranium-235 atom has a mass of 235.043924 amu.

282 Sustainable Nuclear Power

Solution Step 1: Calculate the mass defect using Equation 10-1.

$\Delta m = [Z(m_p + m_e) + (A - Z)m_n] - m_{atom}$

$\Delta m = [92(1.007826\,amu) + (235 - 92)1.008665\,amu] - 235.043924\,amu$

$\Delta m = 1.91517\,amu$

Step 2: Use the mass defect and Equation (10-2) to calculate the binding energy.

$B.E. = \Delta m \left(\frac{931.5\,MeV}{1\,amu}\right)$
$B.E. = 1.91517\,amu \left(\frac{931.5\,MeV}{1\,amu}\right)$
$= 1784\ MeV$

References

1. DOE Fundamentals Handbook, Nuclear Physics and Reactor Theory, Vol. 1 of 2, U.S. Department of Energy, Washington, D.C., DOE-HDBK-1019/1-93, 1993.
2. http://www.aip.org/, May 15, 2002.
3. http://hyperphysics.phy-astr.gsu.edu/hbase/astro/procyc.html, May 15, 2002.
4. "Understanding Radiation." Available at http://www.epa.gov/radiation/understand/.
5. Fact Sheet on Biological Effects of Radiation. Available at http://www.nrc.gov/reading-rm/doc-collections/fact-sheets/bio-effects-radiation.html. National Council on Radiation Protection (NCRP) Report 93, 1987.

CHAPTER 11

Recycling and Waste Handling for Spent Nuclear Fuel

The Nuclear Energy Industry

Providing a sustainable source of energy is a universal goal for all nations. The world population continues to grow and an adequate energy supply must be in the plans to maintain or improve the quality of life. Sustainability is a strategy to meet the energy needs of the present generation and increase our ability to serve the demands of future generations.

There are more than 400 nuclear power plants that currently provide about 16% of the world's electricity. In the United States, about 20% of the electricity is produced with slightly more than 100 nuclear power plants.[1] This technology has been proven reliable and economic on a commercial scale without the environmental impacts of fossil fuel power plants. Fossil fuel power plants (burning coal and natural gas) contribute the major fraction of our electric power, but they are also major sources of the increasing concentration of greenhouse gases in the atmosphere. The sustainable future of nuclear power depends on improving the technology for new energy systems to replace old nuclear plants as they are retired from service.

Recycling and Green Chemistry

Professor Stan Manahan defines green chemistry as "the practice of chemical science and manufacturing that is sustainable, safe, and non-polluting and that consumes minimum amounts of materials and energy while producing little or not waste material". In many instances, green chemistry can increase process profitability. This is typically possible when a process waste stream has higher concentrations of a metal than the concentration in the natural ore that is mined to recover the metal.

The spent fuel rods of commercial reactors are comprised of metal oxides and their cladding. For each 2,000 tons of heavy metals there are 660 tons of cladding and about 269 tons of oxygen. After a 3.4% burn, the uranium content of the waste stream is about 66%, this compares to 0.1% to 0.5% uranium that is typical in mined uranium ore. The fissile material content (U-235 + Pu-239) is about 1.12% as compared to 0.0007% to 0.0036% in the ore.

The reprocessing of spent fuel rods fits, well, the definition of green chemistry. Based on the concentration of the metals in the spent fuel rods as compared to the natural ore, there is the opportunity to use reprocessing to increase the profitability of nuclear power (without even considering the costs of long-term storage of spent fuel rods).

Why Reprocess Spent Nuclear Fuel?

Most of the commercial nuclear power plants in the world use light-water moderated reactors and uranium oxide fuel enriched to 2.6%–4% U-235. The fuel elements remain in the reactor about three years and then are stored in a pool of water on the reactor site as "spent fuel." Figure 11-1 illustrates the process of enriching the natural uranium followed by the nuclear burn.

The spent fuel emits high-energy gamma rays and produces thermal energy as the radioactive fission products decay. The water pool serves as a gamma radiation shield and a heat sink for the decay heat. The average decay rate and the energy release decrease with time but persist for many years, requiring a permanent repository. Such a storage facility is under construction at Yucca Mountain, Nevada, but litigation continues regarding personal safety and material containment in the plant for the very long-term storage. One key technical question concerns the very long-term integrity

Recycling and Waste Handling for Spent Nuclear Fuel 285

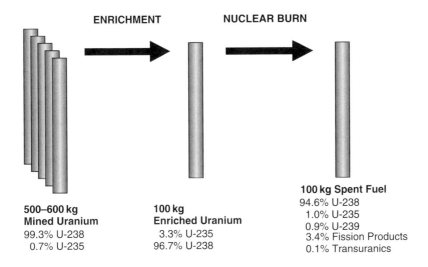

FIGURE 11-1. Mass balance for reprocessing. Mass is mass of heavy metal and fission products of heavy metal.

of the waste canisters in storage. The inventory of spent fuel is now accumulating at the rate of about 2,000 metric tons per year, and the total will exceed the planned capacity of the repository by about 2020.[2]

Current international consensus suggests nuclear energy will be required to ensure future energy security. The challenging technology goals to provide long-term sustainable nuclear energy must focus on resource utilization and waste management. Key issues include economics, safety and reliability, weapons proliferation resistance, and physical protection of plant personnel and the public. These factors apply to the power plants now in operation, and this experience will guide the development of the new generation of technologies that will be the nuclear energy systems.

In the United States, chemical reprocessing of domestic spent nuclear fuel is currently not practiced but is an important goal for long-term nuclear energy systems. The benefits include the following:

- Extending the nuclear fuel supply by recovering fuel values in the current inventory of spent fuel that has produced only 3.4% of the total energy available in the fuel.
- Combining these recycle fuel values with depleted uranium (the U-238 isolated to produce the initial U-235 enriched fuel) extends the fuel supply into future centuries (see Figure 11-2).[3]

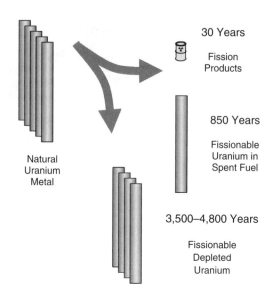

FIGURE 11-2. Mass balance of once-through fuel use as practiced in the U.S. The fission products are the "waste." The years indicate the years of available energy if used at the same rate as used in once-through burns. Both France and England have immobilized the concentrated fission products in glass for long-term storage. (M. L. Adams, *Sustainable Energy from Nuclear Fission Power*. National Academy of Engineering Publications. Volume 32, Number 4, Winter 2002.)

Some of the reactors in Europe and Japan use recycled fuel. Reprocessing technology is available, but the technology could be improved to reduce costs and waste. Reprocessed fuel in France costs 0.90¢/kWh (busbar), while new uranium fuel costs 0.68¢/kWh.

The current management plan in the United States is to place the spent fuel bundles from the "once through" fuel cycle in long-term storage. The spent fuel would be buried and labeled as hazardous. Since the radioactive hazard of this waste will persist for tens of thousands of years, considerable thought has gone into developing a warning/labeling system that would last the test of time.

Alternatively, fuel reprocessing would recover the uranium for fuel. Reprocessing can be broadly categorized into three steps: (1) recovery of unused fuel, (2) waste minimization, and (3) full use of uranium/thorium as fuel.

Recovery of Unused Fuel

Reprocessing options include a process of separating the fuel bundles into the fractions illustrated by Figure 11-3. The goal is to recover the heavy metals for use as nuclear fuel. The heavy metals are the fertile and fissile materials, including U-235, Pu-239, and U-238. The structural metal/ceramic components can be washed to a low level or nonradioactive waste. The fission products are the source of radiation hazard.

Waste Minimization

Recovery of the heavy metals effectively transforms up to 96.6% of the heavy metal in the spent fuel from a "waste" to a fuel for further processing to meet reactor fuel specifications. Of the remaining 3.4%, about 0.4% remains highly radioactive after 30 years of

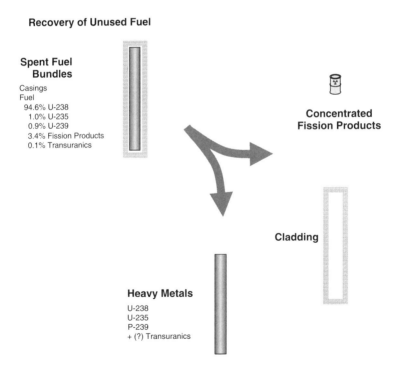

FIGURE 11-3. Recovery of unused fuel is the first phase of fuel reprocessing. Casings are physically separated as the initial step in future nuclear waste management—separation of structural metals in the bundles and recovery of fissionable fuel.

288 *Sustainable Nuclear Power*

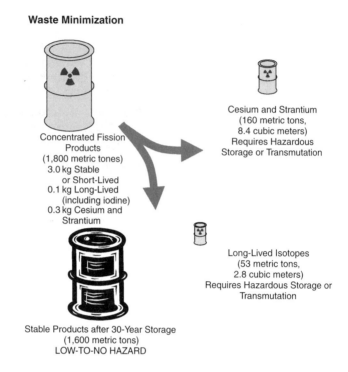

FIGURE 11-4. Schematic of overall process to minimize hazardous waste from nuclear power. The masses in tons represent total estimated U.S. inventory from commercial reactors in 2007—about 50,000 tons. The volumes are based on uranium density—the actual fission product volumes would be about twice the values indicated.

storage. Figure 11-4 illustrates the different fractions into which the 3.4% of fission products could be separated. It is the 0.4% that requires long-term burial or transmutation.

Waste minimization aspects of reprocessing extend the life of a burial site by reducing the quantity of wastes and the troublesome radioactive decay heat. The scientific analysis and demonstration of safe repository performance would be simplified by reducing the radioactive lifetimes of the materials going to geological burial. The storage time for the hazardous materials is significantly reduced, from about 300,000 to less than 1,000 years.

Full Use of Uranium/Thorium

In today's light-water nuclear reactors about 33% to 40% of the energy is provided by the nonfissile U-238 that is in the reactor fuel. New fissile material (Pu-239) is produced that fissions to provide energy. The change from new fuel to spent fuel is represented as follows:

3.0 Parts U-235 + 2.1 Parts U-238

→ 3.4 Parts Fissioned + 0.8 Parts U-235 + 0.9 Parts Pu-239.

Thus, for every 3.0 parts of fertile material that enter the reactor, 1.7 parts of fertile material leave the reactor. Using light-water reactors, insufficient fissile material is available to recycle and recover the centuries of energy value available in the U-238 inventories.

The Generation II light-water reactors are designed for fission propagation with thermal neutrons; this leads to a depletion of fissile material. Most of the Generation IV reactors are designed for fission propagation with fast neutrons. In these reactors much more U-238 is fissioned than U-235. A breeder reactor is designed to produce more fissile material than is fissioned.

Reactors based on fast neutron fission are an important part of reaching the potential of fuel reprocessing. Figure 11-5 illustrates the overall reprocessing cycle including use of fast neutron reactors, referred to as fast-spectrum reactors.

Discovery and Recovery

The 1930s was the decade of discovery for the structure of the atom nucleus. The zero charge neutron with a mass nearly the same as a proton was found to be tightly packed in the nucleus of atoms. Isotopes of many elements were prepared by placing a sample in a stream of neutrons, the nucleus changed with the capture of one or more neutrons (forming an isotope) without altering the balanced charge of the protons and electrons (without changing the chemistry of the atom).

It was during these experiments that irradiation of uranium produced barium. It was proposed that the uranium nucleus splits into two nuclei, one of which was the barium found in the experiment. The new atomic nuclei repelled each other with about 200 MeV gain in kinetic energy. This event also produced a pulse of fast neutrons. The term *nuclear fission* was coined to describe this process.[4]

The huge energy release plus the release of two or three high-energy neutrons suggested a proper quantity and configuration of uranium-235 could sustain a chain reaction. The resulting energy release would produce an explosion much more powerful than any

290   *Sustainable Nuclear Power*

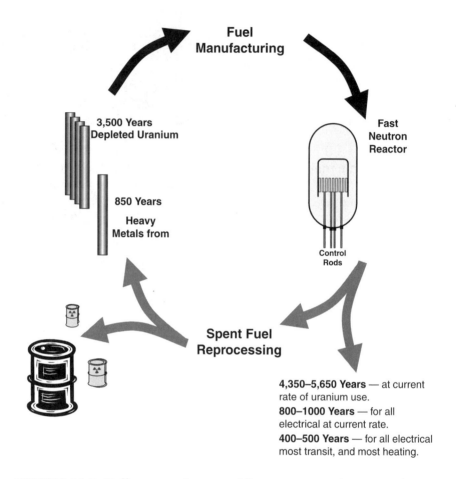

FIGURE 11-5. Full use uranium providing centuries of energy. Thorium is also a fertile fuel that can be used in this closed cycle.

chemical explosive. Lisa Meitner and Otto Hahn (February 1939) authored the paper in *Nature* describing fission just before the beginning of World War II. Shortly thereafter all nuclear research was classified "top secret." The remarkable developments during the 1940s nuclear decade were conducted to produce a nuclear weapon with all of the research and development done under that strict veil of military secrecy.

First Production of Plutonium

Natural uranium is composed of two primary isotopes, 99.27% U-238 and 0.71% U-235, and it was known that the 235 isotope will fission. Isotopic separations are difficult, and producing enough of the fissionable U-235 to make a nuclear weapon was a stiff technological challenge.

In January 1941, Seaborg and coworkers reported the discovery of a new element, plutonium-239, produced by neutron capture of the U-238 isotope.[5] Samples of natural uranium were irradiated with neutrons, and the first separation of plutonium was demonstrated by Cunningham and Werner, who found that lanthanum fluoride, LaF_3, precipitate was an efficient carrier of plutonium to make the separation possible.[6] This separation procedure produced the first few micrograms of Pu-239 metal, used to determine it had a fission cross section about 50% greater than U-235. This made plutonium an attractive alternative to U-235 as a weapons material.

Continued laboratory preparation of plutonium provided about $20\,\mu g$ of metal used to establish its radiological properties and chemical oxidation states and to estimate its physical properties.

The threat that Germany might develop a nuclear weapon convinced President Roosevelt that the Army should be in charge of the new nuclear weapon program. In June 1942, a new unit was formed called the Manhattan District.[7] There were five construction sites under this project. There was little sharing of task information between the groups, and security must have been good. One measure of the size of this secret project is the total workforce on the Hanford site grew to over 40,000 in May 1943.[8]

A group under the direction of Enrico Fermi successfully demonstrated that a nuclear chain reaction could be sustained using a matrix of natural uranium placed in a stack of graphite bricks (a nuclear pile).[9] The pile was initially operated at 0.5 watt and raised to 200 watts thermal energy on December 12, 1942. It was clear that the uranium metal fuel in such a reactor is continuously exposed to the fission neutrons, which then produce plutonium when a neutron was captured by a U-238 atom.

This successful demonstration of the nuclear pile reactor motivated the decision to fund the secret Manhattan District Project to produce plutonium for nuclear weapons. Plans proceeded to build a nuclear pile to produce kilogram quantities of plutonium using known data and a 10^9 engineering scale factor. This plutonium production nuclear reactor would generate about 250 megawatts of thermal energy. The remote site near Hanford in eastern Washington with the nearby Columbia River was selected for this plutonium plant. Construction was underway before the chemical steps for recovering the plutonium from the irradiated uranium fuel were known.

The laboratory recovery methods used ether extraction and produced a gelatinous LaF_3 precipitate, clearly not a good choice for large-scale plutonium production. The explosion hazard using ether and the corrosion problems using aqueous fluorides recommended alternatives. Precipitation remained the method of choice for the separation process because it offered quick development to large scale. Early in 1943, S. G. Thompson showed that $BiPO_4$ precipitate strongly carried Pu^{+4}.[10] The precipitate is crystalline and easily collected by filtration or with a centrifuge. This was the method selected for plutonium production.

Construction of the Hanford plutonium production reactors was underway based on the physics of the nuclear chain reaction and only on limited data indicating that it could be fabricated into a bomb. The scale of the project can be imagined by the Hanford nuclear pile reactors that consisted of 1,200 tons of pure graphite containing about 250 tons of uranium slugs, each slug consisting of a few kilograms of uranium sealed in an aluminum can. The fuel slugs were placed in horizontal aluminum tubes passing through the graphite. Cooling water was pumped through the tubes to remove the thermal energy.

A plutonium production run lasted 100 days, which converted about 1/4,000th of the U-238 atoms to U-239. The uranium slugs were pushed out of the reactor with new fuel to start the next irradiation cycle. The irradiated uranium was stored under water to remove fission product decay heat and provide biological shielding from the gamma radiation the radioactive decay produces.

The sequence to form plutonium from U-238 shown in Figure 11-6 is now well established. The neutron capture of a U-238 nucleus results in the immediate release of gamma rays and a total energy corresponding to the binding energy of the neutron. The new U-239 atom has a half-life of 23.5 minutes and decays releasing a beta particle to form neptunium 239 (Np-239). Np-239 has a half-life of 2.35 days and decays releasing a beta particle to form Pu-239. The plutonium also decays releasing an alpha particle, but the plutonium half-life is 24,000 years making it a relatively stable atom.[11]

The recovery of the plutonium began with removal of the aluminum cans covering the fuel slugs (either mechanically or chemically dissolving). The fuel elements containing uranium, plutonium, and fission products were dissolved in nitric acid. The

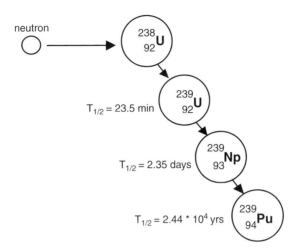

FIGURE 11-6. Conversion of U-238 to Pu-239.

plutonium in solution was reduced to the +4 state and precipitated with $BiPO_4$. The uranium was held in solution using sulfate ion, SO_4^{-2}, to form a soluble uranium ion. This separation step split the very small amount of plutonium from the uranium and most of the fission products. Some of the fission products stayed with the plutonium. Nearly pure plutonium was obtained by dissolving the plutonium-$BiPO_4$ precipitate and oxidizing the plutonium to the +6 state that is soluble in acid solution. The impurities remained insoluble and were separated with the precipitate. The plutonium in solution was again reduced to the +4 state and the precipitation cycle repeated two times with the extracted plutonium ending in a final acid solution. The third $BiPO_4$ precipitation cycle was followed by a cycle using LaF_3 to remove the last traces of fission products.[12]

It is remarkable that following all of these steps, the recovery of plutonium was greater than 95% and the decontamination factor of the plutonium product exceeded 10^7. This might have been considered good luck, since the decision to use the $BiPO_4$ precipitation process was made well before the chemistry of plutonium was known. (Hats off to the insight of the scientists working on the project!)

The "fast-track" military schedule ignored some of the disadvantages of the process. A batch process always requires careful operator attention. The quantity of process chemicals produced a large volume of high-level radioactive liquid waste, which still remains stored in huge tanks on the Hanford site and represents a

WWII legacy waste treatment problem. Nearly all of the uranium remained in the waste stream and additional processing steps were required to separate the uranium from the fission products.

There is an additional difficulty working with irradiated uranium. All of the reactor operations and the chemical treatment must be done behind radiation shielding to protect the plant personnel from the high-energy gamma radiation produced by the decay of the fission products. Mechanical manipulators were designed to provide remote services to operate and maintain the equipment handling irradiate fuel. A 1945 audit indicated that the Hanford plutonium production facility cost over $300 million (1945 dollars).[13] When we include the uranium enrichment plant at Oak Ridge, Tennessee, the weapons development site at Los Alamos, New Mexico, and the metallurgical laboratory in Chicago, the Manhattan Project represents the most ambitious, expensive, successful research and development project in history.

PUREX Process—Cold War Plutonium Production

The signing of the German and Japanese surrender documents was completed in August 1945 and the end of World War II brought a national "sigh of relief." Military personnel quickly returned to private life. This was also true for many of the science and technology people working on the Manhattan Project. The advantage of possessing the most powerful weapon was obvious, but the authority over the program was transferred from the Army (military) to a new civilian committee, the Atomic Energy Commission. Production of plutonium was continued under AEC control.

The first nuclear explosion in the Soviet Union in 1949 came as no big surprise. Winston Churchill's "Iron Curtain" speech in March 1946 certainly gave warning of the aggressive attitude of the Stalin-led Soviet Union toward its WWII Allies. This represents the introduction into the decades-long Cold War.

The Cold War ensured that the military would demand additional plutonium production. Countercurrent extraction was a mature technology when the batch process was selected for recovery and purification of plutonium. The operation of a continuous extraction train is more complicated than a batch process and there was little time available for process development. The bismuth phosphate batch process continued in service until 1951.

The source of plutonium would remain natural uranium irradiated in the graphite moderated reactors at Hanford, Washington, and newer reactors built at Savannah River, South Carolina. The irradiation times remained short with about 1/4,000 for the U-238 atoms converted to plutonium. There were three essential requirements of any proposed separation process:

1. All of the plutonium must be recovered as high purity, weapons grade material.
2. The uranium must be recovered essentially free of radioactive fission products.
3. The mass of the fission product waste stream should be greatly reduced (as small as practical).

The radioactive fission product content of the uranium and the plutonium was set at about the level of natural uranium so these metals could be handled and machined without the cumbersome gamma radiation shielding.

It was known in 1945 that tri-n-butyl phosphate (TBP) could be used as an extraction agent for nuclear fuels.[14] The PUREX (Plutonium-URanium-EXtraction) process was designed to remove the fission products, the source of essentially all of the gamma radiation in the product uranium and plutonium, requiring separation factors of 10^6–10^7. The toxicity of plutonium, inhaled or ingested, required plutonium separation from uranium be set at 10^8. A little uranium in the plutonium was not considered a problem. The recovery of both the uranium for recycle and the plutonium product was to be 99+%. These separation process specifications far exceeded the highest standards in industrial practice at the time.

Since the PUREX process was designed to replace the bismuth phosphate batch process, the feed-irradiated uranium fuel was the same. The first step was to chemically remove the aluminum cans on the uranium fuel slugs and then dissolve the fuel in hot nitric acid. The pH and metal concentration of this feed solution was adjusted to maximize the solubility of the plutonium and uranium in the organic extract phase. This feed solution entered the center of the first extraction train.

The first extraction separated the plutonium and uranium from the fission products. The organic extraction solvent was a nominal 30 (Vol.)% TBP in a paraffinic hydrocarbon (much like kerosene), which gives good flow characteristics in the liquid extraction contact stages. The organic solvent enters one end of the extraction train, with the first contact stage removing all but a trace of the uranium and plutonium from the acid (aqueous) phase containing the fission products. The TBP solution continues to load with U and Pu at each stage until the feed stage. The 2–3 molar nitric acid is fed at the other end of the extraction train and scrubs fission product metals from the TBP phase. The contact time between the TBP and acid phase containing the fission products was kept short to minimize the gamma radiation damage to the organic phase. The acid solution goes to a nitric acid recovery unit that recovers the nitric acid values, removes water, and concentrates the fission product solution for storage.

A second extraction train receives the TBP solvent stream with one end fed dilute nitric acid containing chemicals that reduce the plutonium to the acid soluble Pu^{+3} state. Fresh TBP is fed to the other end of the train to remove traces of uranium stripped into the acid stream that now contains the plutonium. This extraction step completes the separation of the large fraction that is uranium from the small plutonium fraction.

The uranium is released from the TBP with a scrubbing train using dilute acid. The stripped TBP goes to solvent recovery and is cleaned up for recycle. Water is evaporated from the dilute acid containing the uranium. The pH is adjusted, and the uranium extracted with countercurrent TBP-acid scrub to remove traces of plutonium and fission products. The spent nitric acid goes to nitric acid recovery and the fission product waste, including a trace of plutonium, goes to waste concentration.

The uranium is recovered from the TBP with very dilute nitric acid scrub. The TBP is recycled and the uranium solution evaporated. There may be a final uranium "polishing step" before the steps to produce the final uranium product, a uranium nitrate solution or denitration to form UO_3.

The plutonium that was left in the acid solution is oxidized to the Pu^{+4} state and center fed to an extractor that collects the plutonium in the TBP phase and leaves impurities on the acid phase. The TBP phase goes to a dilute acid stripper to recover the product as plutonium nitrate solution. There usually is a product plutonium polishing step to attain the maximum purity of the product. The stripped organic TBP phase and the aqueous acid phase are recycled to recover nitric acid values, renew the organic phase, and reduce the volume of the waste stream.

There are several variations of the general separation steps just described. Mass balance information for the early vintage PUREX plants is given in the References.[15] The first PUREX process plant to produce weapon's-grade plutonium was located at the AEC (Atomic Energy Commission) Savannah River Plant and began production in November 1954.[16] With this successful demonstration, another plant began operation at the Hanford site in January 1956.

Improvements in the operation of the PUREX process continued with operating experience.

The demands for high-purity plutonium and uranium made the development of the PUREX process a real separation technology challenge. Add to this the demands for personnel safety, protection from the toxicity of the heavy metals, and the continuous gamma radiation from the fission products. The engineering task included developing mechanical manipulators to perform all of the process operations and equipment maintenance tasks protected by the radiation shielding. This technology was developed for the military and is the basis for modified PUREX to reprocess domestic spent nuclear fuel.

PUREX Process—Domestic Spent Fuel

The first nuclear reactors designed to produce electricity were installed in nuclear submarines. Such a reactor must provide the electric and thermal power required to sustain the crew underwater, long term, and then to provide the additional variable power required during battle maneuvers. The total mass and the size of the reactor must fit on the submarine and protect the crew from fission product gamma radiation. These reactor designs, the nuclear fuel composition and configuration, were classified "Top Secret" by the military. Civilian contractors built these reactors, and their engineers saw an opportunity, and were encouraged, to extend this technology to civilian electric power production under the Atoms for Peace Initiative.[17]

The nuclear reactors for domestic electric power production were designed to provide baseline power, continuous operation at power plant design capacity for long periods between refueling, and mechanical equipment maintenance shutdowns. The two designs widely adopted and deployed in the United States were the boiling water (BWR) and the pressurized water (PWR) reactors. The fuel for these reactors is uranium oxide slightly enriched to between 2.6% and 4% U-235.

Uranium oxide fuel is commercially manufactured into small, cylindrical pellets about 12–13 millimeters in diameter and the same length. These pellets are loaded into metal tubes (about one centimeter OD), originally stainless steel but soon replaced by Zircaloy (mostly pure zirconium alloyed with tin, nickel, chromium, and iron). The end cap on each tube are welded to isolate the uranium fuel and all the fission products (gases and solids) from the water in the reactor. Often, these tubes are pressurized with helium to improve the heat transfer from the fuel pellets to the tube wall. Zircaloy has a low neutron capture cross section, is corrosion resistant, and quickly became the material of choice for this application.[18]

The fuel for a typical PWR is fixed in a fuel assembly consisting of a 15 × 15 array of fuel tubes fixed in place with space to circulate pressurized water to remove heat and to serve as the neutron moderator (to slow down the neutrons). Such a fuel assembly is about four meters long and weighs about 658 kg. It contains about 523 kg of uranium oxide (461 kg of uranium metal). There is 135 kg of Zircaloy and hardware metal in each fuel assembly.[19] These fuel assemblies contain the spent nuclear fuel that is the feed for fuel reprocessing.

The composition of the spent fuel is determined by the initial composition of the fuel and the radiation history of the fuel assembly. There are three sources of radioisotopes formed during the power cycle of the fuel:

- Fission products formed by the splitting of the fissile elements (initial U-235 and the Pu-239 that is formed by neutron capture of the U-238 in the fuel during the fuel cycle)
- The transuranic elements formed by neutron capture (neptunium, plutonium, americium, and curium)
- The activation products formed by exposure of atoms to the high-radiation field in the reactor

Immediately following reactor shutdown, the fuel will contain more than 350 nuclides, many with very short half-lives that decay in seconds or minutes.[20] These radioactive decay processes produce thermal power and gamma radiation that must be managed to store the spent fuel elements.

For example, consider a reactor that operates with a fuel burnup of about 30,000 megawatt days per tonne (1,000 kg) of uranium metal in the fuel. Immediately after shutdown, these fuel elements will produce nearly 2,000 kW of thermal energy and a nuclide radioactivity of about 2×10^8 curies per tonne. The fuel assemblies are stored in a deep pool of water (containing soluble boron, a neutron absorber) located at each power plant where the thermal energy is removed and the water serves as radiation shielding.[21] The thermal energy release and the gamma radiation decrease with time as each radioactive isotope decays, and after ten years the thermal energy release is about 1.1 kW and the radioactivity is about 3.9×10^5 curies per tonne.[21] The thermal energy release is still much too high to allow isolated underground storage. The gamma radiation requires bulky biological shielding to protect persons transporting spent fuel to any remote storage or reprocessing site.

Reprocessing: Recovery of Unused Fuel

As of the onset of the 21st century, reprocessing technology in the United States has not been used on commercial spent nuclear fuel. An agreement signed during President Carter's administration closed out the option of reprocessing domestic spent fuel in support of the nuclear nonproliferation treaty. Reprocessing would make plutonium more easily attainable to terrorists. If kept in the spent fuel and with the highly radioactive fission products, it is not as available.

The fear of nuclear weapons proliferation is a major obstacle to domestic nuclear fuel reprocessing. Security assurances of excess military plutonium and highly enriched uranium will be important in any decision to proceed to commercial spent fuel reprocessing.

The plutonium in the spent nuclear fuel is less suitable for bombs, even if concentrated, since it contains some Pu-240 isotope.[22] The separation of Pu-240 from the Pu-239 would be more difficult than concentrating of bomb-grade U-235 from natural uranium because the atomic weights of Pu-239 and Pu-240 are more nearly the same than U-235 and U-238.[23]

Since nuclear waste handling is a known significant expense of nuclear power generation, most governments levy a tax on nuclear electrical power to be applied toward disposal. In the United States, this tax is 0.1 cents per kilowatt-hour. Over $20 billion have been collected in this fund. If the revenues collected for handling nuclear waste were used to reprocess spent nuclear fuel, the waste problem could be solved. The British and French have over 35 years of experience in reprocessing spent nuclear fuel. The PUREX process is the primary technology in use.

PUREX Process

Figure 11-7 summarizes the PUREX process steps for reprocessing spent nuclear fuel. The first step for commercial nuclear fuel reprocessing is opening the fuel tubes so the irradiated fuel can be dissolved in nitric acid. Chemical decladding used in military

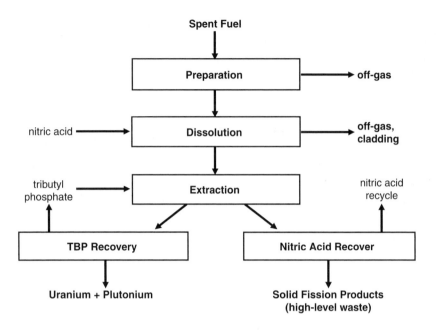

FIGURE 11-7. Block flow diagram of PUREX reprocessing of spent nuclear fuel.

plutonium production to remove the aluminum cladding from the uranium metal slugs is replaced by mechanically shearing the commercial reactor fuel assemblies into short lengths. This releases helium (if helium was filled during fuel manufacture) and the fission product gases (isotopes of krypton, xenon, and tritium are examples) that must be collected. After a reasonable time, these gases decay to stable isotopes and can be released. Radioisotopes, including Kr-84, I-131, and Xe-133, are currently vented to the atmosphere.[24] Other treatment may also be necessary. The long-lived radioactive iodine released during this step is given special attention.[25]

The next step is to dissolve the fuel metal oxides containing fission products, uranium, and plutonium in nitric acid. The stainless steel and Zircaloy pieces from the fuel assemblies do not dissolve and are separated from the nitric acid solution, washed to remove all of the uranium, fission products, and transuranic elements, dried, and packaged as low-level radioactive waste. The nitric acid solution pH is adjusted to ensure that uranium and plutonium are in the most favorable oxidation states for extraction.

Some of the fission products are (or form) metal compounds that exceed solubility limits, and these are filtered before entering the extraction train. For each metric ton of metal in the spent fuel, there will be about 944–946 kg of U-238, 8–11 kg of U-235, 5–9 kg of Pu-239, and 1–9 kg of heavy metal isotopes with atomic numbers greater than uranium (transuranics) in the periodic table. The total mass of the fission product metals (more than 40 elements) is about 34 kg.[26]

A small fraction of the fuel does not dissolve in nitric acid. These residues vary depending on the fuel characteristics, the time the fuel is irradiated, and the procedure used to dissolve the fuel.[27] The acid solution must be clarified before it is fed to the extraction train. These residue solids will be radioactive and a heat source that requires special handling, especially for "young spent fuel" (spent fuel aged less than ten years).

The first extraction train moves the uranium and plutonium to the organic TBP phase and leaves the 43 g of minor actinides and fission product metals in the aqueous phase. The strong gamma radiation of the fission products that cause radiological damage to the TBP phase is essentially all removed in this first extraction step. Extraction steps to strip the remaining fission products and to separate the uranium and plutonium follow, with minor modification of the process for the production of plutonium.

There are other uranium isotopes in this recycle fuel stream that "tag along," and these are neutron absorbers in the recycled fuel for the LWR. These isotopes accumulate each time the fuel is recycled and the different decay routes produce thermal energy and high-energy gamma radiation that may require protective clothing or remote handling during fuel fabrication of recycled uranium metal.[28]

The PUREX process does produce pure plutonium, the source of strong objection from the members of the nuclear weapons nonproliferation group. The proposal to mix plutonium oxide with uranium oxide to form a mixed oxide fuel (MOX) that can be used as fuel for light-water reactors has been used in commercial reactors in Europe. MOX fuels have been successfully used on a limited basis, and this fuel mixture does not present a weapons threat.[28]

Plutonium represents the key ingredient for closing the uranium fuel cycle, and it is an important source of energy for future nuclear power plants. The current fleet of LWRs in the United States produces about 2,000 tons of spent fuel per year containing about 0.5%–0.9% Pu, yielding 10–20 tons of plutonium. The age of the spent fuel (how long it has been in storage) changes the ratio of the isotopes of plutonium and the performance of the plutonium fuel in the LWR reactor. The energy-producing fissile materials in a light-water moderated reactor are U-235 and Pu-239. The other isotopes accumulate with each reprocessing cycle. This disincentive for reprocessing can be overcome using the reprocessed fuel in a new generation of fast-flux (fast-neutron) reactors discussed in Chapter 12.

Further processing can be performed on the mixture of uranium and plutonium, leaving the PUREX process to prepare pure uranium and pure plutonium. The PUREX process produces gaseous effluents and cladding hulls that are not hazardous. The fission products are concentrated into a solid high-level waste. The nitric acid and the solvents (primarily tributylphosphate, TBP) are recycled. The acid and solvents do not add to the volume of waste resulting in a substantial decrease in the total volume of radioactive waste.

The PUREX process separates the uranium and plutonium from the spent fuel, but the fission products and minor actinides that remain in the waste stream represent a very long-term waste storage problem. Figure 11.8 shows a modified PUREX known as URanium EXtraction (UREX) has environmental and antiproliferation advantages over PUREX described in the following excerpt from the DOE 2003 Report to Congress.[29,30]

> In the UREX process, plutonium, other transuranics, and fission products are extracted in a single stream from which transuranics can be extracted for reuse in nuclear fuel. The feature of UREX that makes it much more proliferation resistant than PUREX is the continual presence of minor actinides, the high radioactivity and thermal characteristics of which make these materials relatively unattractive to potential proliferaters.
>
> Because UREX does not place these actinides in the waste stream, there could be a significant reduction in the amount of high-level waste produced. Short-lived radioactive isotopes are separated and may be stored and allowed to decay to harmless elements over several decades.
>
> Further, experiments completed in 2002 have proven UREX to be capable of removing uranium from waste at such a high level of purity that we expect it to be sufficiently free of high-level radioactive contaminants to allow it to be disposed of as low-level waste or reused as reactor fuel. These laboratory-scale UREX tests have proven uranium separation at purity levels of 99.999 percent. If spent fuel were processed in this manner, the potential exists to reduce

Recycling and Waste Handling for Spent Nuclear Fuel 303

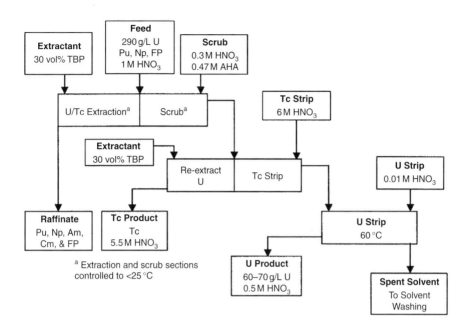

FIGURE 11-8. UREX block flow diagram. (M. S. Thomas, M. A. Norato, G. F. Kessinger, R. A. Pierce, T. S. Rudisill, and J. D. Johnson, "Demonstration of the UREX Solvent Extraction Process with Dresden Reactor Fuel Solution." Report WSRC-TR-2002-00444, U.S. Department of Commerce, NTIS, Springfield, VA, 2002.)

significantly the volume of high-level waste. An additional advantage of UREX is the use of acetohydroxamic acid, which enables the use of chemical processes that are far more environmentally friendly than PUREX.

AFCI (Advanced Fuel Cycle Initiative) Series One research would include the continued development of aqueous chemical treatment technologies including the possible demonstration of UREX at a scale relevant to its eventual commercial use.

An advanced development of UREX, referred to as "UREX+," would be a key element of an AFCI program.[31] This additional research would evaluate different aqueous chemical treatment methods to separate selected actinide and fission product isotopes from the UREX waste stream to minimizes waste. For example, UREX+ would provide mixtures of plutonium and selected minor actinides yielding proliferation-resistant fuels.

Long-lived fission products, iodine-129 and technitium-99, major contributors to the long-term radiotoxicity from spent fuel, could be separated and incorporated into targets for destruction in reactors. A research program must be developed to obtain an understanding of all waste streams, the data needed for a commercial scale treatment facility, and provide the basis for estimating the cost to design, build, and operate such a facility.

If implemented successfully, this treatment technology could significantly reduce the volume of high-level waste from commercial nuclear power. This accomplishment would reduce the cost of the first repository and potentially eliminate the technical requirement for a second.

Key programmatic treatment technology elements for Advanced Fuel Cycle Initiative Series One would include (1) laboratory demonstration of UREX+ using radioactive materials; (2) engineering scale demonstration of UREX+; (3) laboratory demonstration of PYROX (pyrochemical dry treatment) technology using spent LWR fuel; (4) demonstration of PYROX actinide recovery; (5) engineering scale demonstration of PYROX using radioactive materials; (6) demonstration of large-scale metal waste form technology; and (7) treatment facility requirements, costs, and design studies.

The UREX treatment technology, combined with additional processing steps, provides the ability to produce proliferation-resistant transmutation fuels for use in LWRs or gas-cooled reactors.

Successful implementation of this technology would require dealing with several issues such as fabrication and testing of transplutonic-bearing fuels, which would require remote fabrication.

To support this effort, research on transmutation fuels would focus on the development of proliferation-resistant fuel forms, preliminary fuel irradiation testing, and analysis of the resulting transmutation system (including waste streams). In the case of LWR transmutation fuels, several technology options would be considered, including the French CORAIL, Advanced Plutonium Assembly systems, advanced assembly designs, and inert matrix/nonfertile fuel concepts.

Gas-cooled reactors use very small spherical fuel particles which, if manufactured with advanced coating technology, are strong enough to permit much higher burnups than are possible with LWRs and are difficult to reprocess. Very high destruction levels of plutonium (over 90%) have already been demonstrated using pure plutonium fuels; however, the challenge remains to achieve these impressive burnups with proliferation-resistant fuels. Research is needed to address this challenge and will include the development of proliferation-resistant fuel forms, fuel irradiation testing at the High Flux Isotope Reactor at Oak Ridge National Laboratory or the Advanced Test Reactor (ATR) at the Idaho National Engineering and Environmental Laboratory (INEEL), and analysis of the resulting transmutation system performance for gas-cooled reactor fuel.

Advanced Aqueous Separation

More recently, Argonne National Laboratory personnel have developed an advanced aqueous process called UREX+, which has been demonstrated on a laboratory scale.[30] The UREX+ process consists of five solvent extraction steps that separate the dissolved spent nuclear fuel (the PUREX feed) into seven fractions. In the first stage, the uranium and technetium are recovered in separate streams with high total recovery and purity. Next, the cesium and strontium (heat producers, a problem in repository waste storage) are removed. A little feed adjustment allows the plutonium and neptunium to be recovered with impurity levels that allow these actinides to be incorporated into MOX fuel. The fourth step recovers the minor actinides and the rare Earth elements. The final step separates the minor actinides from the rare Earth elements.

The many metal species in spent fuel dissolved in nitric acid form a complex chemical mixture. The UREX+ process demonstration indicated that with additional work to understand this chemistry and to refine the separation parameters, the product streams can yield recycled nuclear fuel, radioactive isotopes that can be formed into targets for transmutation, and a waste product that can meet the demands for long-term geological storage. This work demonstrates an option for the treatment of the inventory of spent fuel accumulating from the current fleet of light-water reactors.

Tables 11-1 and 11-2 provide a summary comparison of these aqueous treatment processes.

Experimental Breeder Reactor II

The nuclear reactors designed to produce military plutonium used uranium metal as fuel. As early as the mid-1940s there were attempts to demonstrate the use of plutonium as a fuel for power production.[32] In 1963, Experimental Breeder Reactor II (EBR-II) was

TABLE 11-1
Summary of separations in PUREX and UREX processes.

PUREX	• Extracts pure U and Pu, free from fission product contamination
	• Minor actinides go to waste along with fission products
	• Separates pure U
	• All transuranics are recovered as a group
	• Cesium and strontium removed to repository heat load
UREX	• Lanthanide fission products can be retained with TRUs if needed to provide limited self-protection radiation barrier
	• Hybrid modification sends the process stream after U, Cs, and Sr removal to a pyrochemical process for separation of TRUs from fission products for fast reactor recycle
UREX+	• Several variants of the UREX process are being studied; all include separation of pure U and removal of Cs and Sr
	• Each variant provides different options for the recovery of transuranics, either as a group or as subgroups, for use in different recycle scenarios in thermal and fast-spectrum systems
	• Provides flexibile response to evolving nuclear systems in the U.S.

TABLE 11-2
Further comparison of aqueous treatment processes.

		Advanced Aqueous Separations	
Process	PUREX	UREX	UREX+
Pure Pu Separation	Yes	No	No
Remote Fuel Fabrication Required	No	Yes	Some Do
Technology Development Completion	Available	2010	2012

the first fast-flux (the high-energy neutron spectrum produced at fission) reactor with the fuel core, the circulation pumps, and the primary heat exchanger submerged in a pool of sodium contained in the reactor vessel.

Sodium has a small neutron cross section that minimizes the neutron slowing-down effect. The primary sodium coolant that is exposed to high-energy radiation in the reactor core becomes radioactive (a portion of the sodium becomes Na-24 with a 15-hour half-life). The secondary liquid sodium is circulated through the heat exchanger and transports the thermal energy to a steam generator to supply steam to a turbine. This higher-pressure secondary sodium would leak into the reactor pool if there was a leak in the sodium heat exchanger preventing radioactive sodium release from the reactor vessel.

The EBR-II operated until 1994, over 30 years without heat exchange problems. The facility was designed to "breed" plutonium to extend the inventory of fissile uranium. Later, the fuel loading was modified to demonstrate it could be a "plutonium burner" (to reduce the inventory of surplus military plutonium). EBR-II provided the test bed for irradiation studies of many proposed metallic and oxide fuels for military and commercial applications.[33]

The EBR-II was designed to be an integral nuclear power plant. This included on site nuclear fuel reprocessing and new fuel production on the reactor site. The fertile atoms (U-238) in the fuel absorb neutrons to form fissile atoms (Pu-239). These would join the fissile metals in the initial fuel to serve as fuel in the fast-neutron flux. The demonstration included the production of steam to drive a turbine to produce electric power to complete the simulation of a commercial power plant.

The engineers on the EBR-II program steadily increased the performance of the fuel for the reactor. Initial problems with the fuel showed the uranium metal expanded during irradiation; mechanically stressing the tubes and limiting service to about 1% of the U-235 in the fuel. The swelling problem was solved using a uranium-plutonium-zirconium alloy fuel that achieved 10% fission of the fuel (compared to 3.4% for commercial LWR metal oxide fueled reactors). The metal fuel was sealed in stainless steel tubes with sodium metal filling the space between the fuel and the tube wall (sodium bonded fuel). The all-metal fuel elements in contact with the liquid sodium pool provided high heat-transfer rates, smaller in core temperature gradients, making possible a small reactor core.[33]

The original EBR-II fuel was enriched uranium to serve as a *driver* (neutron source) for the production of plutonium. The demonstration of "breeding" plutonium in U-238 rods surrounding the reactor core required the recovery and recycling of the new plutonium as fuel. Collaboration between Argonne National Laboratory in Illinois and the Idaho National Laboratory developed a "dry" (no water, no nitric acid) process they called pyropartitioning.[34]

Pyrometallurgical Reprocessing

Unlike the PUREX and UREX processes, pyrometallurgical processing is based on electrochemical separation. Oxide fuels must be reduced to metal prior to electrolysis, as illustrated in Figure 11-9. Electrochemical separation can be compact and relatively simple; however, the process mixes the cladding with fission products, forming more high-level waste than some alternatives.

In the pyropartitioning process, the spent fuel elements are chopped into short pieces and placed in a metal basket in a pool of molten salt. A minimum melting (eutectic) mixture of potassium chloride and lithium chloride (KCl-LiCl) was used as the solvent. An electric current passed from the chopped fuel basket that serves as an anode where the heavy metals in the fuel are oxidized to form metal chlorides dissolved in the molten salt. An inert metal cathode served to collect the reduced metals from the salt separating them as they selectively deposit on the cathode based on the electronegative potential for each metal chloride in the salt. (The voltage for deposition of each metal is like the voltage of the lead-acid battery used in an automobile.) This electrolytic process separates the uranium and plutonium from the fission products and the minor actinides, producing uranium-plutonium metal. The metal deposited on the cathode is harvested, separated

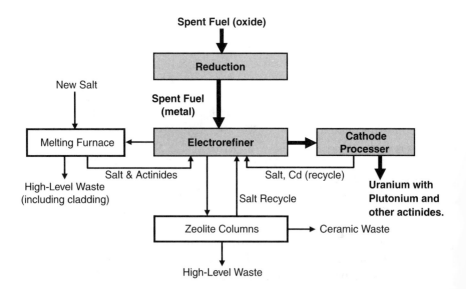

FIGURE 11-9. Pyropartitioning process to recover heavy metals.

from adhering salt, and cast into new metallic fuel pins for fuel or sent to storage. The removal of the uranium and plutonium (about 94%–95% of the mass) leaves the fission products and the transuranic elements in the salt. The metallic sodium bonding agent in the fuel elements forms sodium chloride. The melting temperature of the KCl-LiCl eutectic salt increases as metal chlorides build up. The saturated solvent salt is removed, most of the KCl-LiCl recovered, and the remainder formed into a ceramic waste form suitable for long-term storage.

There are noble metal fission products that remain solid during the electrodeposition and these are combined with the fuel cladding pieces that remain in the anode basket. The solids are stabilized into a metallic waste form and sent to storage.

All of the processing steps must be done in an inert atmosphere (argon) that is essentially free of water, hydrogen, nitrogen, and oxygen. The actinide and rare Earth metals in the spent fuel are chemically active and readily form oxides, nitrides, and hydrides that are insoluble in the molten salt. They collect as solids in the processing equipment. The radioactive fission products in the spent fuel produce high-energy gamma radiation, making gamma shielding necessary.

Remote handling is required for all of the steps in the fuel reprocessing and new fuel fabrication for the EBR-II fuel. Additional protection from the health hazards of inhaling or ingesting the radioactive heavy metals are minimal because the total isolation requirements for the fuel processing protect the workers.

The EBR-II experimental program achievements include the generation of over 2 billion kilowatt-hours of electricity; irradiation of over 30,000 specimens of fuel, structural, and neutron absorber materials; advanced instrumentation testing; a test of inherent safety with demonstration of total loss of coolant flow; and advanced computer technology applied to diagnostics and control.[35] In December 1995, James Toscas (Executive Director of the American Nuclear Society) stated, "EBR-II is arguably the most successful test reactor ever." The technology informing the next generation of fast-flux reactors depends on data collected during the EBR-II experimental run.

The pyropartitioning process can be used to separate the nuclear fuel components from fission products to make fuel for the next generation of fast-flux reactors. All of the LWR spent fuel in the United States, currently more than 40,000 metric tons (expected to exceed 60,000 metric tons by 2010) is ceramic (metal oxides). Anticipating that commercial scale processes will be required to recover fuel values and reduce the mass of radioactive waste going to a repository from this inventory of LWR spent fuel, the Chemical Engineering Division at Argonne National Laboratories has completed a demonstration of an electrochemical process that reduces the oxide fuel to metal.[36] The reduced metal would be fed to the pyropartitioning process to recover the fuel values from the fission products.

The promise and future of nuclear power systems depend on the success of research and development investments to develop and deploy new energy systems. The plans of the international community to cooperate in this project are the subject of Chapter 12.

Mining and Processing

Uranium ore is mined and processed to uranium oxide concentrate (U_3O_8) that is sold as a feedstock for further processing. For use as reactor fuel, the uranium must be enriched in U-235. This is performed in a gas-phase process by forming uranium hexafluoride (UF_6). Enrichment is used to increase the 0.7% natural U-235 concentration to 2.6%–4%. The enriched uranium is processed to uranium dioxide (UO_2) and formed into fuel pellets that are placed in tubes as reactor fuel rods.

Today's world nuclear power production includes about 430 nuclear reactors, with a capacity of 350,000 megawatts. Every year each 1,000 megawatts of power production convert 750 kg of U-235 (and some U-238/Pu-239) into 750 kg of waste fission material mixed with about 30,000 kg of unused fissionable elements (U-235, U-238, and Pu-239).

Proven world uranium reserves are 3.3 million tons,[37] with vast deposits in Australia and Canada.[38] Estimated reserves in addition to the proven reserves include another 10.7 million tons.[39] Recently discovered uranium deposits in Canada are so rich in uranium that they must be mined with robots to avoid exposing miners to their natural radiation. In the United States, 56,000 tons of spent fuel plus 224,000 tons of depleted uranium represent 400 years of fuel for essentially all energy needs (electricity, transportation, and heating). The world estimated reserves are more than 50 times this 400-year supply—at least 20,000 years. Thorium can also be used to fuel nuclear reactors. These are projections because one can hardly predict how technology will provide new ways to meet energy needs in 30 years let alone 500 or 5,000 years in advance.

In the 2005 AFCI Report to Congress[39] an attempt was made to estimate unconventional uranium reserves. These reserves include 180 million metric tons from sandstone, 4,300 from seawater, and 800,000 from phosphate. If 16 million metric tons correspond to 20,000 years of uranium, 804,500 million metric tons correspond to 1 billion years of energy from uranium.

The Report to Congress classifies the massive phosphate reserves as "unconventional," which is an ambiguous term. These reserves are on the edge of commercial viability.

> Uranium from phos-acid is extremely (technically) viable and phosphates represent a major source of U_3O_8. Plants were operated in the 50's in Florida, but with the discovery of uranium in the western United States, the facilities were shut down due to economics.[40]

In the early 1970s interest was revived with the growth of the nuclear power industry and a significant amount of development was done to improve the earlier processes. Much of this work was carried out by Oak Ridge National Laboratory. Private companies also developed processes: United Nuclear, Freeport Chemical, and Westinghouse.

International Minerals and Chemical (IMC) initially worked with United Nuclear to develop a process to recover the highly abundant uranium from phosphates, but their process (after installation by WR Grace) had problems, so they began their own development and eventually worked with Oak Ridge National Laboratory.

Several commercial installations resulted from these developments. Freeport installed a facility at their plant in Louisiana. IMC installed three facilities to extract uranium. Their approach was to install primary extraction facilities at the individual phos-acid plants and produce a concentrated uranium solution (in phos-acid). This material was then trucked to their main refinery for U_3O_8 recovery.

A uranium-from-phosphate plant was also installed at the Gardinier facility in Tampa, Florida. There were also commercial facilities in Europe. In the late 1970s there was in excess of 3 million pounds/year of U_3O_8 recovered from phosphoric acid operations. (As a side note, IMC had about 2.2 MM lbs/yr of capacity and at that time was the fourth largest U_3O_8 producer in the United States.)

Unfortunately Three Mile Island occurred, and nuclear plant activity came to a virtual halt. The long-term U_3O_8 outlook diminished, and existing supply contracts were not renewed as the price of U_3O_8 plummeted. The result was that by the early to mid-1990s all of the phos-acid-based production ended due to low prices for the product.

Based on the earlier economics (adjusted for inflation), U_3O_8 was priced in the $35+ range. In the mid-1970s, $25/lb was the price where interest was generated. Uranium from phos-acid was a commercial industry from the mid-1970s to early 1990s, and the technology is well established to restart this industry.

What is important is that nuclear fuel is available to produce all the electric power needs, and the lessons of U.S. commercial utilization demonstrate that it is safe and has a lower environmental impact than alternatives. Since about 30 years of spent fuel have been stockpiled, nuclear fission can provide abundant energy without additional mining and actually uses more spent fuel than it generates.

Waste Generation from Reprocessing

Vast amounts of nuclear waste were generated in the rush to develop the nuclear bomb and subsequent arms race. The problem of the high-level waste was compounded by landfilling and storing liquids dilute in the actual radioactive materials; concentrated wastes would be much less voluminous. Any spilled liquids can leach through soils and contaminate the ground waters around storage areas. It is this history that leads many to believe that reprocessing will generate more waste volume than untreated waste. Current reprocessing methods are much cleaner than those historically used in the United States.

Table 11-3 summarizes the five reprocessing options put forward by the 2004 U.S. DOE Office of Nuclear Energy, Science and Technology Advanced Fuel Cycle Initiative (AFCI) Comparison Report.[41] These methods are based primarily on the recovery of uranium, plutonium, and (in some cases) other actinides. These technologies reduce the volume of high-level waste by about 80%.

Table 11-4 summarizes the fuel and nonhazardous product from the reprocessing options. The components marked for recycle for future reactors include minor actinide transuranics. The cladding would be washed to low- or zero-level waste levels. The

TABLE 11-3
Estimated quantities in metric tons of chemicals, glass, and salt used in reprocessing waste by different methods. Assume 3.4% of spent fuel is fission products and ends up in waste and that all the glass, salts, and chemicals end up in waste.

Process	Net Chemical Consumption	Net Glass/ Salt Frit Consumption	High Level Waste	Reduction in Waste
Once-Through			2000	0%
PUREX	4.2	420	490	76%
UREX+	7	124	232	88%
UREX/PYRO	5.6	322	280	86%
PYROX	80	500	490	76%
Advanced Aqueous Process	0.8	124	232	88%

Table was compiled from information in U.S. DOE Office of Nuclear Energy, Science and Technology Advanced Fuel Cycle Initiative (AFCI) comparison report.

TABLE 11-4
Fuel and low-level wastes from reprocessing options. Mass in pounds. Mass based on 2000 lbs heavy metal.

		Fissile and Fertile Materials		Low-Level Waste	
Process	Uranium	Recycled Pu, Np, Am, Cm	Recycle for Future Reactors	Cladding	Secondary Waste
PUREX	1892	17.0	0.0	660	2.1
UREX+	1892	18.0	3.2	660	3.5
UREX/PYRO	1892	21.2	21.2	660	4.2
PYROX	1892	21.2	21.2	660	2.1
Advanced Aqueous Process	1892	18.0	3.2	660	1.4

Source: Extracted from Reference 41.

secondary wastes are broken, contaminated equipment and materials contaminated in fuel transport. The cladding could be recycled as cladding on recycled fuel.

The fate of the uranium product stream from reprocessing is not obvious. The amount of U-235 (about 1%) in this uranium is about the same as natural uranium, and has no premium fuel value. The motivation for uranium reprocessing is to keep it from becoming a low-level radioactive waste.

Waste Minimization

The reprocessing technologies of Table 11-3 are based on recovering uranium and plutonium. This is only one of three phases of spent fuel reprocessing:

Phase 1—Recover unspent fuel.
Phase 2—Minimize waste by separating stable fission products from high-level fission products.
Phase 3—Transmute the high-level fission products into nonhazardous materials with total elimination of long-term radioactive waste.

Phases 2 and 3 require technology developments and add costs. Instead of depositing glass-stabilized wastes, the fission products could be placed in temporary storage (30–60 years), with the objective of processing in the future when larger volumes of these materials have accumulated. These wastes could be stored at the reprocessing facility, and the temporary storage could reduce the consumptions of glass.

Waste minimization provides an opportunity to reduce the 3.4 parts of high-level fission product waste to 0.4 part of high-level waste. Due to the different chemical properties of the elements (nuclides) in fission waste, a variety of chemical process steps would provide the desired waste separations. The stable isotopes would not be high-level waste. This would leave the 0.4 part of unstable products that could be placed in temporary storage.

The volume of the unstable products would be small. Technology is available to transmute many of the unstable isotopes—using fast-spectrum reactors and/or accelerators. Storage for centuries would then be unnecessary. The waste would be managed with a program of reduction of the waste that requires long-term storage. Eventually, even long-term storage could be eliminated.

Report to Congress

In January 2003, the U.S. Department of Energy (DOE) prepared the "Report to Congress on the Advanced Fuel Cycle Initiative: The

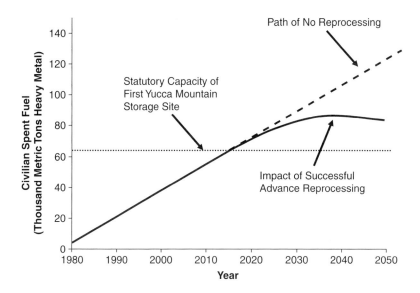

FIGURE 11-10. Impact of advanced nuclear fuel reprocessing.

Future Path for Advanced Spent Fuel Treatment and Transmutation Research."[42] This official document confirms the potential of nuclear technology to meet the U.S. energy needs without continued growth of spent nuclear fuel inventory. This is illustrated by the maximum in civilian spent fuel storage at about 2040, according to Figure 11-10, prepared from this DOE report.

The mass of radioactive waste would decrease if the stable elements were removed and used. For every 3.4 parts of fission products, 3 parts are stable after moderate storage times. The volume of remaining materials that are radioactive hazards is quite small. If the entire fission product inventory (for 30 years) was cast into a metal cube, the volume would be less than the size of a small house. It is an option to deposit this concentrated material for several decades or even a century. Transmutation technology could be developed and used to transform this fraction into benign, stable waste.

References

1. "A Technology Roadmap for Generation IV Nuclear Energy Systems," Issued by the U.S. DOE Nuclear Energy Research Advisory Committee and Generation IV International Forum, December 2002, p. 1.

2. Nuclear Energy Institute—High Level "Nuclear Waste" Is Really Used Nuclear Fuel. Retrieved from http//www.nei.org/index.asp?catnum=1&catid=14.
3. Ibid.
4. J-P. Adloff, and R. Guillaumont, *Fundamentals of Radiochemistry*, Ann Arbor: CRC Press, 1993, pp. 15–20.
5. L. K. Roland, B. G. Jerry, and T. Gary, Eds., *The Plutonium Story: Journals of Professor Glenn T. Seaborg, 1939–1946*. Columbus, OH: Benefiel; Battelle Press, 1994.
6. B. B. Cunningham, and L. B. Werner, "The First Separation of the Synthetic Element 94PU-239." In *The Transuranium Elements, Research Papers*, G. T. Seaborg, J. J. Katz, and W. M. Manningam, Eds. New York: McGraw-Hill Book Company, 1949, pp. 51–78.
7. S. Groueff, *Manhattan Project: The Untold Story of the Making of the Atomic Bomb*. Boston, MA: Little, Brown and Company, 1967, p. 9.
8. H. Thayer, *Management of the Hanford Engineer Works in World War II*. New York: American Society of Civil Engineers Press, 1996, p. 94.
9. S. Glasstone, *Sourcebook on Atomic Energy*, 3rd ed. Princeton, NJ: D. Van Nostrand Company, 1967, pp. 519–523.
10. S. G. Thompson, and G. T. Seaborg, *Progr. in Nuclear Energy*, Ser. III, Vol. 1, 1956, p. 163.
11. Adloff and Guillaumont, p. 20.
12. M. Benedict, T. H. Pigford, and H. W. Levi, *Nuclear Chemical Engineering*, 2nd ed. New York: McGraw-Hill Book Company, 1981, pp. 458–459.
13. H. Thayer, p. 17.
14. T. H. Siddall III, "Solvent Extraction Processes Based on Tri-n-butyl Phosphate." In J F. Flagg, Ed. *Chemical Processing of Reactor Fuels*. New York: Academic Press, 1961, pp. 199–207.
15. Ibid.
16. Benedict, Pigford, and Levi, p. 461.
17. Dwight Eisenhower, Atoms for Peace. Speech delivered to the U.N. General Assembly, Dec. 8, 1953, Idaho National Lab publication 03-GA51120-01.
18. R. G. Wymer, and B. L Vondra, Jr., Eds., *Light Water Reactor Nuclear Fuel Cycle*. Boca Raton, FL: CRC Press, 1981, pp. 39–41.
19. Ibid., p. 65.
20. R. G. Cochran, and N. Tsoulfanidis, *The Nuclear Fuel Cycle: Analysis and Management*, 2nd ed. LaGrange Park, IL: American Nuclear Society, 1990, p. 270.
21. Wymer and Vondra, Jr., pp. 69–72.
22. http://www.uic.com.au/nip09.htm.
23. http://www.uic.com.au/nfc.htm.
24. R. A. Hinrichs, and M. Kleinbach, *Energy—Its Use and the Environment*, 3rd edition, page 475. Brooks/Cole: New York, 2002.
25. Benedict, Pigford, and Levi, p. 475.

26. Wymer and Vondra, Jr., pp. 67–68.
27. Wymer and Vondra, Jr., p. 92.
28. Cochran and Tsoulfanidis, p. 230.
29. "Report to Congress on Advanced Fuel Cycle Initiative: The Future Path for Advanced Spent Fuel Treatment and Transmutation Research." Prepared by the U.S. DOE, January 2003.
30. M. S. Thomas, M. A. Norato, G. F. Kessinger, R. A. Pierce, T. S. Rudisill, and J. D. Johnson, "Demonstration of the UREX Solvent Extraction Process with Dresden Reactor Fuel Solution." Report WSRC-TR-2002-00444, U.S. Department of Commerce, NTIS, Springfield, VA, 2002.
31. G. F. Vandegrift, M. C. Regalbuto, S. B. Asse, H. A. Arafat, A. L. Bekel, D. L. Bowers, J. P. Byrnes, M. C. Clark, J. F. Emery, J. R. Falkenburg, A. V. Gelas, L. D. Hafenrichter, R. A. Leonard, C. Peraira, K. J. Quigley, Y. Tsai, M. H. Vander Pol, and J. J. Laidler, "Lab-Scale Demonstration of the UREX+ Process." WM'04 Conference, February 29–March 4, 2004, Tucson, AZ.
32. A. E. Waltar, and A. B. Reynolds, *Fast Breeder Reactors*. New York: Pergamon Press, 1981, pp. 24–26.
33. History of Nuclear Energy, ANL-W, 1998. Retrieved from http://www.anlw.anl.gov/anlw_history/general_history/gen_hist.html.
34. "A Report to Congress on Electrometallurgical Treatment of Waste Forms." March 2001, pp. 4–6.
35. ANL-W History—Reactors (EBR-II). Retrieved from http://www.anlw.anl.gov/anlw_history/reactors/ebr_ii,html.
36. Electrochemical Process for Spent Fuel Treatment. Retrieved from http://www.cmt.anl.gov/science-technology/nuclear/.
37. Survey of Energy Resources: Part 1 Uranium. World Energy Council. Available at http://www.worldenergy.org/wec-geis/publications/reports/ser/uranium/uranium.asp.
38. http://www.uic.com.au/uran.htm.
39. "Report to Congress Advanced Fuel Cycle Initiative: Objectives, Approach, and Technology Summary." U.S. DOE Office of Nuclear Energy, Science, and Technology, May 2005.
40. Personal conversation with Wes Berry.
41. "U.S. DOE Office of Nuclear Energy, Science and Technology Advanced Fuel Cycle Initiative (AFCI) Comparison Report, FY 2004." Published by the U.S. DOE, September 2004.
42. "Report to Congress on Advanced Fuel Cycle Initiative: The Future Path for Advanced Spent Fuel Treatment and Transmutation Research." Prepared by the U.S. Department of Energy, Office of Nuclear Energy, Science, and Technology, January 2003. Available at http://www.ne.doe.gov/reports/AFCI_CongRpt2003.pdf.

CHAPTER 12

Nuclear Power Plant Design

Nuclear energy has a strong role to play in satisfying our nation's future demands for energy security and environmental quality. The advantages of the desirable environmental, economic, and sustainable attributes of nuclear energy give it a firm place in supplying the growing U.S. and international energy demands. The National Energy Policy issued in May 2001 stated that there should be an expansion of nuclear energy in the United States. This initiative recommended work on advanced nuclear fuel cycles and next generation nuclear reactor technologies. Advanced spent fuel reprocessing and safe disposal of radioactive residues are necessary to make this program long-term sustainable.[1]

The global advances in the standard of living and economic expansion for nations place increasing demands on energy production. The Nuclear Energy Research Initiative (NERI) and its international twin (I-NERI) report on cooperation among nations to finance and conduct the research with the objective to advance sustainable nuclear energy systems.[2] The overall objectives of this program include the development of advanced concepts and scientific breakthroughs, promotion of collaboration with international agencies and research groups, and promotion of nuclear science and engineering capabilities to meet future technical challenges.The plan is to coordinate discussions among governments, industry, and the international research community regarding development of Generation IV Nuclear Energy Systems Initiative (Generation IV), the next generation of nuclear energy systems.

International collaborations depend on government-to-government agreements that remain in place. The current structure of the agreements link together the international partners conducting research on the three mutually complementary programs: Generation IV, the Advanced Fuel Cycle Initiative (AFCI), and the Nuclear Hydrogen Initiative (NHI).[3] The Generation IV International Forum currently has 11 members: Argentina, Brazil, Canada, European Union, France, Japan, Republic of Korea, Republic of South Africa, Switzerland, the United Kingdom, and the United States. Projects are selected on the basis of contribution to the overall objectives of the collaboration and funding for participants provided by each participating country. There will be jointly selected projects that will be cost-shared. The growth of nuclear power will require the United States to be the worldwide leader in the development and demonstration of technical options that do the following:

- Expand the use of nuclear power worldwide
- Effectively manage radioactive waste
- Reduce the threat of nuclear material misuse
- Enhance national security

Generation IV is a step beyond previous energy systems, with planned incremental improvements in economic competitiveness, sustainability (maximize the use of the energy content of uranium while generating much less nuclear waste), development of passively safe systems, and methods that will reduce the routes that lead to nuclear proliferation. The Generation IV Roadmap identifies the six most promising nuclear energy systems for development. The level of active development of each system is based on the technical maturity of the option and the potential to meet program and national goals. With common program goals defined, bilateral and multilateral agreements between member nations provide flexibility to address local needs with the advantage of shared technical data. The agreement structure does seem cumbersome, but open access to technical advances is a way to ensure that nuclear materials are used to produce energy and to close diversion to nuclear weapons.

New-generation nuclear power plants will need to meet the performance standards on safety, low environmental impact, and competitive prices. Since all of these standards are measured per kWh of electrical power produced, improved thermal efficiency is a win-win-win situation and will be discussed first. After a discussion on thermal efficiency, the Generation IV designs will be reviewed. Finally, the lessons history offers will be discussed.

Advances in Thermal Efficiency

As the name "heat engine" implies, heat engines are based on methods of converting high temperature energy (heat) into work and low temperature energy. The illustrative example of Figure 12-1 is for a steam cycle operating at 33% thermal efficiency. For every 100 kW of high temperature heat going into the engine, 33 kW of work is produced and 67 kW of waste heat is rejected into the environment. As illustrated by Equation 12-1, thermal efficiency is defined as the net work divided by the heat input from the high-temperature source. An energy balance on a power plant allows the thermal efficiency to also be written in terms of the heat at the high temperature and heat rejected at the low temperature.

$$\text{Eff}_{thermal} = \text{Net Work Produced}/\text{Heat Input}$$
$$= (Q_H - Q_L)/Q_H = 1 - Q_L/Q_H \quad (12\text{-}1)$$

The majority of advances in heat engine technology targets increasing this thermal efficiency. A French engineer, Nicolas Leonard Sadi Carnot (1796–1832), recognized that the thermal efficiency of a heat engine increased with increasing temperatures of

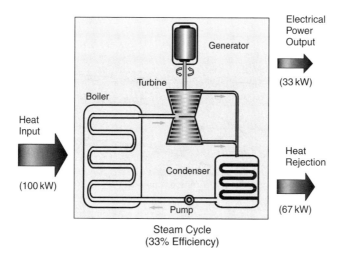

FIGURE 12-1. Steam cycle operating at 33% thermal efficiency.

the heat input and decreasing temperatures of the heat rejection. The best possible efficiency for a given source of heat and reservoir for rejecting heat is with the Carnot cycle. The Carnot cycle is a reversible heat engine operating from hot and cold reservoirs at constant temperatures of T_H and T_L, respectively. Equation 12-2 provides the thermal efficiency of the Carnot cycle where the temperatures are in degrees Kelvin, the absolute temperature scale.

$$Eff_{Carnot} = 1 - T_L/T_H \qquad (12\text{-}2)$$

Equation 12-2 indicates that as the temperature of heat input in a power cycle increases, the thermal efficiency increases. This trend is verified by the historic data of Figure 12-2 that shows the evolution of the steam-power cycle (typically coal-fired).

For large commercial power plants, T_L is set by the environment because it is the only place that is large enough to take in vast amounts of heat without increasing temperature. Heat rejection during warm summers increases T_L in equation 12-2, and this lowers the efficiency. A major equipment component of power plants are the cooling towers that circulate and evaporate water to provide a practical low-temperature place to reject the low-temperature heat from the steam turbine.

Most locations have climates that allow the cooling water heat rejection at 40°C or less year round. Because the heat-rejection temperature is controlled by local climates, the only degree of freedom in the Carnot cycle equation for increasing efficiency is the temperature of a high-temperature energy source (T_H).

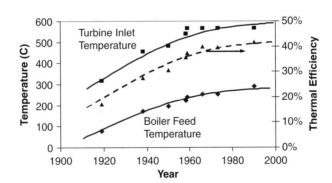

FIGURE 12-2. Evolution of thermal efficiency in a steam cycle. Higher-temperature steam turbine operation was key.

Nuclear Power Plant Design 323

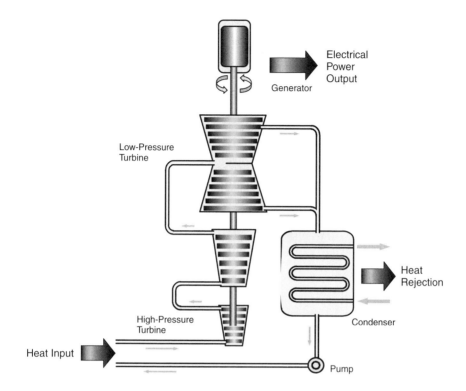

FIGURE 12-3. Staged expansion.

Figure 12-3 illustrates the basic steam cycle. The generator is driven by the turbine and produces electrical power. Steam is condensed by cooling it in the condenser. The flow of cooling water through the condenser results in heat rejection to the surroundings. The heat input occurs in a boiler (not shown). The boiler produces steam that is directed to the turbine(s), and this steam is returned to the boiler in the form of water with a pump to overcome the pressure drop through the turbine(s). The work put into the pump is small in comparison to the power produced by the turbines, resulting in a net production of power.

In practice T_H is the highest temperature that a working fluid reaches in a power cycle. This is usually the temperature of the working fluid as it leaves the heat exchanger producing the steam just prior to expansion in the turbine. In a nuclear reactor, the maximum temperature occurs when the working fluid is in contact with the fuel rods; in a natural gas power plant, it is the combustion temperature; and in a pulverized-coal-fired power plant, it is the temperature of steam as it exits the superheater.

In practice heat input is not at a constant temperature, since the working fluid increases in temperature as it is heated (or as combustion takes place with a gas turbine). The Joule efficiency was defined to take into account that practical engines do not operate with all heat input at a constant temperature. Equation 12-3 defines the Joule efficiency.

$$\text{Eff Joule} = 1 - \text{TLavg}/\text{THavg} \tag{12-3}$$

Because the Joule efficiency does not account for process irreversibilities, further modification is needed to correlate with actual processes. A correlation effective for the historic data of Figure 12-2 applies an overall reversibility factor (f) that indicates that the low-temperature heat rejection increases with increasing irreversibility. This empirical formula is provided by Equation 12-4.

$$\text{Eff}_{\text{Modified Joule}} = 1 - T_{\text{Lavg}}/[f\ T_{\text{Havg}}] \tag{12-4}$$

Figure 12-4 compares the historic data with Equation 12-4, and a reversibility factor of $f = 0.77$ represents the performance of the steam power cycle.

In the correlation of Figure 12-4, T_{Lavg} was taken as 313 K (40°C), and THavg was the arithmetic average of the boiler feed temperature and the turbine inlet temperature. Based on these correlations, the efficiency increases with the average temperature at which heat is received by the working fluid. Implicit in this correlation is that good design practices and efficient turbines/pumps

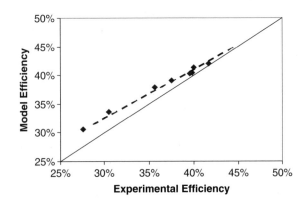

FIGURE 12-4. Accuracy of empirical model for power cycle thermal efficiency.

are used. The reversibility factor of 0.77 is obtained with state of the art turbines and pumps, as well as designs where the difference between the hot and cold streams in the heat exchangers is less than 10°C.

Practical Brayton power cycles fueled with natural gas depend on materials development increasing the temperatures at which the metals of turbine blades can operate and large heat exchangers can be economically manufactured. As illustrated by the trends in boiler feed temperature of Figure 12-2, regenerative heating of the working fluid is just as important as increasing turbine operating temperatures to increase T_{Havg} and the thermal efficiency of the power cycle.

After partial expansion of the steam, some of it is diverted to feed water heaters. This feed water heating uses the lower-quality energy of partially expanded steam rather than heat provided by combustion or the nuclear reactor. Moving from 17% to 42% thermal efficiency, the number of heaters increased from two to eight or more. Higher pressures were necessary to increase the boiler feed temperature above 290°C. The higher pressures required reheating the steam after expansion through the high-pressure turbine. Moving from 17% to 42% thermal efficiency required reheating the steam two times as it moved from the inlet high pressure to the turbine exit pressure.

Figure 12-5 shows a steam cycle with one steam reheat. When steam is produced at higher pressure, a steam reheat is used to keep excessive water from condensing in the turbine. Excessive condensation leads to erosion and failure of the turbine. Reheating the steam before completing expansion in the low-pressure turbine reduces the excessive condensation problem and provides additional high-temperature heat input that increases the thermal efficiency.

Both open- and closed-feed water heaters can be used to preheat the boil feed water. A small amount of condensing steam heats the feed water to the temperature of the steam. Higher-steam pressures, repeated steam reheat, and multiple feed water heaters were all needed in the evolution of the steam cycle to achieve the higher T_{Havg} and convert more heat to work.

Figure 12-6 superimposes the increases in Carnot, Joule ($T_{Lavg} = 313\,K$), and modified Joule ($f = 0.77$) efficiencies as the working fluid (steam) temperature increases. Nuclear and coal-fired power plants closely follow the modified Joule curve, since the reversibility factor of 0.77 is characteristic of current turbine and regenerative heat transfer efficiencies. Natural gas combined

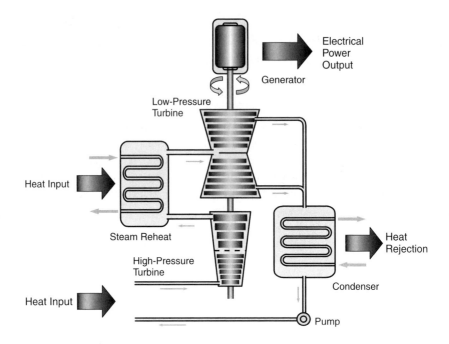

FIGURE 12-5. Steam reheat in power cycle.

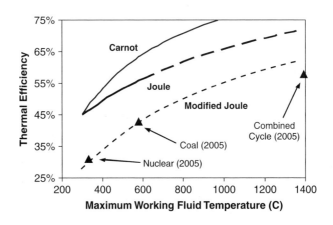

FIGURE 12-6. Comparison of efficiency projections of different models. The Joule and modified Joule models assume a feed temperature of 313 K.

cycles do not follow the modified Joule curve because the lost work associated with air compression and heat transfer to the low-temperatures cycle is inefficiencies of the combined cycle that are not part of the steam cycle.

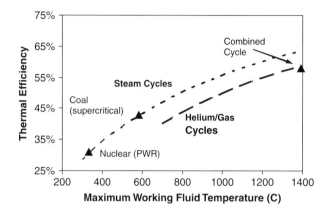

FIGURE 12-7. Projected thermal efficiencies as a function of maximum steam temperature and a low temperature of 313 K.

Figure 12-7 shows (based on the modified Joule equation) the performance potential for the steam cycle and combined cycles. The correlations in Figure 12-7 represent goals for the new Generation IV reactors with higher reactor temperatures that translate to higher-efficiency electricity production from nuclear power. Higher thermal efficiencies reduce both the capital and fuel costs for the nuclear power system.

Steam Cycles in Commercial Operation

The concepts for improved efficiency of heat engines are well known. It is the temperature limitations of current nuclear boiling water reactors (BWR) and pressurized water reactors (PWR) that limit the thermal efficiency of these nuclear power plants.

For comparison, Figure 12-8 illustrates a pulverized coal-fired power plant. Coal is ground into powder so that when introduced into a flame, it burns, similar to a spray of liquid fuel. The hot flue gases rise from the flames to steel pipes in the upper section of the fire box that consists of the boiler, superheater, and reheater components of the power plant. The steel piping contains the liquid, and supercritical fluids passing through the boiler.

The pressures and temperatures of the steam in the boiler and superheater are limited by the extent to which costly materials can be used to fabricate the pipes to withstand the extreme temperature and pressure. Multiple steam reheats (only one shown)

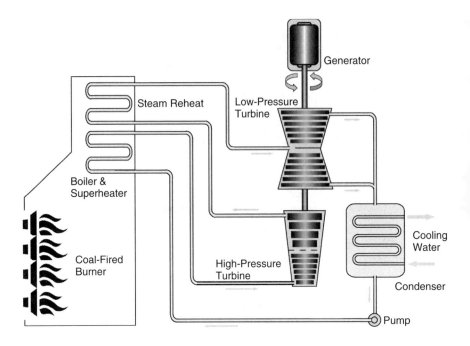

FIGURE 12-8. Boiler, superheater, and steam reheat in a pulverized coal power plant.

can be attained by routing the partially expanded steam back to the reheat section of the boiler.

Boiling Water Reactors

Figure 12-9 shows a nuclear boiling water reactor. Water enters the reactor, preheated by the feed water heaters (about 150°C, not shown). The pressure and temperature in the reactor are maintained near 70 MPa and 286°C.

The water surrounding the fuel rods in the reactor core of the BWR must be maintained as liquid because the core is designed for water to serve as the neutron moderator. The velocity of neutrons slow down through collisions with the water before colliding with the uranium atoms in the nuclear fuel. If water is not present as liquid, an insufficient number of collisions with water occur and the neutrons have too high an energy, the rate of fissions decreases, and the nuclear fission reaction will not be self-sustaining. While this is a desirable feature in case of pump failure, normal operation requires that liquid water should surround the fuel rods. In a BWR,

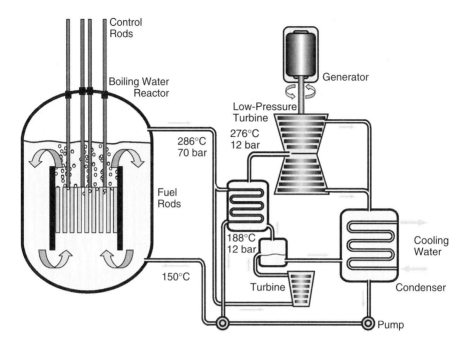

FIGURE 12-9. Boiling water reactor (BWR).

the fraction of vapor in the core can be adjusted by controlling the circulation rate of water through the core; the water circulation rate works to control the nuclear fission processes.

BWR systems employ high-volume jet pumps (not shown) to assist in the circulation of water through the core. Steam is formed, but the high water circulation rate rapidly carries the steam to the top of the reactor, where it separates from the water and flows to the system turbines.

The steam leaving the BWR is saturated steam near 286°C and it expands through a first stage condensing turbine. The liquid water is removed from the turbine continuously and about 10% of the steam is condensed in this stage. As shown in Figure 12-9, the low pressure steam is reheated with some of the 286°C steam from the reactor and then expanded through the low pressure turbine. The T_{Havg} is quite low, about 218°C, but reheating the low pressure steam brings the BWR thermal efficiency to about 33%.

Pressurized Water Reactors

The BWR is an open cycle system with the steam generated in the reactor core going directly to the steam turbine. The PWR is a closed cycle with an isolated, pressurized water loop between the reactor core and heat exchangers that produces steam for the power cycle. This isolates the steam turbine from the reactor core should a fuel element fail. Figure 12-10 illustrates a PWR. Borate is added to the primary water loop to absorb neutrons during the early part of new fuel burn. As fission products build up in the fuel, they absorb neutrons and the borate concentration is reduced to maintain uniform power production. The water in the primary loop remains liquid under pressure and exits the reactor at 315°C and 150 bar above the boiling point pressure, 105.4 bar.

The closed-cycle design of PWR eliminates possible radioactive contamination of the power cycle's working fluid. For this reason, commercial PWRs outnumber commercial BWRs by about 3:1.

A boiler, superheater, and reheat are used with the BWR similar to a coal-fired facility but at lower temperatures and pressures. In principle, the PWR reactor can attain higher efficiencies than the · BWR, but the upper end of the efficiency remains about 33%.

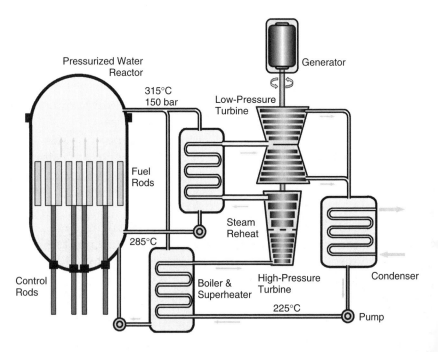

FIGURE 12-10. Pressurized water reactor (PWR).

Generation IV Nuclear Power Plants

The light-water reactors (BWR and PWR) are Generation II reactor designs with the BWR and PWR comprising 90% of the nuclear reactors in the United States and 80% of the nuclear reactors in the world. Table 12-1 lists the most promising Generation IV reactors along with typical maximum temperatures for the power cycles associated with each design.

Table 12-1 shows the lowest anticipated maximum cycle temperatures over 500°C for each of these Generation IV systems. They will attain thermal efficiencies in excess of 40%. Thermal efficiencies up to 50% are possible with the higher temperature systems. This means that a 1 GW power plant becomes a 1.3 to 1.5 GW power plant with the same fuel consumption.

TABLE 12-1
Summary of nuclear reactor designs and operating temperatures.

System	Abbreviation	Typical T_{max}* (°C)	Fast Flux
Generation II			
Boiling Water Reactor	BWR	288	No
Pressurized Water Reactor	PWR	300	No
Generation IV			
Gas-Cooled Fast Reactor System	GFR	850	Yes
Lead-Cooled Fast Reactor System	LFR	540/790	Yes
Molten Salt Reactor System	MSR	680/780	Other, with full actinide recycle
Sodium-Cooled Fast Reactor System	SFR	540	Yes
Supercritical Water-Cooled Reactor System	SCWR	510/540	Option
Very High-Temperature Reactor System	VHTR	990	No

*Temperatures are for working fluid in the power cycle. A 10°C minimum approach temperature is assumed for each heat transfer process for the indirect systems.

Generation IV Reactor Systems

The early or prototype nuclear power reactors built in the 1950s and 1960s are classified as Generation I energy systems. This experience provided the technology for the Generation II light-water moderated reactors deployed in the 1970s that are most of the commercial reactors in the United States today. The evolution of these designs with advances in control, safety, and economics make up the Generation III light-water reactors that have been deployed outside the United States in the 1990s. The success of nuclear reactors deployed in the United States is indicated by these numbers from 2002: They produced 790 billion kilowatt-hours of electricity at an average cost less the 1.70 cents per kilowatt-hour. Three billion tons of air emissions would have been produced by fossil fuel plants producing the same electrical energy in 2002.[4] This historical record drives the plan that future energy sources move toward nuclear energy replacing fossil fuels. The new systems are designated Generation IV nuclear energy systems.

Ten nations have joined to develop technology goals for Generation IV energy systems with sharp focus on four areas: sustainability, economics, safety and reliability, and proliferation resistance and physical protection.[5] Experts working in teams selected six Generation IV energy systems that should be considered as candidates for long-term (30 years) development and deployment. They are listed alphabetically in Table 12-1.

Electricity is the primary product of the current fleet of commercial nuclear power plants. Some of the Generation IV energy systems will be designed to serve the dual role of providing high-temperature thermal energy for chemical processing as well as electricity. The near-term nuclear power system development program for the United States will focus on electric power generation and hydrogen production. Hydrogen will be used as an "environmentally clean" transportation fuel to gradually replace gasoline and diesel fuel, major sources of pollution in high-population-density metropolitan regions. The U.S. DOE is pressing research and development for near-term deployment of the VHTR and the SCWR nuclear energy systems.

Supercritical Water-Cooled Reactor

The mission of the SCWR is the production of low-cost electricity. There are two proven technologies that support the selection o this energy system: Liquid water-moderated reactors are common and therefore provide operating history for development of the SCWR. Coal-fired supercritical water boilers are in operation around the world, so the steam end of this energy system has been developed. The SCWR reactor core, based on the U.S. LWR experience, would be contained in a pressure vessel with the high-temperature, high-pressure, supercritical water expanding directly into the steam turbine. The fuel would be low-enriched uranium oxide, with no need for new fuel development or new fuel reprocessing technology. The increased temperature and pressure will require additional study of the structural material oxidation, corrosion, stress cracking, embrittlement, and creep (dimensional and microscopic stability), all required to ensure the design lifetime of the reactor system. The SCWR design would increase thermal efficiency due to the higher-temperature steam turbine, but there will remain the once-through fuel cycle that characterizes the current LWR systems. Long-term sustainability will require reprocessing of this additional LWR-spent fuel.[6] Figure 12-11 illustrates the supercritical water-cooled reactor. It is much like the BWR schematic in Figure 12-9.

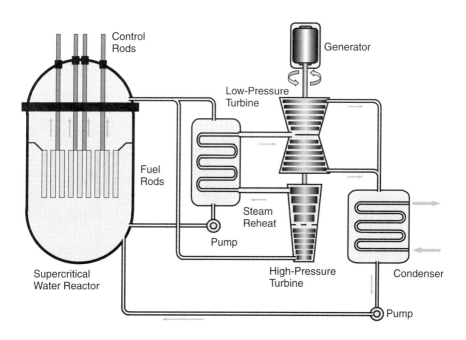

FIGURE 12-11. Supercritical water-cooled reactor.

Very High-Temperature Reactor

The very high-temperature, VHTR, will be designed to produce both electricity and hydrogen.[7] Helium will be circulated through the reactor core at high pressure to pick up thermal energy. Some of the hot helium is passed through a high-temperature heat exchanger to provide process heat. Most of the hot helium will be expanded through a gas turbine to generate electricity and turn the compressors that return the cooled, low-pressure helium to the reactor core pressure. This is an application of the classical Brayton cycle gas-turbine engine for producing work from a hot gas.[8] One proposal uses the electricity to produce hydrogen by electrolysis of high-temperature steam.[9] Figure 12-12 is a schematic of the very high-temperature reactor system.

The design planned for the VHTR will be a graphite-moderated thermal neutron spectrum reactor. The reactor core might be a prismatic graphite block core, or it could be a pebble bed reactor.[10] The fuel pebbles could be uranium metal or oxide particles uniformly distributed in porous graphite surrounded by solid graphite and coated with silicon carbide (tricoated-isotropic (TRISO)-coated gas reactor fuel particles). Each pebble would contain the fission product gases and solids during the irradiation lifetime of the pebble. One proposal is to circulate the pebbles: withdrawing them from the bottom of the reactor vessel and introducing them at the top. The pebbles could then be visually inspected for physical damage and monitored using the emitted gamma radiation to measure fuel burnup. Damaged or spent fuel pebbles would be sent to

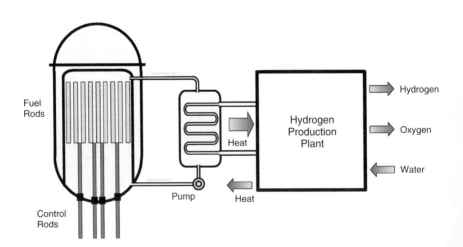

FIGURE 12-12. Very high-temperature reactor.

fuel reprocessing and new fuel added to maintain the pebble fuel inventory.

The prismatic graphite block core would be rigid material that contains the low-enriched uranium fuel and provides a thermal neutron spectrum reactor. The size of this reactor would be much smaller compared to first graphite piles fueled with natural uranium. These next-generation reactors will be designed to use a low-enriched uranium fuel and increase the nuclear fuel burnup beyond that attained with the LWR reactors.[11]

Both of these reactor systems use thermal spectrum neutrons and therefore cannot efficiently fission the minor actinides present in the spent nuclear fuel. A primary objective of the Generation IV nuclear energy program is to develop fast-flux reactors that will fission all of the transuranium elements in recycled spent nuclear fuel. This will reduce the volume and long-term radiotoxicity of the fission product waste stream. Passing from a once-through to a closed-fuel cycle extends the useful energy yield of the world supply of uranium many fold, a long-term energy sustainability objective. The research and development programs on the fast-flux reactor options are designed to select, by the year 1010, the fast-flux energy system(s) for commercial development and deployment by the year 2050.

Gas-Cooled Fast Reactor

The gas-cooled fast reactor offers the advantage of building on the high-temperature fuel technology that will be used in the VHTR. The GFR offers the sustainability feature with reduction of the volume and toxicity of its spent fuel and the added potential to use reprocessed LWR spent fuel that continues to accumulate with the once-through fuel cycle.[12] The GFR fuels and in-core structural components must be shown to survive the high temperatures and the fast-neutron radiation. Since recycled fuel will contain the minor actinides and some fission products, the serviceable life of the fuel will depend on the integrity of these multicomponent fuel elements. Tests must demonstrate the fuel integrity and performance over the irradiation time between refueling. Figure 12-13 illustrates the gas-cooled fast reactor.

Successful deployment of the GFR, a new reactor system, will require detailed safety analysis. The development of computational tools to design the energy system hardware, run simulations of operating transients (example, failure of gas coolant flow), and

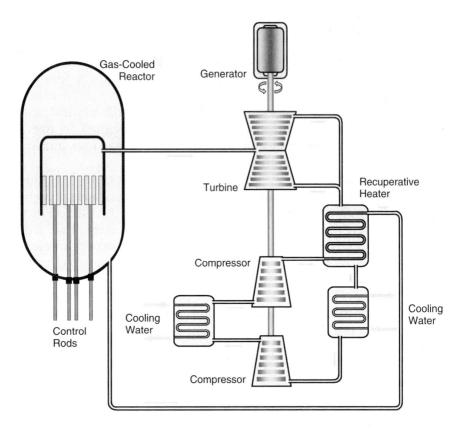

FIGURE 12-13. Gas-cooled fast reactor.

identify data gaps that must be filled with experimental measurements and material qualification data.

Sodium-Cooled Fast Reactor

The sodium-cooled liquid metal energy system features a fast-spectrum reactor and a closed-fuel cycle.[13] Sodium is the reactor core coolant of choice because sodium has a small collision cross section for neutrons, allowing neutrons to pass without slowing down. There has been significant development of the SFR system; the EBR-II program in the United States[14] is the primary source of fast-flux reactor data. This program included on-site reprocessing of the spent fuel to recycle uranium and plutonium. The EBR-II was a pool-type reactor with a low-pressure, inert gas pad above the sodium pool. Figure 12-14 is a schematic of the sodium-cooled fast reactor.

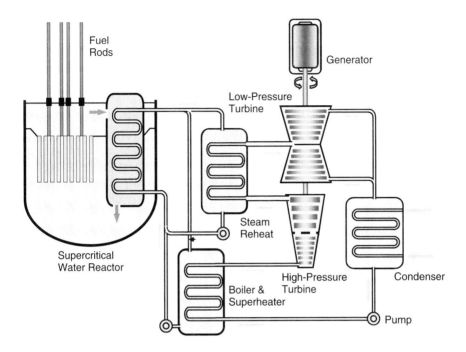

FIGURE 12-14. Sodium-cooled fast reactor.

The French have the most experience with commercial fast-flux reactors. A big jump to the Super Phenix, a commercial sodium-cooled, fast-flux power reactor, was built in France.[15] It operated from 1985 to 1997 when it was shut down due to construction problems and sodium leaks that caused a poor electric power-production record. The decision to proceed to build this 1,200-megawatt (electric) energy system may have been premature, but the commercial failure did produce valuable technical data, operating experience, and identified material problems.

The SFR option includes on-site recycling of the spent fuel. This would close the fuel cycle and provide security assurance that weapon's-grade nuclear material would not be produced. The plutonium would not be separated from the uranium and minor actinides in this process. There would be some fission products in the recycled fuel that would render it radioactive, an additional protection from diversion to weapons. Fission products can be tolerated in fast-spectrum fuels, and reducing the fuel purity makes spent fuel reprocessing much easier. The design and safety characteristics of these recycled fuels will be the focus of the development of the SFR energy system.

Lead-Cooled Fast Reactor

The lead-cooled reactor system proposal seeks to advance all of the Generation IV goals: nonproliferation, sustainability, safety, reliability, and economics.[16] For some time the Russians have been studying the substitution of lead for sodium in a fast-spectrum reactor.[17] The fuel for this reactor might be a mixed oxide with 80% depleted uranium and 20% plutonium. As the plutonium fission occurs, neutrons captured by U-238 would produce replacement plutonium. Since the accumulating fission products do not significantly change the fast-neutron energy spectrum, this fuel might continue in service for ten or more years with burnup to 15%. This would decrease the number of reprocessing cycles required to use all of the uranium to produce energy. Figure 12-15 illustrates the lead-cooled fast-reactor configuration.

The experience with lead-cooled reactors comes from the Russian navy. They built eight reactors to power submarines that used a lead-bismuth eutectic mixture (to lower the melting point of the liquid metal coolant), and there are about 80 years of reactor operation experience from this program.

The plan for this proposed reactor system includes establishing the necessary features of fuel and core materials that will provide a 20-plus-year core life. The benefits of this long core life can

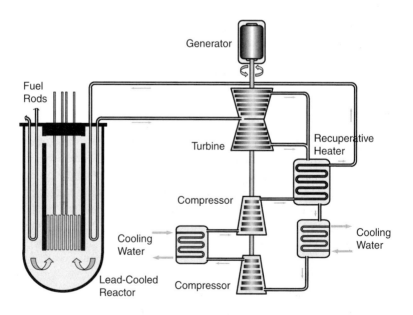

FIGURE 12-15. Lead-cooled fast reactor.

be achieved if the construction materials are developed to resist the corrosive effects of hot lead. The reactor core is set in a lead pool and thermal energy removed from the reactor core by natural convection. Heat exchangers in the upper section of the lead pool transfer the heat to high-pressure gas serving a Brayton cycle or to steam and a conventional steam turbine.

Molten Salt Reactor

Two experimental molten salt reactors were built in the United States during the 1950s and 1960s to study the basic technology of this reactor scheme. These results with the ongoing MSR research in Europe provide the basis to develop an advanced molten salt reactor, with an emphasis on fuel cycles. Figure 12-16 illustrates the molten salt reactor.

Molten salt reactors (MSR) are liquid-fueled reactors that can use actinides as fuel and produce electricity, hydrogen, and fissile fuels. Molten fluoride salt with a 1,400°C boiling point is used as a solvent for the nuclear fuel and fission product metals. This primary salt is circulated through a reactor core that contains a graphite moderator. Fissionable metals fission producing excess neutrons, promoting fertile metals to fissile metals by neutron capture. The heated salt mixture passes to a heat exchanger where the heat is transferred to a secondary molten salt loop, isolating

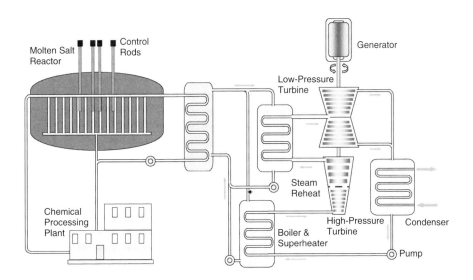

FIGURE 12-16. Molten salt reactor.

the radioactivity in the primary salt. The thermal energy passes to a second heat exchanger to supply a Brayton gas cycle (nitrogen or helium) or a conventional steam cycle turbine-generator unit to produce electricity. The operating temperature of the MSR system can be increased to provide thermally assisted hydrogen production described in the VHTR section.

A portion of the reactor salt is continuously passed to a chemical processing unit. The fission products are removed, and any of the nuclear fuel components can be removed or added to maintain the optimal fuel composition. The reactor salt contains radioactive fission products, so the chemistry steps require gamma ray shielding and remote handling.

The basic technology of the MSR has been demonstrated, but the concept has a low priority for near-term development. The conceptual design for an (advanced) AMSR will provide an understanding of the economic factors for this reactor system. There is promise of using the actinides as AMSR fuel. Disposal of the minor actinides is an important objective of the Generation IV program. Most of the AMSR reprocessing research activities will be performed under higher-priority studies, since the fuel technology of all the proposed energy systems overlap.

Toward the Future

The current fleet of light-water moderated nuclear power plants provides technical and economic data that point toward increased deployment of nuclear energy systems. These reactors use a "once-through" fuel cycle producing spent fuel, a very long-term radiological hazardous material. There is a highly contested proposal to place this spent fuel in the Yucca Mountain, Nevada, geological repository. The capacity of this repository will be exceeded if the reactors operating today shut down when their current operating licenses expire.

Nuclear power systems are large, expensive, and inherently hazardous, which means the evolution to new power systems will be slow and must be accomplished with great care. At present, there is no commercial power system based on a fast-neutron spectrum reactor and plutonium as fuel. The long-term goal of the Generation IV energy systems is to provide high-temperature thermal energy for chemical processing (to produce hydrogen?) and improve the thermal efficiency of electricity production. The next step in this international effort will be to deploy fast-flux breeder reactors, making reprocessing the irradiated fuel necessary. Reprocessing will be greatly simplified, since only the fission products need to be removed. The recycled heavy metals (all the actinide elements) fission or are transmuted to fissionable fuel or extracted as fuel in the next reprocessing cycle.

The commitment of the international community to share in the nuclear energy system project should make it happen. Reprocessing spent nuclear fuel is chemistry of heavy metals. The gamma radiation from the fission products and the toxicity of the heavy metals make remote processing necessary. Chemical analysis of the rocks on Mars is being done today from a control room in Houston, Texas. The challenge of doing remote chemistry has been met, and the experience with nuclear fuel reprocessing provides the basis for future expansion of nuclear energy in the 21st century.

Lessons from History

Nuclear Safety

A nuclear core meltdown is considered the worst-case accident in a nuclear power plant. Both U-235 and Pu-239 must be >90% pure for bomb-grade of applications compared to 3% of the usual enrichment for nuclear reactor fuel. In the diluted forms (<80% U-235 or Pu-239), the fuel cannot produce a powerful nuclear explosion. In the absence of high purity and proper configuration, the initial energy released by a chain reaction will rapidly "splat" the heavy metals too far apart to continue the chain reaction. Worst-case nuclear reactor incidents would potentially release radioactive materials in the form of hot vapors.

This was reported to occur at the Chernobyl nuclear power plant on April 26, 1986. In a meltdown, the chain reaction is not controlled, and reactor fuel temperatures increase until they melt. In addition to the fuel rods melting, the heat passes to the water in the reactor; this generates high pressures. If the hot uranium contacts water, it can react to form hydrogen. The steam pressures and/or the explosion of the hydrogen can rupture the reactor vessel and allow radioactive vapors to escape. The radioactive vapors settle to the Earth and result in radiation poisoning. The incident in Chernobyl occurred when technologists decided to use the commercial reactor to run an experiment and the experiment went wrong.

The Chernobyl Catastrophe (1986)

The Chernobyl accident is an example of failure to follow operating procedures. The plant managers were attempting to produce electric power as the reactor was being shut down for refueling. The reactor was operating at very low power. The control rods were almost fully withdrawn to allow fission to occur, even though there was significant decrease in fission due to the fission products in the spent fuel. Operators were not observing safety precautions that led to the dangerous situation. They had not informed the reactor safety group that they were running this experiment.

The Chernobyl reactor used graphite to slow down the neutrons to improve the fission probability of the U-235 in the fuel. This graphite contained tubes with the uranium oxide fuel elements in them. Channels between the fuel elements and the graphite had circulating water to remove the heat produced by the fission reaction. This reactor design made it unstable and susceptible to loss of control when cooling water flow is lost. The nuclear chain reaction and the power output increase. During the experiment, cooling water flow was lost, and there was a power surge. Some of the fuel elements ruptured, and the hot fuel reacted with the water, making hydrogen and causing an explosion. This lifted a 1,000-ton cover from the reactor, rupturing most of the remaining tubes and causing a second explosion. The reactor core was now completely exposed to the atmosphere.

The graphite in the core caught fire, and this very hot fire vaporized or produced aerosol particles of the reactor core materials. These materials included radioactive fission products and reactor fuel that were scattered as aerosols rather than as shrapnel during the explosions. The radioactive cloud from this accident spread for thousands of miles. It is easily the worst industrial nuclear accident in history. There were deaths of those who fought the fire due to radiation poisoning. There were many cases of radiation sickness and increased incidents of cancer in those exposed to the radioactive cloud. The whole region is still a laboratory for the study of radiation poisoning of the land and vegetation. The shadow of this accident will always fall across the best efforts to use nuclear reactors to produce electricity.

The Three Mile Island Accident

The Three Mile Island incident was the worst commercial nuclear disaster in U.S. history. In this incident, the reactor operators did not follow emergency procedures, misread water-level indicators in the control room, and turned off water pumps that would have cooled the reactor core. Some of the core melted, with the melt being contained in the lower part of the reactor pressure vessel. High temperatures caused steam to be released with some radiation release to the environment. The most significant health-related effects were due to the psychological stress on the individuals living in the area.[18] Scientists still disagree whether the radiation vented during the event was enough to affect the health of those who lived near the plant.[19]

The Three Mile Island accident occurred in March 1979 in a reactor located near Harrisburg, Pennsylvania. This was a pressurized water reactor that had been brought to full power late in 1978. The accident began when the feed water pumps to the steam generator stopped. The pressure in the vessel containing the reactor core increased. This caused a relief valve on the reactor pressure vessel to open and drop the control rods that stopped the neutron chain reaction. The fission product decay produced about 200 megawatts of heat immediately following reactor shutdown. The pressure in the reactor vessel continued to drop when the pressure relief valve failed to close and the water level in the reactor core continued to drop as water evaporated and vented as steam through the relief valve that was stuck open.

The reactor operators did not know that the vent valve was stuck open, since the indicator on the control panel showed it was closed. They did not replace the water in the reactor vessel. There were emergency feed water pumps that should have been running, but the reactor was now being operated in violation of safety rules. The dry fuel rods melted, and the hot fuel pellets reacted with the water to form hydrogen. This high-pressure hydrogen bubble prevented water from covering the reactor core for several days. The core meltdown did release fission products, but they were held in the reactor containment structure. A small amount of the volatile fission products did escape the containment structure to the environment.

This accident did alert the nuclear power industry to the possibility of a core meltdown. Operator training and emergency responses have been put in place with emphasis on loss of cooling accidents. Operators receive extensive training to respond properly to any emergency. Engineering changes have been made that ensure that the control system will respond automatically to emergency shutdown conditions.

Lessons Learned

Valuable lessons were learned from these two nuclear incidents. The most important is how these incidents were the result of actions of one or a few individuals who overrode the safety system and protocols designed to assure safe emergency reactor shutdown. Reactor design and operator training can overcome major operator errors and prevent emergency shutdowns from becoming disasters.

Improved reactor design, improved operating procedures, improved operator-override protocols, and location in unpopulated areas provide nuclear power that is safer than coal or natural gas power. If recent amendments to NOx and particulate matter emission standards are an indication that these emissions lead to environmental and health risks, the safety record of nuclear power generation in the United States is better than natural gas or coal.

Challenges in Nuclear Power Plant Design

Nuclear power plants have unique challenges and design features for the following reasons:

- Radioactive leaks can lead to long-term environmental contamination.
- Wind can spread radioactive aerosols over great distances.
- Nuclear fission products produce delayed heat release, unlike fire, which releases heat instantaneously.
- Radiation energy transfer mechanisms can pass through walls that present high resistance to thermal conduction.
- High thermal efficiencies associated with high-temperature heat transfer require improved high-temperature materials technology.

These aspects of nuclear processes must be addressed when designing nuclear reactors and represent the challenges that are usually covered outside nuclear science and engineering.

Chernobyl versus Three Mile Island

While both the Chernobyl and Three Mile Island incidents were the result of errors that resulted in nuclear reactor meltdowns, their similarities end there. Three Mile Island was a testament to good design and the safety of the U.S. commercial nuclear power industry. Chernobyl was an attempt to balance risk with benefit and risk won.

The Chernobyl design/incident:

- it was known that the graphite core design becomes unstable in the event of coolant loss,
- there was no secondary containment building, and
- a commercial facility without proper instruction was used to run experiments to recover additional energy during shutdown for refueling.

When coolant was lost in the Chernobyl reactor liquid water was replaced with vapor, the rate of nuclear fission increased.

Alternatively, in light-water reactors the absence of water (loss of coolant or uncontrolled heating of coolant) between the fuel rods results in fewer thermal neutrons and more fast neutrons. The fast neutrons have a much lower fission cross section, and thus are about 20 times more likely to escape from the core. The fission rate drops which leads to a reactor shutdown. Automatic shutdown when heat is not removed fast enough is known as a passive safety design feature.

Radioactive isotope decay heat continues after shutdown of the core. This energy release is manageable because there is less heat released and the cooling water flow is maintained.

Commercial nuclear reactors in the United States include standby pumps and emergency power to make sure that coolant flow is not lost. The reactor containment structure is designed to hold any radioactive material that leaves the reactor. The reactor and the power production are lost but the power plant remains and the public is protected.

Simplified plant designs provide passive safety by reducing the operation, maintenance, and testing requirements for new plant designs. Systems are designed to use gravity, natural circulation, and compressed gas rather than pumps or fans to provide emergency cooling. "Fail safe" valves are used that go to the safe position (open or closed) in the event of total plant power failure.

Possible design features include passive water injection, residual heat removal, and containment cooling. Passive safety features can be put in place so operators cannot override their operation (operator error caused both the Chernobyl and Three Mile Island incidents). All passive and active safety features can be installed for application to the primary core containment. A secondary containment is an additional passive safety design feature.

In the boiler, superheater, and the reheat piping of a coal-fired power plant, all heat must pass through the pipe walls to the steam or water by thermal conduction. In a nuclear power plant, 167 of the 200 MeV nuclear binding energy released on fission is converted to heat in the fuel tubes and is transferred to the water by heat conduction. The remaining 33 MeV passes through the tube walls as β-particles, γ-rays, and neutrons (see Chapter 10) and by-passes heat conduction through the fuel tube wall. This energy is released by interaction of the radiation energy with the cooling water. This parallel heat transfer mechanism allows higher temperature in the water and lower temperature of the fuel than is possible when all energy is transferred by heat conduction through the fuel tube walls.

High Reactor Core Temperatures

A limiting factor in power plant design is the materials available for the structural parts that can tolerate the high temperature and pressure. It is these high temperatures and fluid pressures that improve the thermal efficiencies discussed in Chapter 13. Higher thermal efficiencies are important to improved fuel economy.

Supercritical steam cycles operate at pressures above 221 bar, and this increases the equipment and maintenance costs. Fortunately, higher efficiencies depend on the nuclear reactor temperature and the pressure is secondary. There are several working fluids that operate at moderate pressure and high temperature that can be used in the design of nuclear reactors to efficiently move the thermal energy from the reactor. A second heat exchanger produces steam that drives the generator to produce electricity.

Figure 12-17 is an assessment of materials capabilities as summarized by an independent review group for Idaho National Energy and Environmental Laboratory (INEEL).[20]

At maximum turbine temperatures near 300°C, light-water reactors have thermal efficiencies from 30% to 33%. Under normal operation in these reactors, the average temperature of the uranium oxide fuel is 1,094°C, with the Zircaloy cladding cooled by the water at 343°C.[21] The INEEL report indicates that temperatures of 850°C are readily attainable (assuming corrosion is not a problem), these temperatures would bring efficiencies of 47% to 53%. These higher efficiencies can reduce both capital and fuel costs and make nuclear power a sustainable low-cost option for producing electrical power.

Nuclear Power Plant Design 347

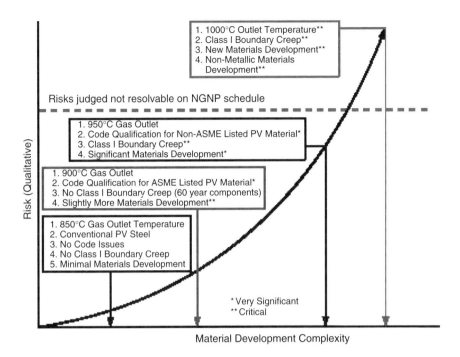

FIGURE 12-17. Level of risk as the level of materials development complexity increases.

Implementation Strategies and Priorities

The U.S. nuclear technology entered the 21st century with the industry paralyzed by regulations and cheap natural gas prices. Once the indisputable leader in nuclear power production, over two decades passed without the start of new construction. By 2005 China initiated efforts to construct more than 40 new nuclear power plants, and the French had a reprocessing program supplying 80% of electrical power needs with closed-cycle nuclear power.

Now a more favorable outlook for the U.S. nuclear industry has emerged. The convergence of possibly losing a technological leadership position, the huge $50 billion price tag for Yucca Mountain, and growing concerns about greenhouse gas emissions led to a series of Reports to Congress. These factors provided the stimulus to revisit nuclear power as an option, but it is the impressive U.S. safety, environmental, and economic history that maintain the momentum. At an average of 1.68 ¢/kWh, nuclear power (excluding capital depreciation) is second only to low-cost hydroelectric power in price. In 2003, the U.S. DOE published the U.S. Generation IV Implementation Strategy[22] consisting of the following three steps designed to a sustainable U.S. nuclear power production program:

- Deploy Generation III nuclear energy systems in the United States. This is undertaken in the Nuclear Power 2010 program announced by Secretary Abraham in February 2002. This program seeks to reduce the regulatory, economic, and technical uncertainties associated with the licensing and construction of new nuclear power plants. Nuclear Power 2010 is an industry cost-shared effort to identify sites for new nuclear plants, develop advanced nuclear plant technologies, evaluate the nuclear business case, and demonstrate untested regulatory processes, leading to an industry decision by 2005 to order a new nuclear plant for deployment in the 2010 timeframe.
- Develop separations and transmutation technology for reducing the volume and radiotoxicity of accumulated spent nuclear fuel. This is undertaken in the Advanced Fuel Cycle Initiative (AFCI). This initiative addresses the intermediate-term issues associated with spent nuclear fuel, specifically reducing the volume of material requiring geologic disposition by extracting the uranium (which represents 96% of the constituents of spent nuclear fuel), and reducing the proliferation risk through the destruction of significant quantities of plutonium contained in spent nuclear fuel. The AFCI also addresses long-term issues associated with spent nuclear fuel, specifically the development of fuel cycle technologies that could sharply reduce the long-term radiotoxicity and long-term heat load of high-level waste sent to a geologic repository.
- Develop fourth-generation nuclear energy systems for the long term that employ AFCI fuel cycle technologies. This is undertaken in the Generation IV Program. Successful development and deployment of Generation IV systems would provide a very long-term, sustainable fuel supply for the expanded use of nuclear energy. Systems developed under Generation IV and deployed in the United States would complement the existing fleet of reactors, and all would use fuel cycle technologies developed under AFCI.

This path forward sends a signal to the U.S. nuclear industry that the government wants to work with industry to overcome the regulatory paralysis and proceed to a sustainable program. However, it stops short of the commitment to make it happen. Serious flaws in this approach include:

- Generation III systems may benefit from reduced regulatory and licensing uncertainties, but low thermal efficiencies lead to marginal economics relative to coal-fired power plants.
- Recalling that reprocessing technology has been available for decades, it is the responsibility of this generation to reprocess all spent fuel that is rapidly accumulating. The spent fuel contains fuel for energy production for many years that can be recovered. If it was deposited in Yucca Mountain, it would be a radioactive hazard for 300,000 years. Recovering fissionable/fertile materials would provide nuclear fuel, remove stable fission products for nonhazardous disposal, and temporary storage of high-level fission products for future treatment would minimize the burden to future generations—it would also be sustainable.
- The lessons of history are that programs with 35-year timeframes (2040 implementation of Generation IV systems) require continuous industrial and political support. A 35-year timeline provides ample opportunity for opponents to delay or end the program.

Technologies for reprocessing, fast-spectrum reactors, and high thermal efficiency combine to form an energy source that is environmentally and economically sustainable. Especially when considering how plug-in HEV technology could increase the demand for sustainable electrical power to sustainable transportation, these advances would be a great technology feat. It is here where the 35-year timeline for implementation of fast-spectrum reactors is troubling.

History shows that great technological feats occur on 2- to 8-, not 35-, year timeframes. From the commitment to the mission as a national priority to its realization, the following were achieved in less than 8 years:

- Landing on the moon
- Panama Canal
- Atomic bomb
- Converting atomic bomb technology to nuclear power production in submarines (and then to commercial production)
- Erie Canal (regional effort, not national)
- Hoover Dam

- Space shuttle
- National Internet network
- British Channel Tunnel
- Human genome project

While the interstate highway system appears to have a long timeframe, substantial interstate networks were established and in use in less than 8 years.

Two of the most recent national initiatives announced in the United States have been on the hydrogen economy and sustainable nuclear power. The first phase of the hydrogen economy initiative was the 2001 to 2015 timeframe for demonstrating fuel cell and hydrogen storage technologies. The DOE's implementation strategy for fast-spectrum reactors was 2003–2040. The lessons of history show that longer timeframes are effective for obtaining funding for national laboratories but not effective for implementation of the technology. Changing the energy infrastructure is a long-term project that connects directly to the national economy. Federal funding for energy research is critical. Industrial leaders are not "economic risk takers," so government subsidy for implementing new technology must "walk" the new commercial nuclear power plants into production.

Recommended Reading

Advanced Fuel Cycle Initiative (AFCI) Comparison Report, FY 2003, October 2003. Available at http://www.ne.doe.gov/reports/reports.html.

References

1. *Generation IV Nuclear Energy Systems Ten-Year Program Plan, Fiscal Year 2005*, Vol. 1. Office of Advanced Nuclear Research, DOE Office of Nuclear Energy, Science and Technology. Released March 2005, Executive Summary.
2. *International Nuclear Energy Research Initiative, 2004 Annual Report*. U.S. Department of Energy, Office of Nuclear Energy, Science and Technology.
3. Generation IV, pp. 23–25.
4. Ibid., p. 1.

5. *A Technology Roadmap for Generation IV Nuclear Energy Systems, December 2002.* Issued by the U.S. DOE Nuclear Energy Research Advisory Committee and the Generation IV International Forum, p. 11.
6. Generation IV, pp. 28–31.
7. Ibid, pp. 23–28.
8. J. R. Howell, and R. O. Buckius, *Fundamentals of Engineering Thermodynamics*, 2nd ed. New York: McGraw-Hill, Inc., 1992, pp. 425–436.
9. Generation IV, pp. 23–28.
10. Ibid, pp. 23–28.
11. Ibid, pp. 23–28.
12. Generation IV, pp. 31–34.
13. Ibid, pp. 39–41.
14. ANL-W History—Reactors (EBR-II). Retrieved from http://www.anlw.anl.gov/anlw_history/reactors/ebr_ii.html.
15. R. L. Garwin, and G. Charpak, *Megawatts and Megatons*. New York: Alfred A. Knopf, 2001, pp. 131–135.
16. Generation IV, pp. 35–38.
17. Garwin, and Charpak, p. 164.
18. http://www.libraries.psu.edu/crsweb/tmi/accidnt.htm.
19. http://www.washingtonpost.com/wp-srv/national/longterm/tmi/tmi.htm.
20. Design Features and Technology Uncertainties for the Next Generation. Nuclear Plant, June 30 2004. Available at http://www.ne.doe.gov/reports/reports.html.
21. R. A. Hinrichs, and M. Kleinbach, *Energy—Its Use and the Environment*, 3rd ed. New York: Brooks/Cole, 2002, p. 477.
22. *The U.S. Generation IV Implementation Strategy.* U.S. DOE Publication 03-GA50439-06. September 2003.

CHAPTER 13

For-Profit Industrial Drivers

Profitability is the primary driver for commercialization. Profitability projections provided the motivation for the expansion of the U.S. nuclear industry in the 1970s, and the negative impact on profitability of the high costs to meet new regulations led to the end of expansion in the 1980s. This chapter reviews the fuel costs, capital costs, and operating/maintenance costs that impact the profitability projections of future nuclear power plants. The social and profitability factors for state governments will be considered as approaches to influence corporate decisions on electrical power generation options.

Levelized Cost Approach

To allow cost comparisons for capital, fuel, and operation/maintenance, a levelized cost formula provides a basis for comparing annual contributions of each components with Equation 13-1.

$$LC_{total} = LC_{capital} + LC_{fuel} + LC_{O\&M} \quad (13\text{-}1)$$

where LC is the levelized cost in $/kWh of electricity produced, often referred to as the busbar cost.

In a levelized cost comparison, the cost of providing electrical power to the grid allows the capital costs paid prior to startup to be directly compared to the fuel, operating, and maintenance costs paid during power production.

354 Sustainable Nuclear Power

TABLE 13-1
Summary of U.S. busbar costs. All costs are in ¢/kWh for base load power generation for a 2003 basis. Values are comparable to values from the report Ilten except the price of natural gas has been adjusted to reflect trends since 2003. A 25-year plant life and 10% discount factor are assumed.

Case	Plant Type	Capacity Factor*	Capital	O&M	Fuel**	Total
Nuclear	Open Cycle	80%	3.55	1.03	**0.68**	5.26
Natural Gas	Combined Cycle	80%	**0.93**	0.32	3.94	5.19
Coal	Pulverized	80%	2.53	0.60	1.05	4.18

*A capacity factor of 80% was used.
**For natural gas, a combined cycle plant efficiency of 50% is assumed. For case 1, this translates to 3.94¢/kWh on a natural gas (not busbar) basis. Here, $0.0394/kWh-e × 0.53 kWh-e/kWh-f × 2.778 kWh/0.009486 MBtu results in a fuel cost of $6.12/MBtu for natural gas.

The recent comparison of electrical costs by Ilten[1] provides a basis for discussion and sensitivity analyses. Table 13-1 compares a nuclear base case to base cases for coal and natural gas. The busbar costs are similar to those reported by Ilten. The fuel costs for each option are given in Table 13-2.

Table 13-1 show the factors that, on the average, give nuclear, natural gas, and coal their unique competitive advantages in the United States. Natural gas has an advantage due to reduced capital costs. When natural gas prices are more than about $4/MBtu, coal is the better option for base load power generation. For seasonal power plants, natural gas can be a better option than coal even at gas prices higher than $4/MBtu due to the low capacity factors

TABLE 13-2
Fuel cost basis for Table 13-1.

Case	Fuel Cost ($/MBtu)	Efficiency (Fuel to Busbar)	Busbar Fuel Cost (¢/kWh)
Nuclear	0.62	31%	0.68
Natural Gas	6.11	53%	3.94
Coal	1.29	42%	1.05

of these facilities. A low capacity factor occurs when the plant operates part of the day or well below designed power output.

Coal has lower fuel costs than natural gas and lower capital costs than nuclear. This combination makes coal a better long-term investment than natural gas and a lower capital risk than nuclear. In recent years, the ability of coal to compete with natural gas for new power plant construction has changed with increasing prices of natural gas. Natural gas prices tend to fluctuate wildly, but coal prices are more stable near $1.29/MBtu. With recent increases in natural gas prices (from $3 to $6 and even $12 per MBtu), coal is often favored over natural gas for new power plant construction.

Nuclear power's advantages are the abundance and low cost of uranium fuel. The advantage of near-zero greenhouse gas emissions strengthens the case for nuclear. To realize its potential, the levelized capital costs of nuclear will need to decrease from the $5.26 per kWh to values comparable to coal at $4.18 per kWh. Additional cost reductions can be realized by increasing power cycle efficiency, which can reduce both fuel and capital costs.

Capital Costs

Base Case Assumptions

The levelized capital costs of Equation 13-1 and Table 13-1 take into account the actual dollars spent to build the power plant facilities (K in dollars, overnight capital), the discount factor (r in percent, used in place of the borrowing) used to represent costs in terms of a reference year, and the construction time (c in years). Reductions in each of these factors can reduce the capital costs.

By selecting the reference year as the first year of electrical power production (year 1), the levelized capital cost is calculated by converting the overnight capital to a capital cost discounted to year 1 (I in dollars) and dividing this by lifetime electrical power generation of the plant discounted to year 1.

Equation 13-2 provides the method of calculating I, assuming an even expenditure of overnight capital for each year of construction.

$$I = \sum_{t=1}^{} S_t K(1+r)^{T-t+1} \approx K \sum_{t=1}^{} (1+r)^{T-t+1}/T \qquad (13\text{-}2)$$

The summation is from 1 to T is the construction period in years, and S_t is the percent of overnight capital spent each year. This approximation assumes that the overnight capital expenditure is the same each year.

Equation 13-3 relates the levelized capital cost ($LC_{capital}$ in \$/kWh) to the total overnight capital costs (K, in dollars).

$$LC_{capital} = I/[FE \sum_{t=1}(1+r)^{t-1}] \qquad (13\text{-}3)$$

The summation is over the number of years of electrical power production, F is the capacity factor, and E is the theoretical maximum amount of power that can be produced with the power plant. For a 1 GW by facility, E is 8.76E9 kWh (E = 1 GW × 365 day × 24 hours).

Ilten's estimate of $LC_{capital}$ for the once-through nuclear facility is 3.55 ¢/kWh (see Table 13-1). This is consistent with F = 80% (capacity factor), r = 10%, 25-year production life, and a construction time of 7 years. In equation form, this is \$0.0355/kWh = 1.491 K/(9.89 × 0.80 × 8.76E9) or K = \$1.65 billion for a 1 GW facility. With these assumptions the $LC_{capital}$ of 3.55 ¢/kWh corresponds to an overnight capital cost of \$1,650/kW.

Table 13-3 provides the overnight capital costs and assumptions for the levelized capital costs provided in Table 13-1. Values of the total overnight capital costs reported from other sources are included and show that the values are reasonable and consistent.

TABLE 13-3
Summary of overnight capital costs (K) with assumptions that link them to levelized capital costs ($LC_{capital}$). Assumes a 25-year production life and 10% discount factor.

Case	$LC_{capital}$ (/kWh)	T (yrs)*	F**	K (/kW)	Comparative K (/kW)*
Nuclear-OC	3.55 ¢	7	80%	\$1,650	\$1,600
Natural Gas-CC	0.93 ¢	3	80%	\$531	\$590
Coal-P	2.53 ¢	4	80%	\$1,374	\$1,350

*Construction times were taken from Table 3.5 of reference. Overnight capital costs are also reported in reference, as listed in this column. References could be redone to cite original.
**A capacity factor of 80% was used rather than 70% to be more internally consistent. Overnight capital costs estimated later in this chapter.

The assumptions and values listed in Table 13-3 provide the base case and the basis for discussions of factors that impact the $LC_{capital}$ portion of projected nuclear busbar costs.

Parameters Impacting Capital Cost

In the competition for baseload power generation, natural gas has a price disadvantage compared to coal and is not a good fuel choice. Due to the use of low-capacity factors (80% rather than 90%) and the low natural gas price ($6/MBtu rather than the current market $12/MBtu), the cost of natural gas in Table 13-1 is low.

Comparing coal to nuclear, variations in the years of production, discount rate, and capacity factors would be similar for the coal and nuclear options. Therefore, nuclear power would gain only an incremental advantage over coal—too small to detect with this calculation. The greatest impact in a sensitivity analysis is those parameters that uniquely impact nuclear power. Table 13-4 summarizes factors unique to nuclear over coal that were used in the economic sensitivity analysis.

Standardized Designs

Standardized and preapproved nuclear power plant designs could reduce the construction times from 7 to 4 years. The value of 4 years compares with 3.4 years, the minimum construction time for existing nuclear power facilities in the United States, and 5.3 years for the average construction times of facilities since 1993. The summation term of the approximation of Equation 13-2 gives the effect of this change in time. For a 10% discount factor

TABLE 13-4
Summary of parameters to be varied in sensitivity study.

Parameter	Base Case	Parametric Study
Reductions in Construction Time (years)	7	4
Decreases in Overnight Capital Costs ($/kW)	$1,650	$1,365
Selective Loan Guarantees (% interest)	10%	5%
Increases in Fuel Efficiency	31%	47%
Improved Reprocessing Technology ($/kWh)	0.68	0.40
State Subsidies ($/kWh)	0	tax neutral

(interest rate), this term changes from 1.491 to 1.276, a 14.4% decrease in the levelized capital costs, or a 0.51 ¢/kWh decrease.

Improved Reactor Designs

Ilten summarizes estimated capital costs for three alternative nuclear reactor designs as estimated by SAIC, Scully, and EIA. In each case at least one design option (pebble bed, advanced technology, modular helium reactor) gave a capital cost at or below $1,365/kW while maintaining the nuclear fuel cost. A decrease from $1,650 to $1,365 in the overnight price of a new reactor design represents a 17.3% decrease in the levelized capital cost, or a 0.614 ¢/kWh decrease.

Guaranteed Loans

Providing guaranteed loans can be a good option for a federal government to stimulate the development of a first-of-a-kind nuclear power facility. It is the goal of the federal government to promote actions that benefit the public. For the corporation building the facility, the loan guarantees would offset the risk associated with new technology and the increased construction time building a new design. A loan guarantee would correspond to the corporation qualifying for a low interest rate. Decreasing interest from 10% to 5% during construction, the Equation 13-2 summation changes from 1.491 to 1.221 for an 18.1% decrease in the levelized capital costs, or a 0.64 ¢/kWh decrease.

Efficiency

The low efficiency of nuclear power generation is just above peak power gas turbines in the mix of electrical power technologies. The source of the inefficiency is the low maximum operating temperatures of the pressurized water reactors (about 340°C) that limits the efficiency to about 33%. An increase in this efficiency to 48% is possible with a corresponding 33.3% reduction in fuel consumption. This would reduce the fuel cost from 0.68 ¢/kWh to 0.46 ¢/kWh or a decrease of 0.22 ¢/kWh.

Improved reactor designs with higher operating temperature more efficiently produce power. The new reactors are smaller but provide the same power output. Most of the 0.614 ¢/kWh decrease in cost associated with improved design is due to the smaller reactor core made possible by increased thermal efficiency.

Reprocessing

Reprocessing spent nuclear fuel is desirable because it reduces the mass of high-level radioactive materials for long-term storage by about 96%. It is assumed that all fission products must be handled as radioactive waste and other "contaminated" materials can either be reused or disposed of as low-level waste at lower disposal cost. Reprocessing consists of several physical and chemical processing steps. The French do reprocess fuel in its closed cycle nuclear processes where fuel costs are 0.90¢/kWh compared to once-through technology in the United States at 0.68¢/kWh. This once-through technology does not include the current undetermined cost of spent fuel disposal.

Despite the years of experience in France with reprocessing, the cost is not near the bottom of the cost curve. One option not explored is relaxing the metal purity of the reprocessed fuel. Rather than removing essentially all fission products and actinides, the fuel could be processed to remove only fission products and concentrate the fissile materials. Generation IV reactors would be designed to use reprocessed fuel that contains some of the fission products. This approach would reduce fuel costs about 0.28¢/kWh, but it does require new fast-flux reactors.

Reprocessing performed at the nuclear power plant site benefits the state. The 0.68¢/kWh to provide new uranium fuel would be spent to bring fuel from another state.

State Incentives

States often provide incentives for electrical power generation technologies that produce state tax revenue and provide quality jobs. A case can be made that building and operating a nuclear power plant alternative to a natural gas power plant could be a job-creation equivalent to attracting an automobile manufacturer to the state. A 1-GW power plant produces about 7.9E9 kWh of electrical power per year (90% capacity factor). At a natural gas busbar fuel cost of $0.0394/kWh, this represents $310 million per year in cash flow from the state for each natural gas combined cycle power plant producing base load electrical power.

A nuclear power plant with the same capacity would pay $54 million per year for uranium fuel. The $256 ($= 310 - 54$) million reduction in fuel cost represents savings to the state economy with a nuclear power plant rather than a natural gas plant.

Capital investments in a state also benefit the state. The investment in the construction of a new power plant comes from corporate funds that can be invested anywhere. When used for construction in state, about one-third of the capital costs are for site labor and site materials.[2] The cash flow from new construction and the money paid for fuel both have an impact on the state economy.

Table 13-5 compares the different electrical power generation options showing the cents per kilowatt-hour of electrical power produced that leave the state.

Considering the total impact, a 1-GW coal-fired power plant places about 1.87 ¢/kWh (= 4.56 − 2.69) more money into the state economy than a 1-GW natural gas combined cycle facility, natural gas assumed to be $6.11/MBtu. This represents $148 million per year reduction in fuel cost. If the 2005 natural gas prices of $12.22/MBtu were used, this difference increases to $458 million per year. The cost of electricity from a natural gas power plant is very sensitive to the price of natural gas.

These calculations indicate that building a power plant that has a lower fuel cost component in the total per kWh busbar cost is about the equivalent of attracting a large manufacturer to the

TABLE 13-5
Impacts of power plant options on state cash flow from a state that imports natural gas, coal, and uranium. All are busbar in ¢/kWh. The state impact numbers assume 40% of the capital expenditures and 80% of the O&M expenditures stay in the state during plant construction.

Case	Total & (Imported) Capital	Total & (Imported) O&M	Total & (Imported) Fuel*	Total State Import in ¢/kWh
Nuclear	3.55 (−2.13)	1.03 (−0.206)	0.68 (−0.68)	−3.02
Natural Gas	0.93 (−0.558)	0.32 (−0.064)	3.94 (−3.94)	−4.56
Coal	2.53 (−1.518)	0.60 (−0.12)	1.05 (−1.05)	−2.69
Reprocessed Nuclear	4.05 (−2.43)	1.43 (−0.286)	0 0	−2.72

*For natural gas, a combined cycle plant efficiency of 50% is assumed. For case 1, this translates to 3.94 ¢/kWh on a natural gas (not busbar) basis. Here, $0.0394/kWh-e × 0.5 kWh-e/kWh-f × 2.778 kWh/0.009486 MBtu results in a fuel cost of $6.12/MBtu for natural gas.

state. The cost of producing electric power represents a significant economic factor for the community and state.

The data in Table 13-5 include the option of a nuclear facility with and without on-site fuel reprocessing. On-site reprocessing reduces state cash export by about 0.3 ¢/kWh (= 3.02 − 2.72) representing a $24 million per year for a 1-GW facility.

Because the busbar costs of nuclear power in Table 13-5 are more than for coal, nuclear tends to export more dollars than coal even though the fuel costs for nuclear are less. Table 13-6 shows an analysis for proposed nuclear technologies with busbar cost parity with coal. In this comparison, nuclear reduces the export of cash flow relative to coal—a direct annual impact of $25 to $59 million on the state cash flow.

Multiplying factors from 4 to 6 are often used to represent the true economic impact of avoided trade deficits on a local economy. Multiplying factors and reduced trade deficits typically do not produce corporate incentives to make investments that are good for the state. One approach that improves corporate incentives is for states to pass incremental tax revenues to the corporations in the form of subsidies. An estimate of the incremental difference in state personal income tax revenues resulting from the different power plant options is 10% of the differential flows of cash from

TABLE 13-6
Impacts of power plant options on state cash flow for targeted nuclear processes. All are busbar cost in ¢/kWh. The state impact numbers assume 40% of the capital expenditures and 80% of the O&M expenditures stay in the state.

Targeted Process	Total & (Imported) Capital	Total & (Imported) O&M	Total & (Imported) Fuel*	Total State Import in ¢/kWh
Targeted Nuclear	2.47 (−1.482)	1.03 (−0.206)	0.68 (−0.68)	−2.37
Targeted Reprocessed Nuclear	2.75 (−1.65)	1.43 (−0.286)	0 0	−1.94

*For natural gas, a combined cycle plant efficiency of 50% is assumed. For case 1, this translates to 3.94¢/kWh on a natural gas (not busbar) basis. Here, $0.0394/kWh-e × 0.53 kWh-e/kWh-f × 2.778 kWh/0.009486 MBtu results in a fuel cost of $6.12/MBtu for natural gas.

the states. It is both reasonable and within the legislative power of states to pass tax incentives to corporations. For once-through nuclear versus coal, these translate to about 0.025 to 0.059 ¢/kWh.

For all of the estimates of cash flow from states, it is assumed that 60% of the capital costs go out of state. These include an average of 37.3% factory equipment costs.[3] Most of the remaining 22.7% is cost of administration by the primary construction contractor and out-of-state workers. For those states with commercial power plant construction, nuclear fuel reprocessing, or power plant equipment, the 60% of capital costs assumed to leave the state can be reduced. An additional 1 ¢/kWh or more could remain in the state for fuel reprocessing or power plant construction if the firms were located in the state.

Sensitivity Analysis

While coal has cost advantages relative to traditional nuclear power, there is potential for technology and regulatory changes to reduce the costs for nuclear power plants. Table 13-7 summarizes the various parameters on the levelized cost of production for nuclear power. The parameters apply to nuclear power over coal.

The greatest reduction in busbar costs would come from new technology that reduces the cost of a nuclear power plant by increasing the energy efficiency of the facility. Increased efficiency decreases the capital costs (at constant electrical power output)

TABLE 13-7
Summary of parameters to be varied in sensitivity study of nuclear versus coal.

Parameter	Decrease in LC_{total} (¢/kWh)
Improved Thermal Efficiency with Improved Design That Also Decreases Capital Costs	0.61 + 0.22
Loan Guarantees (Reduce Construction Times)	0.64 (0.51)
Decreases in Overnight Capital Costs Only	0.61
Increases in Fuel Efficiency (Impact on Fuel Costs)	0.22
Improved Reprocessing Technology	0.28
State Subsidies (Upper Value)	0.10

and decreases the levelized fuel costs. Combined, these can reduce the cost of nuclear power by about 0.83 ¢/kWh. These savings are cumulative and add to the cost savings from reducing the plant construction time.

The second greatest reduction can be achieved by reducing construction costs with precertified designs. The first-of-a-kind facilities with longer construction times could be covered by guaranteed loans. The estimated savings are 0.51 to 0.64 ¢/kWh. These construction costs will decrease as more precertified power plants are built.

There are additional incremental gains for nuclear power of 0.10 to 0.28 ¢/kWh through states returning incremental tax revenues to corporations and with improved fuel reprocessing technology.

Case Studies

Scenario 1: 40-Year Production Life with 4-Year Construction Times

A shift from a 25-year plant life to a 40-year plant life is reasonable based on operating experience by nuclear power plants currently in use. With a 4-year construction time for both nuclear and coal facilities, Table 13-8 was prepared to compare the busbar costs. For nuclear, the $LC_{capital} = 1.276 \times \$1.65E9/(10.67 \times 0.80 \times 8.76E9)$ or $LC_{capital} = 2.82$ ¢/kWh for a 1-GW facility. For coal, $LC_{capital} = 1.276 \times \$1.65E9/(10.67 \times 0.80 \times 8.76E9)$ or $LC_{capital} = 2.35$ ¢/kWh. Reducing construction times to 4 years and assuming a 40-year plant life reduced the busbar price difference between nuclear and coal by about half with coal still holding an advantage.

TABLE 13-8
Scenario 1 busbar costs for 40-year plant lives, 10% discount factor, and 4-year construction times for both nuclear and coal. Values are in ¢/kWh.

Case	Plant Type	Capacity Factor	Capital	O&M	Fuel	Total
Nuclear	Open Cycle	80%	2.82	1.03	0.68	4.53
Coal	Pulverized	80%	2.35	0.60	1.05	4.00

TABLE 13-9
Scenario 2 busbar costs for 40-year plant lives, 10% discount factor, and 4-year construction times for both nuclear and coal. New generation nuclear power plant has overnight capital cost similar to coal with 47% thermal efficiency. Values are in ¢/kWh.

Case	Plant Type	Capacity Factor	Capital	O&M	Fuel	Total
New Nuclear	Open Cycle	80%	2.35	1.03	0.46	3.84
State Import			−1.41	−0.206	−0.46	−2.08
Coal	Pulverized	80%	2.35	0.60	1.05	4.00
State Import			−1.41	−0.12	−1.05	−2.58
New Nuclear	Closed Cycle	80%	2.75	1.23	0	3.98
State Import			−1.65	−0.246	0	−1.90

Scenario 2: A $1,365/kW Nuclear Power Plant at 47% Thermal Efficiency

New generation nuclear power plants are projected to have overnight capital costs similar to coal plants with thermal efficiencies of about 47%, which is slightly better than coal. Table 13-9 summarizes the levelized costs for this scenario. Nuclear power has a cost advantage that is independent of plant life and discount factor.

The busbar cost of 3.84¢/kWh is consistent with a series of nuclear options reported by Ilten[4] with costs between 3.6 and 4.1¢/kWh. Ilten's lowest projected cost for coal facilities is 3.7¢/kWh.

The results in Table 13-9 indicate that nuclear power can be less costly than coal. In addition, up to 0.68¢/kWh (= 2.58 − 1.90) in cash flow leaving the state can be eliminated with nuclear power versus coal. This represents a direct $54 million per year per 1-GW facility.

Costs of Reprocessing

A relatively firm estimate of the cost difference between reprocessed nuclear fuel and new fuel is 0.22¢/kWh. This is the difference between the cost of once-through U.S. fuel at 68¢/kWh and closed cycle French fuel at 90¢/kWh.

TABLE 13-10
Estimates of capital costs for fuel reprocessing and fuel fabrication. Numbers are billions of dollars and are for 2,000 metric tons per year (spent fuel generation rate in 2005). Data from U.S. DOE Office of Nuclear Energy, Science and Technology Advanced Fuel Cycle Initiative (AFCI) Comparison Report.

Process	Reprocessing Facility	Fuel Fab. Facility
PUREX	8.0	2.0
UREX+	6.0	2.0
UREX/PYRO	6.0	3.0
PYROX	7.0	3.0
Advanced Aqueous Process	4.0	2.0

Table 13-10 lists reprocessing facility and fuel fabrication capital costs available from the U.S. DOE to supplement this estimate.

The combined reprocessing plus capital cost estimates range from $6 to $10 billion for capacities to handle all the U.S. commercial spent fuel. If the policy is to store the spent fuel for 30 years before reprocessing, the 2,000 metric tons per year capacity is enough to handle U.S. commercial waste for the next 30 years.

In 2002, 103 operating nuclear power plants produced 790 billion kWh of electric power. Assuming overnight construction of the reprocessing and fuel fabrication facilities, the capital costs produce levelized busbar costs of 0.76 to 1.270, and 0.03 to 0.05 ¢/kWh based on the power produced generating the electricity. Since the U-235 content of the reprocessed fuel is about the same as natural uranium, it offers no premium fuel value. The premium fuel value is for the plutonium present at about 0.9% in the spent fuel. The adjusted levelized cost of the plutonium from reprocessing is 0.092 to 0.154 ¢/kWh.

The costs for 25 years of operation are summarized in Table 13-11. Assuming a 0% discount factor, these costs are 0.17–0.072 ¢/kWh on the basis of electricity produced generating the waste or 0.21–0.49 on the use of reprocessed Pu-239 to enrich uranium fuel for power production. These costs are less than the price of new fuel at 0.68 ¢/kWh. However, in the U.S. Generation IV Implementation Strategy, FY2003 only provides a $12 billion fuel sale credit versus the $15.2 to $23.0 billion for operating and D&D.

The reactor fuels fabricated from reprocessing facilities would be mixed-oxide fuels. The fissile material content of the

TABLE 13-11
Estimated costs for operating reprocessing and fuel fabrication facilities for 25 years. Data from the U.S. Generation IV Implementation Strategy, FY2003.

Process	D&D	Operating
PUREX	3.0	20.0
UREX+	2.4	14.0
UREX/PYRO	2.7	12.5
PYROX	3.0	14.0
Advanced Aqueous Process	1.8	12.5

"U.S. DOE Office of Nuclear Energy, Science and Technology Advanced Fuel Cycle Initiative (AFCI) Comparison Report." Published by U.S. DOE, October 2003.

U-235 + Pu-239 from the reprocessing could be increased using excess weapon's-grade plutonium or highly enriched U-235.

The HEU (highly enriched uranium extracted from nuclear weapons) agreement signed between the United States and the Russian Federation (1993) provides for the U.S. purchase of 500 metric tons of highly enriched uranium between 1993 and 2013.[5] If this uranium (greater than 80% U-235) was available and used to enrich the 0.9% Pu-239 and 0.8%–1.1% U-235 in the heavy metals of reprocessed fuel (to reach 3.3% fissile material), it would be sufficient to enrich 29,000–36,000 metric tons of recycled fuel. At 2,000 tons per year, this would provide fuel for up to 18 years. No new uranium would need to be mined or enriched to prepare this fuel.

During this time, 0.1¢/kWh has been collected anticipating the cost of waste disposal. Thirty years times 790 billion kWh/yr times 0.1¢/kWh produced $23.7 billion. The expenses for development of the Yucca Mountain Repository, including attorney fees, have been spent from this account. There is available more than $10 billion that might be invested in reprocessing technology and recycled fuel fabrication. This would require spending authorization, but is in the spirit of the original nuclear waste disposal plan.

If these funds were used to build recycle facilities, the reprocessed fuel cost to cover the operating and distribution would be about 0.2¢/kWh. Reprocessed fuel would be lower cost than new fuel. Strategies exist to avoid very long-term storage by separating the unstable fission products from the unstable isotopes and eventually transmuting these materials. All the objectives of waste management would be met.

The present nuclear power program based on once-through fuel would fill the Yucca Mountain repository by 2015. The estimated life cycle cost for a second repository similar to Yucca Mountain is estimated to be $50 billion. The 2003 U.S. Generation IV Implementation Strategy places the cost of the once-through repository approach to be $33 to $43.2 billion more than initiating spent fuel reprocessing.

Advocates for Nuclear Power

The end of new nuclear power plant construction in the United States began in 1977 when President Carter issued an Executive Order to end reprocessing to demonstrate compliance with the nonproliferation treaty. New regulations increased construction times, construction costs, and maintenance costs. Since 1979, no new reactors have been ordered in the United States.

The problem appears to be lack of political advocates for nuclear power. The coal, natural gas, and petroleum lobbyists are numerous and powerful. There are lobbying groups opposing nuclear energy expansion.

Historically, the scientific/academic communities have been advocates for nuclear power. Environmentalist groups shared the same campuses and social circles as the scientific groups that advocated nuclear. When nuclear became unpopular in these social circles, the scientific/academic advocates gradually became silent.

In 2005, the academic community perspective was changing. The American Chemical Society Green Chemistry Conference is a premier U.S. conference each year that specifically advocates sustainable and environmentally friendly alternatives to petroleum. In 2002, mention of nuclear as a power source complementary to green chemistry met with visible scorn. By 2005, nuclear power was mentioned favorably by at least three different presentations and without visible scorn.

Nuclear power is now recognized as one of the technologies with ready solutions to global warming. The remarkable safety record of U.S. commercial nuclear power plants can no longer be ignored. The safety and low environmental impact of nuclear power plants surpass alternatives in the United States throughout its history—including the Three Mile Island accident.

Nuclear power is seen as part of a solution to global warming and may soon be recognized as the best way to replace imported petroleum. The lobbying efforts of the scientific/academic communities have deteriorated since the 1970s, and greater advocacy will be required to make changes. International programs have been set up with forward planning extending at least 30 years.

Nuclear power advocates in state governments and among community leaders need to be heard. The elected officials and their staff people responsible for energy legislation, the targets for energy lobbyists, must also become advocates for nuclear power if the advantages of nuclear power over fossil fuels are to be fully realized.

In Scenario 2, nuclear power keeps about 0.68 ¢/kWh more in the state than coal. This becomes about $54 million per year and provides many jobs for each nuclear power plant. The economic multiplying factor and a 40-year life for the facility generate $8 billion per 1-GW power plant that is nuclear rather than coal. For a natural gas facility with gas at $12/MBtu, the reduction in state imports is $520 million per year for a 1-GW nuclear power plant rather than a 1-GW natural gas power plant or a state impact of $83 billion.[i]

The advocacy role taken by non-coal-producing states and their elected officials will be important for the future of nuclear. These states have much to gain from the commercialization of the next generation of nuclear power plants shown by the busbar power costs summarized in Table 13-9. These states should advocate nuclear power and move to develop the commercial infrastructure to support construction of both nuclear power plants and nuclear fuel reprocessing plants.

When the projected overnight capital costs, the 4-year construction times, and higher thermal efficiencies are attained, the industry will be a sustainable low-cost power production option. Hundreds of years of fuel have been mined and stored as spent fuel at nuclear power plants and there are large stores of depleted uranium at DOD and DOE facilities.

Expansion of this industry does not put states in competition with each other if each state works to supply its own electrical power needs. It is in the interests of every state to form partnerships and advance the technology. The next generation of efficient, closed-cycle nuclear power plants should be developed soon and deployed no later than the current plan date of 2035.

There is an economic advantage to provide incentives to assure coal or nuclear power plants are built rather than natural gas. The

---

[i] A levelized state impact in ¢/kWh of $-4.56 - 3.94 + 1.90 = -6.2$ or $520 million per year. Applying a multiplying factor of 4- and 40-year life, this translates to $83 billion.

TABLE 13-12
Examples of new-generation nuclear reactor designs being advocated by proponents. NOAK is nth-of-a-kind.

Parameter	Target	Indirect Cycle Prismatic	Direct Cycle PBMR
Plant Design Life (years)	60	60	40/60
Thermal Efficiency	>50%	46%	45%/55%
Operating Costs			
Staffing Level per Plant (FTEs)	<250	~250	131
O&M ($/MW-h)	<5	~6	5.3
Fuel Cost ($/MW-h)	5.0	6.4	4.4–5.2
Misc.			1.3
Fuel Enrichment (%)	<20	19.9	9.6/19.5
Fuel Burnup (thousands of MWd/ton)	>100	120	92/200
Fuel Cycle (months)	24	18	continuous
NOAK Construction Costs ($/kWe)	<1,000	1,300	1,100–1,225

"U.S. DOE Office of Nuclear Energy, Science and Technology Advanced Fuel Cycle Initiative (AFCI) Comparison Report." Published by U.S. DOE, October 2003.
Design Features and Technology Uncertainties for the Next Generation. Nuclear Plant, June 30, 2004. Available at http://www.ne.doe.gov/reports/reports.html.

customer pays for the energy, not the state, but dollars spent on imported natural gas are lost from the state economy.

INEEL Next Generation Design Targets

Table 13-12 summarizes two new-generation nuclear reactor designs advanced by nuclear power planners. At 45% thermal efficiency the pebble bed reactor (PBMR) is on target with the economics presented earlier in this chapter. The estimated capital cost of $1,100–$1,225/kW is low enough to give an optimistic development future.

Transportation and Nuclear Power

The abundance of fuels to produce electrical power combined with the electrical grid distribution system makes electrical power the most likely candidate to replace petroleum as a transportation fuel. An emphasis is on replacing petroleum with ethanol, improving energy efficiency, and using biodiesel to incrementally reduce petroleum consumption. New plug-in HEV approaches using electrical power are now competitive (although development vehicles are not available to consumers) with conventional vehicles with the potential to replace about 80% of the petroleum fuel with domestic electrical power.

If nuclear power is used to replace petroleum, each of the U.S. states might realize favorable economic balances and the creation of jobs. Using battery storage, about 64% of the power delivered to an automobile goes to the powertrain. Table 13-13 includes a line for powertrain nuclear costs compared to busbar nuclear based on this transportation energy option.

Assuming an energy efficiency of 30% for a gasoline engine to a powertrain, a powertrain petroleum cost can be compared to a nuclear energy powertrain cost. At $1.50 per gallon for gasoline (0.119 MBtu/gallon), 0.009486 MBtu/2.778 kWh, and 30% efficiency, the cost of powertrain petroleum is about 14.3 ¢/kWh. At $3.00 per gallon, the powertrain cost is 28.6 ¢/kWh. This compares to 6 ¢/kWh for nuclear.

New-generation nuclear power plants are projected to have overnight capital costs similar to coal and thermal efficiencies of 47%, which is slightly better than coal. Table 13-13 summarizes the levelized costs for this comparison. Nuclear power has a cost advantage that is independent of plant life and capital discount factor.

Trade balances for a state would increase by 11 ¢/kWh (see Table 13-13, 14.3-3.3 ¢/kWh) by replacing imported gasoline with

TABLE 13-13
Scenario 3 levelized costs at an automobiles powertrain including the impact of each item on a state's cash flow. Values are in ¢/kWh. The state impact numbers assume 33% of the capital expenditures and 80% of the O&M expenditures stay in the state of construction.

Case	Capital	O&M	Fuel	Total
Nuclear Busbar	2.35	1.03	0.46	3.84
Nuclear Powertrain	3.67	1.61	0.72	6.00
State Impact, Nuclear Busbar	−1.41	−0.206	−0.46	−2.08
State Impact, Nuclear Powertrain				−3.3
State Impact, Petroleum Powertrain			−14.3	−14.3

domestic nuclear power. For a 1-GW power plant that is assumed to be used just for transportation, this represents $554 million[ii] per year improvement in the state's balance of trade. The per-mile costs for grid-powered transit are less than petroleum costs, and the price of grid power is much more stable than petroleum. These estimates are based on $1.50 per gallon gasoline ($1.92 less $0.42 taxes). A $2.00 price for gasoline translates to a much greater economic advantage.

A more rigorous analysis of petroleum replacement with electrical power and PHEV technology shows that about 60% of the petroleum imports are replaced with PHEV components like batteries. Table 13-14 summarizes the impacts of using electrical power rather than natural gas or petroleum. The basis is 90% utilization of the capacity of a 1-GW nuclear power plant. This analysis shows that the impact on a state economy would be about $554 million per year for replacement of petroleum at the rate of 1 GW if the state were to use PHEV vehicles powered by nuclear power and to produce the batteries and other hybrid components for the vehicle in the state.

TABLE 13-14
Impact of 1-GW capacity nuclear power plant and related technologies on state economy based on reducing flow of cash out of state. A 0.9 capacity factor is assumed with a 1-GW power plant providing 7.88E9 kWh of electricity in a year.

Transition	Busbar (¢/kWh)			Year (in millions)
	Before	After	Impact	Impact
Nuclear Rather Than Natural Gas	−4.56	−2.08	2.48	$196
On-Site Reprocessed Fuel	−2.08	−1.90	0.18	$14
40% of Electrical Power Rather Than Imported Petroleum	−9.12	−1.90	7.22	$228
In-State Hybrid Vehicle Upgrades*	−9	−4.5	4.5	$326–$355

*Based on Figure 9-5, annualized vehicle modification costs are ~1.5 the annual fuel costs (9 = 1.5 × 6¢/kWh). Fifty percent of the hybrid component (mostly batteries) costs are assumed to stay in state.

[ii] $867 million is 10.3¢/kWh × 0.9 Capacity Factor ×1,000,000 kW × 365 days × 24 hr/day × 0.64 (busbar/powertrain).

The potential demand for electric power to replace petroleum for transportation is about three-fourths of the total current demand for electricity. Considering the supplies of nuclear fuel that are stockpiled (in the forms of spent fuel and depleted uranium), these energy demands could be met with nuclear. The potential increase in electrical demand presents a rare opportunity. A 75% increase in electric power demand would require a redesigned and expanded electrical grid infrastructure. The expansion of the energy industry would occur over time and include:

- Measures could be taken to reduce daily and annual fluctuations in electrical power demand. For example, overnight charging of electric vehicles combined with flexibility in charging times could be used to level and maximize baseload power demands. This would lower average electrical costs and increase the efficiency for the entire electrical power grid.
- Measures could be taken to increase the efficiency of electrical power distribution. For example, electrical superconductor corridors might be implemented.
- Improved use of cogeneration (use of lower-quality heat from power plants by industry and communities) could also be implemented to increase the overall energy efficiency.

It is important in transportation applications to note that PHEV technology represents a replacement for gasoline transportation applications. PHEV technology is not an alternative to diesel used in tractor-trailers and farm tractors. Increased use of trains for long-haul shipping containers is an alternative to much of the diesel consumed for tractor-trailer freight transport. Low-till farming and use of perennial crops would reduce demand for diesel fuel for farm tractors.

Expanded Use of Nuclear Power in Residence and Commercial Applications

Increased use of electrical power for space and water heating could reduce the use of natural gas and heating oil in this application as indicated by Figure 13-1. The per-kWh impact is about the same as that for transportation—about a $554 million per year improvement in the trade balance per 1 GW of capacity. The total GW impact for space and water heating is about one-third of that projected for transportation.

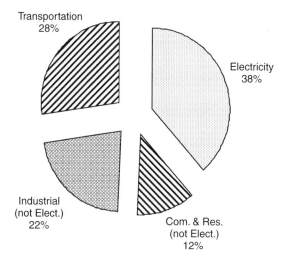

FIGURE 13-1. Distribution of energy consumption in the U.S. by sectors with electricity separated as its own sector.

Approaches to Long-Term Handling of Spent Nuclear Fuel

The natural resources available within states vary considerably. The majority does not have significant reserves of natural gas and coal that can be used at competitive prices. The stockpile of spent nuclear fuel represents such an energy resource. Figure 13-2 shows the spent fuel in storage on the nuclear power plant sites after 30 years of operation. The mass and volume of the fission products in the spent fuel are shown. If the uranium in the spent fuel is all recycled, it would fuel the power plant for about 850 years. The depleted uranium left from the production of the initial fuel used during the past 30 years would give the power plant fuel for 3,500 years (based on 4 parts depleted uranium per part enriched uranium produced). Adams[6] reports this ratio is 5.7 parts, which would translate to 5,000 years rather than 3,500 years. These estimates are based on total recovery of fuel values. It is obvious that even a fraction of the fuel values in the current inventory of spent fuel represents a huge energy asset.

The data of Figure 13-2 can be used to assess the impact on the economy of a state holding the reserves. On average, a state with one commercial nuclear power plant has enough uranium stored at the commercial reactor site to provide all that state's

374 *Sustainable Nuclear Power*

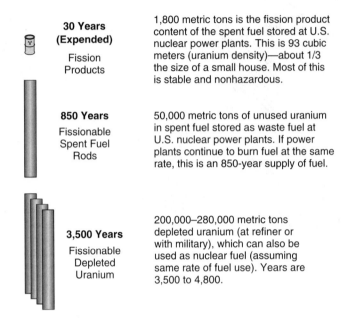

FIGURE 13-2. Energy reserves available as a direct and indirect result of 30 years of operation of nuclear power plants in the United States. The mass is all commercial-heavy metal for 30 years of operation. The years are the period a power plant could continue to operate using the uranium that was mined to provide the first 30 years of operation.

electrical power needs for the next 200 years[iii] (assuming the states have stockpiles rather than the nation as a whole). Using established nuclear reprocessing technology, the spent fuel could be reprocessed at a busbar cost of about 0.60¢/kWh (assuming the 0.90¢/kWh closed cycle fuel cost in France and a 50% increase in thermal efficiency of new generation nuclear power plants) compared to the current busbar rate for coal at 1.05¢/kWh. If the fuel is reprocessed in state, the benefits remain in the state from job creation and cash flow passing into the local economy.

The U.S. government relies on commercial nuclear power plants to store spent fuel on-site and this is the reason these

---

[iii] Nuclear represents 18%–19% of electrical power produced in the United States. Not all states have nuclear power plants. A value of 200 years is 850 years (Figure 2) divided by 23.5% where 23.5% is a rough estimate of the fraction of power provided by nuclear for the average state that has at least one nuclear power plant.

resources are available locally. These stockpiles will have accumulated an average of 30 years by 2012. About 3.5% of the spent fuel is fission products—the remainder is fuel.

Plans call for separation and transmutation technology to be developed to reduce the volume of the fission products to smaller quantities that require long-term storage. The easiest storage option is to let the commercial power plants store the spent nuclear fuel. With these fuel rods accumulating for 30 years, a new power plant on the same site could reprocess the fuel and store the 3.5% that is fission products for the 40-year life of a newly commissioned reactor. The storage facilities and expertise are on-site at this facility. Figure 13-3 illustrates this process—the amount of stored spent fuel on-site will remain about constant.

The goal would be to develop technology by the end of this 40-year cycle that would use or transmute the majority of the fission products that is produced. With new technology, time, and new plant commissioning, the goal would be to have the fission products processed to decrease the quantity of fission products on-site using improved methods for final, long-term disposal.

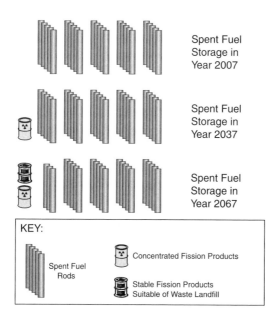

FIGURE 13-3. Spent nuclear fuel stored on-site at a nuclear power plant if reprocessing started in 2007.

By the second or third reprocessing cycle, most of the "aged" fission products could be removed from the power plant site for use or long-term storage. Nuclear power plants do not release fission product gases.[7] Nuclear fuel reprocessing plants would collect fission product gases that are released when spent fuel is reprocessed.

A benefit of this approach is that funds set aside for long-term nuclear waste storage (tens of billions of dollars) could be reauthorized to include support for building fuel reprocessing plants (especially first-of-a-kind facilities) and reactor technologies that produce less waste per kWh generated.

Technology and funds are available to proceed on this path. Recognize the spent nuclear fuel inventories as assets and new technologies will be developed. Handling the waste will not continue to be a financial burden but a way to use an available energy resource.

Fuel Costs and Energy Options

The focus has been on comparing nuclear to coal for supplying base load electric power. Table 13-15 summarizes and ranks feedstock costs by their representative prices at the end of the 20th century. The "Liquid Fuel" column gives the feedstock costs for the energy equivalent of one gallon of gasoline. The last column estimates the feedstock cost for producing 1 kWh of electricity.

The data of Table 13-15 show that fuels like natural gas, biomass, and petroleum cost much more than nuclear. These are fuel costs and do not include capital costs, operating costs, and maintenance costs. Both MSW and spent nuclear fuel use waste disposal costs to offset new fuel costs when the materials are used as fuel rather than placed in storage. MSW and spent nuclear fuel show the greatest potential for low-cost electrical power. There is not enough MSW to meet growing energy needs. An MSW energy program does provide a way to dispose of the waste and produce electricity.

Burns & McDonnell[8] report prices of delivered coal, $1.31/MBtu; natural gas, $7.00/MBtu; and biomass $6.48/MBtu. These are consistent with those given in Table 13-15 and used in the economic estimates.

Under the assumptions (1) 30 years of stockpiled spent nuclear fuel, (2) 3.5% consumption of fissionable metal, (3) meeting 18%

TABLE 13-15
Summary of feedstock costs for commonly used and considered fuels.

Fuel	Price ($/MMBtu)	AVG. ($/kWh)	Liquid Fuel* ($/gasoline gallon equ.)	El. Conv. Efficiency** (%)	Electricity Cost ($/kWh)
Municipal Solid Waste (MSW)	$(−2.00) to $(−4.00)	−0.0102	−0.36	20%–35%	−0.037
Spent Nuclear Fuel	$(0.08)	−0.0003	−0.01	25%–45%	−0.001
Full Uranium	$0.08	0.0003	0.01	25%–45%	0.001
Conventional Uranium	$0.62				0.0068
Coal	$1.20–$1.40	0.0044	0.145	25%–45%	0.011
Oil Sands ($10–$15/barrel)	$2.00–$3.00	0.0086	0.30	28%–53%	0.021
Natural Gas	$6.00	0.0205	0.68	53%–53%	0.039
Biomass	$2.10–$4.20–$6.80	0.0149	0.52	20%–45%	0.044
Petroleum ($45–$75/barrel)	$9.00–$15.00	0.0411	1.44	28%–53%	0.135

*Assuming 0.119 MMBtu per gallon of gasoline.
**Electrical conversion efficiency is the thermal efficiency of the cycle. The price of the feedstock fuel is divided by the thermal efficiency to estimate the cost of fuel consumed to generate 1 kWh of electricity.

of the nuclear power needs during 30 years, and (4) 100% availability of the spent fuel, the use of spent nuclear fuel could provide 100% of electrical power needs for the next 150 years.[iv] This 30 years of stockpiled spent fuel will be reached by 2007 with the linear projection shown in Figure 13-4.

The calculation does not account for increased power demand, but increased power plant efficiency can partially account for increases in power demand. The U.S. total inventory of spent nuclear fuel represents about the same energy content of the U.S. reserves for recoverable coal.

[iv] 850 years times 18%.

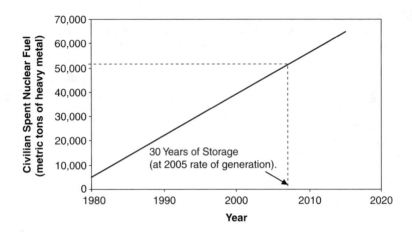

FIGURE 13-4. Extrapolation of recent rates of spent fuel accumulation from U.S. commercial facilities. Based on the rate of generation between 1990 and 2002, 30 years of stored spent nuclear fuel will be reached in 2007. ("Report to Congress on Advanced Fuel Cycle Initiative: The Future Path for Advanced Spent Fuel Treatment and Transmutation Research." U.S. DOE, January 2003.)

Comparison to Other Studies on Economics of Nuclear Power

Table 13-16 summarizes estimated costs of nuclear, coal, and natural gas electrical power production provided by Ilten.[9] The analyses in this chapter attempted to be consistent with this work. However, the natural gas prices ($3.05/MBtu) in that 2003 study are

TABLE 13-16
Summary from Ilten on French and U.S. busbar costs. All costs are in ¢/kWh.

Country	Plant Type	Capital	O&M	Fuel	Total
France	Nuclear Closed Cycle	3.91	0.69	0.90	5.50
U.S.	Nuclear Open Cycle	3.55	1.03	0.68	5.26
France	Gas Combined Cycle	1.50	0.42	4.09	6.01
U.S.	Gas Combined Cycle	0.93	0.32	1.97	3.22
France	Pulverized Coal	3.14	0.95	2.65	6.74
U.S.	Pulverized Coal	2.53	0.60	1.05	4.18

neither consistent with recent high prices of ($12.00/MBtu) nor projected prices ($6.11/MBtu).

The data in Table 13-16 show why France uses nuclear energy to meet 80% of its electrical demands—it is the high prices for coal and natural gas in France.

The low natural gas prices in 2003 led Ilten to project natural gas as the lowest cost option in the United States. This is not true based on updated natural gas prices. In addition, no previous studies evaluated the impact of importing natural gas or coal on the local trade deficit. Ilten did project lower capital costs for nuclear power consistent with costs reported in this chapter.

It is in the interest of non-coal-producing states (France, Japan) to advocate the development of lower-cost nuclear options. Advocates for nuclear power should promote spent nuclear fuel as a cost-effective way to expand nuclear power and to use nuclear energy to address the demand for transportation fuels.

Concluding Comments

Coal generally provides the low cost option for generating electricity in the United States today. Standardizing new nuclear power plant designs with decreased construction times from 7 to 4 years can cut the energy cost difference between coal and nuclear in half, but coal will still have an economic advantage. There are new regulations limiting coal plant emissions that will add to the cost of coal power. The cost of nuclear power will have to decrease to economically compete with coal.

Nuclear power does have an advantage over coal on the basis of fuel cost. Uranium fuel costs about $0.62 per delivered MBtu of thermal energy (using current nuclear reactor technology) compared to coal at $1.29. The fleet of nuclear power plants in operation today run at 30% to 33% thermal efficiency. The proposed new nuclear facilities will operate at higher temperatures that will allow operation at thermal efficiencies up to 45%. Higher thermal efficiencies will reduce both the capital cost and the fuel cost. These improvements are attainable and would make nuclear power sustainable and less expensive than coal.

Advocates for nuclear power must present a plan to achieve the potential nuclear energy offers in the present political climate. The technology can be developed and procedures put in place to reduce the cost of power and meet the regulations to build and certify the safety of nuclear plants. Those with the most to gain from a transition to nuclear power are non-coal-producing states. One new 1-GW nuclear power plant can reduce local energy costs by several hundred million dollars per year when nuclear energy replaces imported coal, natural gas, or petroleum.

The potential of nuclear technology leading to improved economics rests on the promise of abundant, relatively inexpensive uranium fuel and suggests that nuclear energy can become a sustainable power source for the long term. The inventory of stockpiled uranium available in the United States is under control of the Department of Defense and the Department of Energy. Releasing this uranium for domestic energy production would satisfy most energy demands for several hundred years. There are vast untapped sources of uranium in phosphate deposits increasing the available fuel supply. Nuclear energy technology essentially eliminates air pollution and substantially reduces greenhouse gas emissions. The problems of spent fuel management have not been addressed but the chemical science is known, so the pathway to spent fuel management technology is available.

Those familiar with nuclear technology recognize the clear lessons of history: Commercial nuclear reactors operating under the safety protocols of the United States, Europe, and Japan are the safest energy sources available with the lowest environmental impact. Reprocessing the spent nuclear fuel recovers remaining energy values and addresses nuclear waste disposal problems. Figure 13-5 illustrates reprocessing in a sustainable mode, with a small quantity of fission products formed from a large quantity of

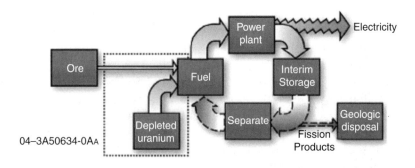

FIGURE 13-5. Uranium/plutonium in its "sustainable" phase. ("Report to Congress Advanced Fuel Cycle Initiative: Objectives, Approach, and Technology Summary." U.S. DOE Office of Nuclear Energy, Science, and Technology, May 2005.)

energy produced. Not shown in the figure is the partitioning of the spent fuel fission products into stable isotopes separated from high-level radioactive waste. This partitioning is the responsible course of action even without the sustainability plan. The stable isotopes can be released for sale or stored without concern for radiological exposure to the public. The long-lived radioactive fission products can be stored in a repository or made into transmutation targets, changing them to nonradioactive chemical isotopes.

Spent nuclear fuel also contains the very long-lived heavy metal isotopes: neptunium, plutonium, americium, and curium. They become nuclear fuel in the next generation, fast neutron flux reactors. When they are combined with uranium to form nuclear fuel, the spent fuel in the next reprocessing cycle produces the familiar fission products and more nuclear fuel.

The citizens of the United States have a huge appetite for energy that will be served. The global petroleum market makes the imported supply unreliable and increasingly expensive. Natural gas, the premium fossil fuel, is increasingly imported with a demonstrated unstable price. Domestic coal supplies can serve but carry emission control demands that weigh in to increase its cost. The carbon dioxide from fossil fuel combustion is a leading source adding to the atmospheric greenhouse gas inventory. Wind, solar, and bio-fuels are poised to contribute, but each is limited by raw material supply, public acceptance, or technology limitations.

Nuclear power plants are the demonstrated technology available to produce electricity in quantity to meet increased national demand. New light-water reactors are under construction overseas. These designs are ready for deployment. Sustainable nuclear energy follows by completing the chemistry of reprocessing the inventory of spent nuclear fuel and learning the technology of Generation IV high performance reactors. The supply of nuclear fuel extends centuries into the future. Public acceptance of the nuclear power option would enable centuries of sustainable nuclear energy.

References

1. N. Ilten, "International Comparisons of Electricity Generation by Types and Costs." August 28, 2003. http://people.cs.uchicago.edu/~nilten/docs/final.pdf#search='Busbar%20Cost%20of%20Nuclear%20Power.
2. "The Economic Future of Nuclear Power." Table 3-2. A study conducted by the University of Chicago, August 2004.

3. "Economic Future." Table 3-2.
4. "Economic Future." Table 1-1.
5. "Report on the Effect the Low-Enriched Uranium Delivery Under the HEU Agreement Between the Government of the United States and the Government of the Russian Federation Has on the Domestic Uranium Mining, Conversion, and Enrichment Industries, and the Operation of the Gaseous Diffusion Plant," December 31, 2003.
6. M. S. Adams, "Sustainable Energy from Nuclear Fission Power." *National Academy of Engineering Publications*, Volume 32, Number 4, Winter 2002.
7. R. A. Hinrichs, and M. Kleinbach, *Energy—Its Use and the Environment*, 3rd ed. New York: Brooks/Cole, 2002.
8. "Analysis of Baseload Generation Alternatives, Big Stone Unit II." Burns & McDonnell, September 2005.
9. Ilten.

Index

Absolute temperature scale-130
Absorption cross section-255
Accelerator driven systems-273
Accessible fossil fuel-34
Actinides-267
Activation products-298
Acute radiation syndrome-279
ADS, see Accelerator driven systems
Advanced fuel cycle initiative-271, 303
Advanced Spent Fuel Treatment-315
Advanced Test Reactor-305
AEC-see Atomic Energy Commission
AFCI, see Advanced fuel cycle initiative
Air conditioner-204–210
Air Pollution Research Act-26
Alpha decay-258
Alpha particles-253, 277
Alternating current generation-193
Alternative fuels-168
Amber-185
American Society for Testing Materials-166
Ammonia-38
Amoco Cadiz-28
amu, see atomic mass unit
Antitrust law-67, 89
Antitrust tax-90
Apollo Program-147
ARB, see California Air Research Board
Argonne National Laboratory-307
ASTM, see American Society for Testing Materials
Atomic Energy Commission-296
Atomic mass number-43, 255
Atomic mass unit-249
Atomic mass-44
Atomic number-250
Atomic stability-251
Atoms for Peace-139
Atoms-249
ATR-see Advanced Test Reactor
Axial flow gas compressor-143

383

384　*Index*

Barber, John-142
Barriers to commertialization-87
Base load power-146
Baseload demand-213
Baseload power generation-357
Basic steam cycle-323
Battery charger-228
Battery energy storage-232
Battery pack-183
Battery powered automobiles-226
Battery powered electric vehicle-229
Battery Technical Advisory Panel-233
Becquerel-259
Benjamin Franklin-15
Beta emissions-257
Beta particles-253, 277
BEV, see Battery powered electric vehicle
Binding energy-45, 257, 264
Biodiesel-62, 65, 176
Biological effects of radiation-278–280
Biomass-61, 173
Biomass feedstocks-224
$BiPO_4$-see Bismuth phosphate
Bismuth phosphate process-294
Bismuth phosphate-294
Black liquor-100
Boiler failure-134
Boiling water reactor-139, 297, 328–329
Bond funding-95
Boron-51
Brayton cycle gas turbine-334
Brayton gas cycle-340
Brayton power cycle-325
Breeder reactor-56, 263, 289

British Intelligence-17
BTAP, see Battery Technical Advisory Panel
Burns & McDonnell-376
Busbar cost-353
BWR-see Boiling water reactor

CAA, see Clean Air Act
CAFÉ, see Corporate Average Fuel Economy
California Air Research Board-233
California free market model-103
Canadian National Oil Policy-87
Canadian oil sand-68
Capacity factor-356
Capital costs-355
Capture cross section-266
Carbon dioxide emissions-115
Carbon dioxide-10, 170
Carbon monoxide-38, 170, 171
Carnot, Nicolas Leonard Sadi-321
Carter, Jimmy-299
Case studies-363
Catalytic converter-180
Catalytic reforming-167
Cellulose-174–175
Cellulose-to-ethanol-65
Cetane number-167
Charge of electricity-185
Chemical binding energy-48
Chernobyl-279, 342–346
Chilled water storage-107, 215
Chinese tallow tree-67
Churchill, Winston-294
Clarence Kemp-58
Classical thermodynamics-120

Clean Air Act Amendments 1990-113
Clean Air Act-26
Climax Solar Water Heater-58
Coal gasification-170
Coal-12
Coefficient of performance-240
Cogeneration-109
Cold start-171
Columbia river-291
Combined cycle-146
Combustion-124
Commercial coke-170
Compression ignition engine-166
Compression ratio-165
Conductor-186
Consumption tax-68, 88, 92
Control rods-51
Converter reactor-263
COP, see Coefficient of performance
Corn-62
Corporate Average Fuel Economy-177
Corporate income tax-91
Corporate lobbying-99
Corporate profitability-93
Cost of fuel fabrication facilities-364
Cost of reprocessing facilities-364
Critical energy-262
Critical mass-51
Crude oil-20
Cunningham, B. B.-291
Curie, Marie-259
Curie, Pierre-259
Curtis, Charles C.-142

Decay chain-259
Decontamination factor-293
Deforestation-25
DeLaval, Carl G.-135
Department of Health and Human Services-278
Depleted uranium-54
Deuterium-275
Diesel engines-179, 139
Diesel fuel-20
Diesel, Dr. Rudoph-163
Diffuse thermal energy-124
Direct exposure-277
Disintegrations-277
Distillers grain-62
Diversity-102
Dividend tax-92
Domestic oil production-21
Donora, Pennsylvania-26
Double action piston-133
Drake, Colonel-162
DuPont-17
Dynamo-188

E85 fuel-230
EBR II-see Experimental Breeder Reactor II
Economic comparisons-378
Economic incentive-152
Edison, Thomas A.-192
Electrochemistry-187
Electric cars-110
Electric lightbulb-192
Electric motor-192
Electric power generator-189
Electric power industry-196
Electric Power Research Institute-230
Electric space heating-239
Electric water heating-239
Electrical current-187
Electrical distribution-193
Electrical power generation efficiency-199

Electrical water heating-245
Electricity-6, 185
Electromagnetic radiation-258
Electromagnetic spectrum-276
Electromagnets-188
Electron volt-252
Electrons-249
Electrostatic attraction-185
Emerging energy
    technology-76
Energy conservation
    principle-120
Energy consumption-224
Energy guide labels-206–207
Energy levels of atoms-257
Energy Policy Act
    1975-113, 176
Energy sources-21
Energy-efficient building-201
Engine crankshaft-133
Engine thermal efficiency-179
Environmental impact-25
Environmental Protection
    Agency-26, 75
EPA fuel requirements 2007-81
EPRI, see Electric Power
    Research Institute
Estimated energy
    reserves-36, 40
Ethanol-62
Ethyl Corporation-17, 27
Excitation energy-258
Excited states-252
Expanded use of electric
    power-225
Experimental Breeder Reactor
    II-304–307, 309
Extraction train-301
Exxon Valdez-28

Faraday, Michael-188
Farm commodities-112

Fast neutron fission-289
Fast neutron reactors-270
Fast neutrons-267
Fast spectrum reactors-268
Federal Insurance
    Contribution Act-92
Federal subsidy-64
Federal tax incentive-68
Feed water heating-325
Fermi, Enrico-291
Fertile material-263, 287
FICA, see Federal Insurance
    Contribution Act
Fischer-Tropsch fuels-68
Fischer-Tropsch synthesis-171
Fischer-Tropsch technology-80
Fischer-Tropsch-5
Fissile material-261, 284, 287
Fission cross section-266
Fission product gases-300
Fission product metals-296
Fission products-270, 293, 298
Fission-12
Fissionable materials-262
Fort Jackson Army
    Installation-215
Fossil fuel-33
Franklin, Benjamin-186
Freeport Chemical-311
French fast-spectrum
    reactor-272
Friction-122
Frictional force-123
Fuel cell
    commercialiization-234
Fuel cell emission benefits-155
Fuel cell schematic
    diagram-150
Fuel cell system costs-238
Fuel cell technology-234–238
Fuel cell vehicle-111
Fuel cell-29, 147
Fuel cycle efficiency-148

Fuel oil-20
Furnaces-211

Gamma ray-258, 277
Gas turbine fuels-168
Gas turbine generator-145
Gas turbine-4, 142
Gas-Cooled Fast Reactor-335
Gas-cooled reactors-304
Gasification technology-81
General Motors-17
Generation II light water reactors-289
Generation IV nuclear reactors-8
Generation IV reactors-269, 289, 331–340
Generation-IV Nuclear Energy Systems Inititative-273
Genetic effects-280
Geothermal energy-56
Geothermal heating-57
GFR-see Gas-Cooled Fast Reactor
Global warming-29, 114, 218
Glycerin-176
Grass-to-ethanol-65
Gravity-force-121
Green Chemistry Conference-367
Green chemistry-284
Greenhouse gas emissions-114
Grid electricity-183
Grid-based electrical power-109
Gross profit analysis-100
Gross profits-77
Ground source heat pump-242–244
Ground state-252
Guaranteed loans-358
Gulf War 1991-29

Hahn, Otto-290
Half-life-255, 259
Halide, Andrew-160
Hanford nuclear pile-292
Hanford, Washington-291
Heat converted to work-125
Heat engine cycle-128
Heat engine-124, 127, 321
Heat pumps-116, 204, 240–245
Heating oil-20
Heating season performance factor-211
Heating Ventilation, and Air Conditioning-201
Heavy metals-287
Heavy oil reserves-37
Heavy oils-36
Helium-3, 44, 275
Henry, Joseph-188
HEU-see Highly enriched uranium
HEV, see hybrid electric vehicle
Hidden costs-109
High Flux Isotope Reactor-305
High-dose exposure-279
High-energy gamma-294
High-energy proton-273
High-level radioactive waste-270
Highly enriched uranium-366
Hiroshima-52, 280
Honda Accord Hybrid-181
Hot cells-271
Hot water heaters-203
HSPF, see Heating season performance factor
HVAC, see Heating Ventilation, and Air Conditioning
Hybrid electric vehicle-181
Hybrid heat pump system-244

Hybrid vehicles-88,
    108, 180
Hydrocarbons-36
Hydroelectric power-105
Hydroelectric-60
Hydrogen bomb-275
Hydrogen economy-110, 350
Hydrogen fuel cell
    efficency-153
Hydrogen-38, 170
Hyundai-94

I. G. Farbin-17
Ice storage-107, 215
Iceland-57
Idaho National Energy and
    Environmental
    Laboratory-346
Idaho National Laboratory-307
Ideal engine cycle-129
INEEL-see Idaho National
    Energy and Environmental
    Laboratory
Ingestion exposure-278
Inhalation exposure-277
Insulator-186, 187
Intangible costs-82, 86
Intangible risk-89
Integral nuclear power
    plant-307
International Minerals and
    Chemical-311
Interstate power grid-195
Investment rate of return-83
Investment threshold-93
Iodine-129–303
Ionizing radiation-277
Iran-21
Iraq-21
Iron Curtain speech-294
IRR, see Investment rate of
    return

Irradiated uranium-292
Isooctane-167
Isotopes-49, 250

Jet aircraft engine-144
Jet aircraft fuel-163
Jet aircraft-18
Jet fuel-20
Joule efficiency-324

Kinetic energy-48, 123
Kyoto Protocol-29

Lanthanum fluoride-291
Latent period-279
Lawrence Berkeley National
    Laboratory-48
Lead poisoning-25
Lead-Cooled Fast Reactor-338
Leaded gasoline-27
Levelized cost-353, 355
LiF3-see Lanthanum fluoride
Light water reactor-284, 301
Lignin-175
Linear no-threshold dose-279
Liquid fuels from coal-169
Liquid fuels-161, 169
Liquifaction-169
Lithium-44
LNT, see Linear no-threshold
    dose
Local economy-361
Lodestone-186
London smog-26
Low freezing point-168
Low temperature fuel cells-236
Low volatility-168
Low-enriched uranium
    fuel-335
LWR-see Light water reactor

Magnetic field-187
Manhattan District-291
Manhattan Project-53
Mass defect-257
Maxwell theory of electromagnetism-188
Maxwell, James Clerk-188
Meitner, Lisa-290
Mercury poisoning-25
Methane hydrates-39
Methane-39
Methanol fuel cells-154
Methanol fuel-171
Methyl tertiary butyl ether-173
Miles per gallon-178
Minnesota BD Use-113
Minor actinides-301
Mixed oxide fuel-301
Molten Salt Reactor-339
Monopolies-18
Morse code-189
Morse, Samuel F. B.-188
MOX-see mixed oxide fuel
mpg, see Miles per gallon
MSR-see Molten Salt Reactor
MTEB, see Methyl tertiary butyl ether
Municipal solid waste-77

Nagasaki-52, 280
National Cancer Institute-279
National Energy Policy-May 2001-319
Natural fraction processes-164
Natural gas combined cycle-325
Natural gas pipeline infrarstructure-172
Natural gas-6, 172
Natural nuclear reactor-13, 50
Natural radioactive decay-251
Natural radioactivity-259
NERI-see Nuclear Energy Research Initiative
Net present value-229
Neutron absorption cross section-268
Neutron capture-289
Neutron capture-292
Neutron decay-258
Neutron flux-269
Neutron induced fission-251
Neutrons-249
New Energy Policy Act-113
New generation power plants-370
New York subway-160
Newcomen, Thomas-131
Newton, Sir Isaac-186
Next generation design targets-369
Next Generation Vehicle Program-181
NHI-see Nuclear Hydrogen Initiative
Nickel metal hydride batteries-233
Nitrogen oxide-26
NPV, see net present value
Nuclear bomb-18, 119
Nuclear chain reaction-51, 291
Nuclear cross sections-265
Nuclear Energy Research Initiative-319
Nuclear energy-6
Nuclear fission-48, 250, 289
Nuclear fusion-10, 48, 119, 274
Nuclear Hydrogen Initiative-320
Nuclear nonproliferation treaty-299

Nuclear pile-291
Nuclear power advocates-368
Nuclear power-139
Nuclear reactions-45
Nuclear reprocessing
   technology cost-374
Nuclear safety-341–343
Nuclear submarines-297
Nuclear transitions-259
Nuclear waste-31, 312
Nuclear weapons
   nonproliferation
   group-301
Nucleus stability-250
Nucleon binding energy-42
Nucleons-258
Nuclide chart-254
Nuclide radioactivity-298
Nuclide-250

Oak Ridge National
   Laboratory-311
Occupational accidents-31
Octane number-165
Octane scale-167
Oil crops-67
Oil imports-21
Oil sand reserves-38
Oil sands-5, 19, 36
Oil shale-36
Oils-175
Once through fuel
   cycle-286, 340
On-site fuel reprocessing-361
On-site reprocessing-336
Ørsted, Hans Christian-187
OPEC, see Organization of
   Petroleum Exporting
   Countries
Open circuit voltage-152
Operation Desert Storm-29

Organization of Petroleum
   Exporting Countries-90
Over night capital costs-368
Overnight capital-355

PAFC, see Polymer electrolyte
   membrane fuel cells
Palm trees-114
Parsons, Charles-135
Particle accelerator-48, 272
Partners for a New Generation
   of Vehicles-181
Passive safety design-345
PCM, see Phase change
   materials
Peak demand-105, 106
Peak load generation-146
Peak load shifting
   strategy-203, 204,
   211, 214
Pebble bed reactor-334
PEM fuel cells-70
PEM, see proton exchange
   membrane
Permanence scale-42
Personal income tax-91
Petroleum displacement
   technology-225
Petroleum fuel fractions-162
Petroleum refining
   technology-164
Petroleum reserves-82
Phase change materials-107,
   216
PHENIX, see French
   fast-spectrum reactor
PHEV technology-371
PHEV, see Plug-in hybrid
   electric vehicle
Phosphoric acid fuel
   cells-236
Photonic emissions-253

Photosynthesis-10, 11
Photovoltaic cells-58
Phytoplankton-12
Pith balls-185
Plastics industry-18
Plug-in hybrid electric vehicle-6, 68, 70, 88, 111, 116, 183, 227–233
Plutonium based fuels-271
Plutonium production-290
Plutonium recovery-292
PNGV, see Partners for a New Generation of Vehicles
Polymer electrolyte membrane fuel cells-236
Potential energy-123
Powertrain-370
Precertified designs-363
Pressurized water reactor-50, 139, 297, 330
Priestly, Joseph-186
Prismatic graphite block-334
Process steam-137
Profitability-353
Programmable thermostat-241
Property tax-92
Proton exchange membrane-151
Protons-43, 249
Pu-240–299
Pumped water storage-105
Pumping natural gas-145
PUREX process-294–297, 299–302
PWR fuel-298
PWR-see Pressurized water reactor
Pyrochemical dry treatment-304
Pyrolysis-170
Pyrometallurgical reprocessing-308

Pyropartitioning-307
PYROX-see pyrochemical dry treatment

Radar-18
Radiation hazard-287
Radiation shielding-294
Radiation sickness-279
Radiation-276
Radioactive nuclides-250
Radioactivity-277
Radiological properties-291
Radiological toxicology-276–280
Radon-277
Rare earth metals-309
Rechargeable fuel cell-29, 110
Recoverable fuels-40
Regenerative braking-227
Remote natural gas-172
Reprocessed fuel-286
Reprocessing spent nuclear fuel-7, 359
Reprocessing technology-101
Return on investment-83
Reversibility factor-325
Rockefeller, John D.-16, 17
ROI, see Return on investment

Sales tax-92
San Francisco cable car-160
Saturated fat-67
Saudi Arabia-21
Savannah River, South Carolina-294
Savery, Thomas-129
Scatter cross section-266
Seaborg, Glenn-291
Seasonal Energy Efficiency Ratio-208
Secondary wastes-313

SEER, see Seasonal Energy Efficiency Ratio
Sensitivity analysis-354
Sensitivity analysis-362
SFR-see Sodium-Cooled Fast Reactor
Shipingsport, Ohio-53
Silicon carbide-334
Site labor-360
Site materials-360
Smith, Adam-91
Sodium coolant-306
Sodium-Cooled Fast Reactor-336–337
Solar energy-58
Solar wall-110
Solar water heating-58
Solid biomass-62
Solid State Energy Conversion Alliance-236
Sorghum-174
Soybeans-65
Space heating fuels-213
Spallation process-273
Spark ignition fuels-162, 164
Spent fuel inventories-55
Spent fuel-284
Stable isotopes-255
Stable nuclides-255
Standard Oil monopoly-16
Standardized design-357
Standardized fuel specifications-162
Starch-175
State tax revenue-359
Static electricity-186
Steam boiler-134
Steam engine-15, 127
Steam reheat-325
Steam turbine power cycle-138
Steam turbine-4, 134
Subcritical reactor-272
Submarine-139

Sugars-175
Sulfur oxide-26
Super PHENIX-337
Supercritical Water-Cooled Reactors-332
Supernova-47
Sustainability-108, 223
Sustained nuclear reaction-50
Syncrude-5
Synthesis gas pipeline-37
Synthesis gas-170
Synthetic fuel-5
Synthetic oil-18

Tariff-free trade-92
Tax credits-64, 112
TBP-see Tri-n-butyl phosphate
Technitium-99-303
Technology Innovation-18
Telegraph-15, 188
Tetraethyl lead-166
Thales-185
Thermal efficiency-321
Thermal efficiency-37, 138, 197
Thermal energy efficiency-141
Thermal energy storage-106, 107, 204
Thermal energy-298
Thermal neutron cross section-255
Thermal neutrons-255, 261, 267, 270
Thermodynamics first law-120
Thermostat setting-241
Thompson, S. G.-292
Thorium-259, 264, 288, 310
Three Mile Island accident-342–343
Three-phase power-195

Toscas, James-309
Total cross section-266
Transformer-194
Transistor-18
Transmutation fuels-304
Transmutation processes-263
Transmutation research-315
Transmutation-263, 270
Transportation infrastructure-160
Transuranic elements-269, 298
Tricoated-isotropic reactor fuel particles-334
Tri-n-butyl phosphate-295
TRISO-see Tricoated-isotropic reactor fuel particles
Truman, Harry S.-17
Turbine blade-143
Turinia-135
TWA flight 800-168

Unemployment tax-92
United Nuclear-311
Unstable nucleous-259
Upside potential-89
Uranium Extration-302
Uranium hexafluoride-310
Uranium ore-310
Uranium oxide fuel-297
Uranium reserves-54, 310–312
Uranium-12, 259
Uranium-from-phosphate-311
UREX+-303–305
UREX-see Uranium Extraction

Value added tax-93
Vapor compression air conditioners-208

VAT, see value added tax
Vegetable oil-70
Vehicle fuel-173
Very High-Temperature Reactor-334
VHTR-see Very High-Temperature Reactor
Vickers Viscount turboprop-144
Visible light-252
Volta, Count Alessandro-187
VW Beetle-180

Waste heat-196
Waste minimization-288, 314
Water vapor-124
Watt, James-133
Weapons grade plutonium-79
Weapons grade U-235-53, 79
Werner, L. B.-291
Westinghouse, George-136
Westinghouse-311
Wind energy-60
Wind farms-61
Wind power-61
Windmill-134
Wood-62
Wood-to-ethanol-65
Work-definition-121
Working fluid-126
World War II-19
Wyoming coal-34

Yellowstone Park-57
Yucca Mountain-366
Yucca Mountain-284
Yucca Mountain-78, 272

Zircaloy-298